AF794880

Mobile Edge Computing and Communications

Mobile Edge Computing and Communications

Driving Forces, Technology Foundation, and Application Areas

Aaron Yi Ding
Delft University of Technology
Netherlands

Chamitha De Alwis
University of Bedfordshire
United Kingdom

Madhusanka Liyanage
University College Dublin
Ireland

WILEY

IEEE Press

This edition first published 2025
© 2025 John Wiley & Sons Ltd.

All rights reserved, including rights for text and data mining and training of artificial technologies or similar technologies. No part of this publication may be reproduced, stored in a retrieval system, or transmitted, in any form or by any means, electronic, mechanical, photocopying, recording or otherwise, except as permitted by law. Advice on how to obtain permission to reuse material from this title is available at http://www.wiley.com/go/permissions.

The right of Aaron Yi Ding, Chamitha De Alwis, and Madhusanka Liyanage to be identified as the authors of this work has been asserted in accordance with law.

Registered Offices
John Wiley & Sons, Inc., 111 River Street, Hoboken, NJ 07030, USA
John Wiley & Sons Ltd, New Era House, 8 Oldlands Way, Bognor Regis, West Sussex, PO22 9NQ, UK

For details of our global editorial offices, customer services, and more information about Wiley products visit us at www.wiley.com.

Wiley also publishes its books in a variety of electronic formats and by print-on-demand. Some content that appears in standard print versions of this book may not be available in other formats.

Trademarks: Wiley and the Wiley logo are trademarks or registered trademarks of John Wiley & Sons, Inc. and/or its affiliates in the United States and other countries and may not be used without written permission. All other trademarks are the property of their respective owners. John Wiley & Sons, Inc. is not associated with any product or vendor mentioned in this book.

Limit of Liability/Disclaimer of Warranty
In view of ongoing research, equipment modifications, changes in governmental regulations, and the constant flow of information relating to the use of experimental reagents, equipment, and devices, the reader is urged to review and evaluate the information provided in the package insert or instructions for each chemical, piece of equipment, reagent, or device for, among other things, any changes in the instructions or indication of usage and for added warnings and precautions. While the publisher and authors have used their best efforts in preparing this work, they make no representations or warranties with respect to the accuracy or completeness of the contents of this work and specifically disclaim all warranties, including without limitation any implied warranties of merchantability or fitness for a particular purpose. No warranty may be created or extended by sales representatives, written sales materials or promotional statements for this work. The fact that an organization, website, or product is referred to in this work as a citation and/or potential source of further information does not mean that the publisher and authors endorse the information or services the organization, website, or product may provide or recommendations it may make. This work is sold with the understanding that the publisher is not engaged in rendering professional services. The advice and strategies contained herein may not be suitable for your situation. You should consult with a specialist where appropriate. Further, readers should be aware that websites listed in this work may have changed or disappeared between when this work was written and when it is read. Neither the publisher nor authors shall be liable for any loss of profit or any other commercial damages, including but not limited to special, incidental, consequential, or other damages.

Library of Congress Cataloging-in-Publication Data Applied for:

HardBack ISBN: 9781119611523

Cover Design: Wiley
Cover Image: © buffaloboy/Shutterstock

Set in 9.5/12.5pt STIXTwoText by Straive, Chennai, India
Printed and bound by CPI Group (UK) Ltd, Croydon, CR0 4YY

C9781119611523_030225

The manufacturer's authorized representative according to the EU General Product Safety Regulation is Wiley-VCH GmbH, Boschstr. 12, 69469 Weinheim, Germany, e-mail: Product_Safety@wiley.com.

Contents

About the Authors *xiii*
Preface *xv*
To Our Readers *xvii*
Acknowledgments *xix*
Acronyms *xxi*

Part I Introduction and Edge Applications *1*

1 Introduction *3*
1.1 Focus of the Book *3*
1.2 The Vision of Edge *4*
1.3 Stakeholders and Related Paradigms *6*

2 Case for Multi-access Edge Computing *9*
2.1 Case for Multi-access Edge Computing *9*
2.1.1 What Is MEC? *9*
2.1.2 Why We Need MEC? *9*
2.1.3 Evolution of MEC *11*
2.2 MEC Reference Architecture *12*
2.2.1 Mobile Edge System Level *13*
2.2.1.1 User Equipment (UE) and Applications *13*
2.2.1.2 Customer Facing Service Portal (CFSP) *13*
2.2.1.3 Operation Support System (OSS) *14*
2.2.1.4 User Application Lifecycle Management Proxy (UALCMP) *14*
2.2.1.5 Mobile Edge Orchestrator (MEO) *14*
2.2.2 Mobile Edge Host Level *15*
2.2.2.1 Mobile Edge Host (MEH) *15*
2.2.2.2 Mobile Edge Platform Manager (MEPM) *16*
2.2.2.3 Virtualization Infrastructure Manager (VIM) *16*
2.2.3 Mobile Edge Services and Interfaces *17*
2.2.3.1 Mp—MEC Platform Interfaces *17*
2.2.3.2 Mm—Management Interfaces *17*

2.2.3.3 Mx—External Interfaces *18*
2.3 MEC Standardization *18*
2.4 MEC Use Cases and Applications *19*
2.4.1 Critical Infrastructure *19*
2.4.2 UHD Video Streaming and Analytics *21*
2.4.3 Massive Machine Type Communication (mMTC) Applications *22*
2.4.4 Intelligent Transportation System (ITS) *23*
2.4.5 Augmented Reality (AR), Virtual Reality (VR), and Mixed Reality (MR) *25*
2.4.6 Unmanned Aerial Vehicles (UAVs) *26*
2.5 Related Computing Paradigms to MEC *26*
2.5.1 Mobile Cloud Computing (MCC) *27*
2.5.2 Fog Computing *27*
2.5.3 Cloudlets *28*
2.6 Comparison with Related Computing Paradigms *28*
2.6.1 MEC and MCC *28*
2.6.2 MEC and Fog Computing *29*
2.6.3 MEC and Cloudlets *30*
2.7 Summary *31*

3 Edge-Based Video Analytics and Mobile Augmented Reality *33*
3.1 Video Analytics Overview *33*
3.2 Edge Video Analytics *35*
3.2.1 Definition and Enabling Technologies *35*
3.2.2 Edge Video Analytics in Practice *37*
3.3 Edge-Enabled Augmented Reality (AR) *40*
3.3.1 AR Overview *40*
3.3.2 Recent Advancement Towards MAR *40*
3.3.3 MAR in 5G and Beyond *42*
3.3.4 MEC for MAR *42*
3.4 Architecture Options for Mobile Augmented Reality *43*
3.4.1 Cloud Based Architecture *43*
3.4.1.1 Advantages and Disadvantages of Cloud Based Architecture *44*
3.4.2 MEC-Based Architecture *45*
3.4.2.1 Advantages of MEC-Based Architecture *46*
3.4.2.2 Disadvantages of MEC-Based Architecture *46*
3.4.3 Localized Architecture *46*
3.4.3.1 Advantages of Localized Architectures *47*
3.4.3.2 Disadvantages of Localized Architectures *47*
3.4.4 Hybrid Architecture *48*
3.4.4.1 Advantages of Hybrid Architecture *49*
3.4.4.2 Disadvantages of Hybrid Architecture *49*
3.4.5 Summary of Architecture Options *49*
3.5 Applications of MEC-Based MAR *51*
3.5.1 IIoT *51*
3.5.2 Healthcare *51*

3.5.3 Travel and Tourism *52*
3.5.4 Smart City and Home *52*
3.5.5 Automotive *53*
3.6 Technical Aspects of MEC-Based MAR *53*
3.7 Summary and Outlook *56*

4 *MEC*^{2} for eHealthcare: Empowering Next-Generation Healthcare Systems *57*
4.1 Introduction *57*
4.2 MEC Enabled Use Cases *59*
4.2.1 Use Case 1: Augmented Reality (AR) Based Remote Robotic Monitoring *59*
4.2.2 Use Case 2: Smart Medical Data Aggregation/Patient Monitoring and Treatment *60*
4.2.3 Use Case 3: AR Assisted Remote Surgery *62*
4.3 Realizing Use Cases with MEC *62*
4.3.1 AR-Based Remote Robotic Monitoring with MEC *62*
4.3.2 MEC Assisted Smart Medical Data Aggregation/Patient Monitoring and Treatment *64*
4.3.3 AR Assisted Remote Surgery with MEC *64*
4.4 Deployment Challenges and Solutions *65*
4.4.1 Security and Privacy Issues *65*
4.4.2 Communication Issues *65*
4.4.3 Scalability Limitation *66*
4.4.4 Legal Adaptation and Ethics *66*
4.5 Conclusion *66*

Part II Technology Foundation *69*

5 Edge AI *71*
5.1 Overview of Edge AI *71*
5.2 Training and Inference in Edge AI *73*
5.2.1 Edge Training *73*
5.2.2 Edge Inference *76*
5.3 Edge AI in Practice *79*
5.3.1 Energy-Efficient Edge AI Using Approximate Computing *79*
5.3.2 Applications at the Edge *79*
5.4 Summary and Outlook *80*

6 Edge Offloading *83*
6.1 Motivation and Fundamental Concept *83*
6.1.1 Rationale for Edge Offloading *83*
6.1.2 Data Offloading *84*
6.1.3 Computation Offloading *87*
6.2 Edge Offloading *89*

6.2.1 Edge Data Offloading *89*
6.2.2 Edge Computation Offloading *90*
6.2.3 Enabling Technologies for Edge Offloading *92*
6.3 Edge Offloading in Practice *92*
6.3.1 Use Cases of Edge Offloading *92*
6.3.2 AI-Driven Edge Offloading *93*
6.4 Summary and Outlook *94*

7 Edge Storage and Caching *97*
7.1 Edge Storage *98*
7.1.1 Edge Storage Design *99*
7.1.2 Data Robustness in Edge Storage Systems *100*
7.1.3 Data Security and Privacy in Edge Storage Systems *101*
7.2 Edge Caching *102*
7.2.1 Introduction to Caching *102*
7.2.2 How Edge Caching Works *102*
7.3 Edge Caching Towards Information-Centric Networking *103*
7.3.1 Interest Packets and Data Packets *104*
7.3.2 Named Content *104*
7.3.3 In-Network Caching *105*
7.3.4 Data Structures *105*
7.4 Conclusion *105*

8 Edge Communication, Hardware and Enabling Frameworks *107*
8.1 Advancement of Communication Technologies *107*
8.1.1 THz Communication *108*
8.1.2 Zero Touch Network and Service Management *108*
8.1.3 Blockchain *108*
8.1.4 Federated Learning *109*
8.1.5 Large Intelligent Surfaces *110*
8.1.6 Visible Light Communication *110*
8.1.7 Quantum Computing and Communication *110*
8.2 Evolution of Communication Networks *110*
8.2.1 Further Enhanced Mobile Broadband *112*
8.2.2 Extremely Reliable Low-Latency Communication *113*
8.2.3 Extremely Low-Power Communication *113*
8.2.4 High Spectrum Efficiency *113*
8.2.5 High Area Traffic Capacity *114*
8.2.6 Mobile Broadband and Low-Latency *114*
8.2.7 AI-Assistive Extreme Communications *114*
8.3 Edge Computing Landscape *115*
8.3.1 Software Development Timelines and Landscape *116*
8.3.1.1 Early Stages of Edge Computing (2000s–2010s) *116*
8.3.1.2 Emergence of IoT and Edge Computing (2010s–2020s) *116*
8.3.1.3 Latest Edge Software Trends (2020s–Present) *117*

8.3.1.4 Future of Edge Software Development *117*
8.3.2 Hardware Development Timeline and Landscape *117*
8.3.2.1 Early Development of Edge Hardware (2000s–2010s) *118*
8.3.2.2 Rise of IoT and Edge Computing (2010s–2020s) *118*
8.3.2.3 Latest Edge Hardware Development (2020s–Present) *118*
8.3.2.4 Future of Edge Software Development *119*
8.4 Edge Computing System Solutions *119*
8.4.1 Legacy Systems *119*
8.4.2 Application Oriented Systems *120*
8.4.2.1 Scalys *120*
8.4.2.2 Enclave Device Blueprint *120*
8.4.2.3 Nokia *121*
8.4.2.4 Amazon *123*
8.4.2.5 IBM *124*
8.5 Edge Hardware and Devices *125*
8.5.1 The Jetson Family—GPU Oriented *126*
8.5.2 Coral Platforms *127*
8.5.3 Raspberry-Pi Series *127*
8.5.4 Intel Devices *128*
8.5.5 Beelink *129*
8.5.6 NXP *129*
8.6 Edge Accelerators *130*
8.6.1 FPGA *130*
8.6.2 Tensor Processing Units *131*
8.6.3 ASIC *132*
8.6.4 Key Takeaway: The Future of Edge Acceleration *133*
8.7 Industry Frameworks *133*
8.7.1 KubeEdge *133*
8.7.2 Edgent *134*
8.7.3 GreenGrass *135*
8.7.4 Azure IoT *135*
8.7.5 Private Edge Service Providers *135*
8.8 Concluding Remarks *136*

Part III Integration and Driving Forces *137*

9 Driving Forces for Edge Integration *139*
9.1 Internet of Things Toward Internet of Everything *139*
9.2 The Growth of Small Data *140*
9.3 AI/ML Based Communication Technologies *140*
9.4 Novel Technologies and Applications *141*
9.4.1 Industry 5.0 *141*
9.4.2 Smart and Intelligent Healthcare *142*
9.4.3 Unmanned Aerial Vehicles *142*

9.4.4 Extended Reality Toward Holographic Teleportation *142*
9.4.5 Connected and Autonomous Vehicles *143*
9.4.6 Collaborative Robots *143*
9.5 Conclusion *143*

10 Edge Security *145*
10.1 Introduction *145*
10.1.1 Importance of MEC Security *145*
10.2 Security Threats in MEC *146*
10.3 Threat Vectors Related to the Access Network *146*
10.3.1 A1: Link Between the User Equipment and a Base Station *147*
10.3.2 A2: Ad Hoc Connectivity Between User Equipment *148*
10.3.3 A3: User Equipment (UE) *149*
10.4 Threat Vectors Related to the Mobile Edge Network (MEN) *150*
10.4.1 E1: Mobile Edge Platform Manager (MEPM) *150*
10.4.2 E2: Virtualization Infrastructure Manager (VIM) *151*
10.4.3 E3: Mobile Edge Host (MEH) *151*
10.4.4 E4: Connectivity Between Mobile Edge Hosts *152*
10.4.5 E5: MEC Platform Connectivity Between the Edge and the Core *152*
10.4.6 E6: Connectivity Between Mobile Edge Apps Operated Under Mobile Edge Hosts at Different Base Stations *153*
10.4.7 E7: Connectivity with the Mobile Edge Host and the Cloud Servers *153*
10.5 Threat Vectors Related to the Core Network *153*
10.5.1 C1: User Application Life-Cycle Management Proxy (UALCMP) *154*
10.5.2 C2: Operation Support System (OSS) *154*
10.5.3 C3: Mobile Edge Orchestrator (MEO) *154*
10.5.4 C4: Customer Facing Service Portal (CFSP) *155*
10.5.5 C5: Connectivity of the Mobile Edge Orchestrator (MEO) and the 5G Core Network *155*
10.5.6 C6: 5G Core Network *156*
10.6 Architectural Threat Vectors *156*
10.6.1 AR 1: Network Slicing (NS) *157*
10.6.2 AR 2: Traffic Steering *157*
10.6.3 AR 3: Service Migration *158*
10.6.4 AR4: Mobility Management *159*
10.7 Security Requirements of MEC *160*
10.7.1 Classical Security Requirements of MEC *160*
10.7.1.1 Confidentiality *160*
10.7.1.2 Integrity *161*
10.7.1.3 Availability *161*
10.7.1.4 Authentication and Authorization *162*
10.7.2 MEC Specific Security Requirements *163*
10.7.2.1 Secure Charging and Billing *163*
10.7.2.2 Mitigating Service Impeding Threats *163*
10.7.2.3 Secure Mobile Offloading Schemes *164*

10.7.2.4 Robustness of MEO *164*
10.7.2.5 Secure Virtualized Environment and Orchestration at the Edge *164*
10.8 Possible Security Solutions for MEC *165*
10.8.1 Security as a Service *165*
10.8.2 Artificial Intelligence with Machine Learning and Deep Learning Methods *166*
10.8.3 Network Slicing (NS) *166*
10.8.4 Context Aware Security *167*
10.8.5 Physical Unclonable Functions (PUFs) *167*
10.8.6 Blockchain *167*
10.8.7 Modern Authentication and Key Agreement Schemes for MEC Infrastructure *168*
10.8.8 Virtual Machine Introspection (VMI) *168*
10.8.9 Virtual Private Local Area Network Service (VPLS) *169*

11 Edge Privacy *171*
11.1 Introduction *171*
11.2 Privacy in 5G and Beyond *172*
11.2.1 Data Privacy *173*
11.2.2 Location Privacy *173*
11.2.3 Identity Privacy *174*
11.3 MEC Privacy *174*
11.3.1 Impact on Data Privacy *176*
11.3.2 Impact on Location Privacy *176*
11.3.3 Impact on Identity Privacy *176*
11.3.4 Other Privacy Issues *177*
11.3.4.1 Authorized and Curious Adversaries *177*
11.3.4.2 Computational Offloading *177*
11.3.4.3 Service Migration *177*
11.3.5 GDPR on MEC *177*
11.4 Objectives for Privacy Preservation in MEC *178*
11.4.1 Global Compliance for Privacy Policies *178*
11.4.2 Responsibility of MEC Service Providers and Consumers *179*
11.4.3 Privacy Compliance on Integrating Technologies *179*
11.4.4 Data Portability *179*
11.4.5 Accountability and Transparency of Data Handling *179*
11.4.6 Declaring Minimum Specification Requisites of UE for Subscribing MES *180*
11.4.7 Optimal Utilization of UE Resources with Embedded Privacy Enhancing Mechanisms *180*
11.5 Privacy Protection Mechanisms for MEC *180*
11.5.1 Regulatory Approaches *181*
11.5.1.1 Governmental or Continental Regulations *181*
11.5.1.2 Industry Self-regulation *181*
11.5.1.3 Consumer or Market Regulation *181*
11.5.2 Privacy Preservation in Mobile Offloading *182*

11.5.3 Strong Encryption and Strong Authentication *182*
11.5.4 Privacy by Design *183*
11.5.5 Software Defined Privacy Approach *184*
11.6 Future Directions *184*
11.6.1 Hybrid Cloud Approach *185*
11.6.2 Privacy Protection as a Service *185*
11.6.3 Privacy Preserving Data Aggregation *185*
11.6.4 Novel Privacy Preserving Techniques *186*

12 Telecom and Cloud Integration *189*
12.1 Introduction *189*
12.2 Network Softwarization in 5G Networks *190*
12.3 MEC Integration in 5G Backhaul *191*
12.3.1 Why MEC Needed in 5G *191*
12.3.2 Challenges Related to MEC in 5G Backhaul Networks *193*
12.4 Key Technologies for MEC 5G Integration *193*
12.4.1 Network Function Virtualization *194*
12.4.1.1 Mutual Benefits of NFV MEC Integration *195*
12.4.1.2 Challenges and Obstacles of NFV MEC Integration *197*
12.4.2 Software Defined Networking *197*
12.4.2.1 Mutual Benefits of SDN MEC Integration *198*
12.4.2.2 Challenges for SDN MEC Integration *198*
12.4.3 Information Centric Networking *199*
12.4.3.1 Mutual Benefits of ICN MEC Integration *200*
12.4.3.2 Challenges and Obstacles of ICN MEC Integration *201*
12.4.4 Network Slicing *201*
12.4.4.1 Mutual Benefits of Slicing MEC Integration *202*
12.4.4.2 Challenges and Obstacles of Slicing MEC Integration *203*
12.5 Conclusion *203*

Part IV Outlook *205*

13 Conclusion and Outlook for *MEC*^{2} *207*

References *209*
Index *245*

About the Authors

Aaron Yi Ding leads the Cyber Physical Intelligence (CPI) Lab as tenured Associate Professor at TU Delft and Associate Professor (permanent) in Computer Science at the University of Helsinki. He has over 17 years of R&D practices in EU, Switzerland, UK and USA. Prior to TU Delft, he has worked at TU Munich with Jörg Ott, at the University of Cambridge with Jon Crowcroft, at Columbia University with Henning Schulzrinne, and on sabbatical at ETH Zurich with Adrian Perrig. His PhD is with Sasu Tarkoma (Helsinki) and Jon Crowcroft (Cambridge). Funded by the Nokia Foundation, vital part of his PhD programme is completed at the University of Cambridge and Columbia University in New York. He has supervised 100+ students and won EU research grants as Consortium Director and Principal Investigator (PI) in tight collaboration with industrial companies of Nokia, Ericsson, Deutsche Telekom, Broadcom, Telia, NEC, Telefonica and WithSecure. His research focuses on edge computing, edge AI, and data-driven IoT services. For contributing to mobile edge computing with over 90 peer-reviewed publications, he received best paper awards and recognition from ACM SIGCOMM, ACM EdgeSys, ACM SenSys CCIoT, ACM IoT, IEEE INFOCOM, IEEE CVPR MAT, and TU Delft. Being an active member of ACM, IEEE, IETF, he is the founder of ACM EdgeSys, Associate Editor for ACM TIOT, IEEE OJ-ITS and Springer Nature Computing. He has served on the chairing and programme committees for numerous prestigious conferences including ACM SIGCOMM, ACM MobiCom, ACM UbiComp, ACM WWW, ACM CoNEXT, ACM MobiSys, ACM/IEEE SEC, and IEEE INFOCOM. He is a recipient of the esteemed Nokia Foundation Scholarship. More info: https://aaronyiding.github.io.

Chamitha De Alwis, a Senior Member of the IEEE, currently holds the position of Senior Lecturer in the School of Computer Science and Technology at the University of Bedfordshire, United Kingdom. His academic journey includes the completion of a PhD in Electronic Engineering from the University of Surrey, United Kingdom, in 2014, as well as a distinguished BSc with First Class Honors in Electronic and Telecommunication Engineering from the University of Moratuwa, Sri Lanka, in 2009. Chamitha is recognized for his leadership and academic contributions as the Founder Head of the Department of Electrical and Electronic Engineering at the University of Sri Jayewardenepura, Sri Lanka, where he concurrently served as a Senior Lecturer. Beyond academia, he has demonstrated his expertise in roles as a consultant and network engineer in the dynamic field of telecommunications. Maintaining an active research profile, he boasts a commendable track record of scholarly publications and has been the recipient of several competitive grants. Notably, he has played a pivotal role in numerous research projects, including those funded by the EU FP7. His research pursuits are centered around critical areas of inquiry, with a primary focus on network security, the evolution of 5G/6G technologies, and the transformative landscape of blockchain.

Madhusanka Liyanage an Assistant Professor/Ad Astra Fellow and Director of Graduate Research at the School of Computer Science, University College Dublin, Ireland. He is also a Adjunct Professor at the University of Oulu, Finland, University of Ruhuna, Sri Lanka, and University of Sri Jarawardhenepura, Sri Lanka. He received his Doctor of Technology degree in communication engineering from the University of Oulu, Finland, in 2016. He also received the prestigious Marie Skłodowska-Curie Actions Individual Fellowship and the Government of Ireland Postdoctoral Fellowship during 2018–2020. During 2015–2018, he has been a Visiting Research Fellow at the CSIRO, Australia, the Infolabs21, Lancaster University, UK, Computer Science and Engineering, The University of New South Wales, Australia, School of IT, University of Sydney, Australia, LIP6, Sorbonne University, France, and Computer Science and Engineering, The University of Oxford, UK. He is also a senior member of IEEE. In 2020, he received the "2020 IEEE ComSoc Outstanding Young Researcher" award by IEEE ComSoc EMEA. In 2021 and 2022, he was ranked among the World's Top 2% Scientists (2020 and 2021) in the List prepared by Elsevier BV, Stanford University, USA. Also, he was awarded an Irish Research Council (IRC) Research Ally Prize as part of the IRC Researcher of the Year 2021 awards for his positive impact as a supervisor. In 2022, he received "2022 The Tom Brazil Excellence in Research Award" by SFI CONNECT Center. Moreover, Madhusanka received a Special Commendation for IRC Early Career Researcher of 2022 by Irish Research Council, Ireland. He is also an expert consultant at European Union Agency for Cybersecurity (ENISA). In 2021, Liyanage was elevated as a Funded Investigator of the Science Foundation Ireland CONNECT Research Centre, Ireland. Moreover, he is an expert reviewer at different funding agencies in France, Qatar, UAE, Sri Lanka, and Kazakhstan. His research interest includes 5G/6G, Blockchain, Network security, Artificial Intelligence (AI), Explainable AI, Federated Learning (FL), Network Slicing, Internet of Things (IoT), Multi-access Edge Computing (MEC) and Cyber-Physical Systems (CPS). More info: www.madhusanka.com.

Preface

This book entails a special journey that is accomplished throughout the pandemics of 2020–2023. As the adoption of information and communication technologies has been accelerated by the growing trend of remote working and learning, the book is developed in constant reflections on the opportunities and challenges behind this rapid transition.

Recently, we have witnessed a remarkable proliferation of new applications and services in the context of Internet-of-Things (IoT) and mobile computing. With the mission of being smart, the various concepts under the umbrella of IoT, including smart factory, smart cities, and autonomous driving, have introduced new ways for people to interact with the physical world. At the same time, mobile computing has evolved to new dimensions with more advanced paradigms such as virtual reality/augmented reality (VR/AR). On the one hand, these applications and services are expected to contribute significantly to the economy and human life quality. On the other hand, a lot of new challenges have been imposed or magnified, among which real-time big data processing and analysis become prominent. Despite the fact that mobile devices are becoming more and more powerful nowadays, handling the constantly generated data flows in real time on those devices for providing application intelligence is still undesirable or non-optimal, one of the reasons being the limited battery lifetime of the mobile device.

With the help of cloud computing, Mobile Cloud Computing (MCC) targets the processing capability gap. Unfortunately, the uncertainty of the network connection to the remote cloud renders most of the MCC solutions impractical, especially for the aforementioned applications. Meanwhile, wireless communication has advanced to a large extent in the past decade. Under the 5G regime, various technologies have been developed that can achieve properties of ultra-high bandwidth, ultra-low latency, and high reliability. Benefiting from these new communication technologies, MCC seems to become applicable but the reality is crucial as the Internet—being part of the remote connection, comes with no such nice properties of 5G, which does not seem to change soon. In the end, only the last hop connectivity can be expected to embrace the new opportunities brought by the 5G adoption.

The situation described above naturally leads to the advent of edge computing—a technology concept that aims to provide intelligence from the edge of the networks. In this book, we deliberately use the term **MEC^2** in a more general manner for the essential technologies of mobile edge computing and communications. The idea of *MEC^2* is to generalize the traditional cloud computing into more distributed environments with close proximity (even in a single hop) to the mobile and IoT devices. Thanks to the 5G advancement, the computing and storage resources at the network edge now can facilitate the real-time data processing for emerging applications with performance guarantees. Note that such computing/storage resources are already partially available on the infrastructure as a result of the recent network campaign called Network Function Virtualization (NFV), where network functions are virtualized and run on general-purpose servers deployed within the edge network. In certain sense, edge computing can be regarded as an extension of NFV from the core network to mobile and IoT applications.

Many enabling technologies, such as Software Defined Networking (SDN), network slicing, Information Centric Networking (ICN), and their synergies have also been extensively explored by both the academia and industry. From the computing perspective, new distributed computing infrastructures that feature locality-awareness have been proposed, adopting new programming/management concepts such as microservice or serverless computing. In addition, multiple communication technologies (e.g., those under 5G, WiFi, VLC, and LoRa) have also been examined for their potentials of supporting different applications in MEC^2. The integration of both computing and communication parts is complex and becomes unprecedentedly challenging, despite that the aforementioned enabling technologies can provide significant simplification to system management and control. With all the research and development efforts, MEC^2 is expected to tremendously benefit a large variety of emerging mobile applications and services and will enable a completely new paradigm for future computing and communications.

Following the Top-Down approach to start from the "layer" of applications and dive further into the underlying technologies and integration, this book is divided into four parts:

- **Part I:** The first part of the book introduces emerging "edge native" applications for readers to comprehend the practical usage of edge computing.
- **Part II:** The second part dives deeper into the technology foundation that supports the edge applications, including system frameworks, enabling technologies and edge hardware.
- **Part III:** The third part covers the technology driving forces and the integration with cloud and mobile ecosystems. Security and privacy aspects of MEC^2 are discussed.
- **Part IV:** The final part illustrates open challenges and future directions for MEC^2.

Instead of being a competing paradigm, the future success of MEC^2 is bundled with the cloud and mobile ecosystems by benefiting from the mature cloud and mobile access infrastructures. In a nutshell, MEC^2 is an exciting domain that is growing fast and fostering numerous opportunities. We wish you enjoy this book and have fun in reading it.

Delft, 2024

Aaron Yi Ding
Chamitha De Alwis
Madhusanka Liyanage

To Our Readers

This book covers mobile edge computing and communications, an emerging computing-communication paradigm that augments today's centralized cloud by distributing processing and storage capabilities into mobile/access networks, close to where end users connect. Drawing influence from cloud and cellular domains, this computing-communication paradigm is a fast growing phase. Standardization efforts have been made by the European Telecommunications Standards Institute (ETSI) Industry Specification Group (ISG) in terms of mobile edge computing/multi-access edge computing.

In this book, we deliberately use the term $\mathbf{MEC^2}$ in a more generic manner to address the essential technologies for mobile edge computing and communications. This book provides a walk-through of applications, technology foundation and driving forces for MEC^2.

We take a "Top-Down" approach in this book. As a reader, you will first learn the emerging "edge native" applications to comprehend the practical usage of MEC^2 (Part I). Then, you can dive deeper into the technology foundation (Part II), the integration and driving forces (Part III). Finally, the book concludes with open challenges and outlook on future computing directions (Part IV).

The content of this book appeals to a wide range of audience in the ICT domain, including:

- **ICT and IoT Service Developers:** As computing requirements and user demands can outpace the traditional tools available, this book provides practical information on the frameworks and technical enablers of MEC^2, which can facilitate developers to create innovative ICT applications catering to future demands.
- **Mobile Network Operators (MNOs):** MNOs are currently looking to adopt new technologies such as edge computing to offer responsive services to their customers. This book offers the guidelines, methods, tools and mechanisms that can benefit their development and deployment in the next generation mobile infrastructure, i.e., 5G and beyond.
- **Academic Researchers:** Mobile edge computing has already been an area of active research and study for major educational institutions across the world. This book meets the growing demand from academics who can benefit from the latest literature reference and use cases, especially in their daily teaching and research on this critical domain of interest.
- **Computing Professionals and Industrial Experts:** Edge computing is going to cross the traditional borders and is going to have visible impact to enterprises and organizations who are planning to transform into IoT-driven business. It would be critical for

computing professionals and industrial experts such as IT architects to start aligning their technology adoption and architectures to the future needs of computing and standards. This books offers resources to design and build an edge-driven architecture and maintain it.

- **Advanced Undergraduate and Graduate Students:** For educational institutes offering computing and networking lectures, this book provides an essential source of up-to-date information for fundamental technologies in MEC^2. The use cases and practical examples shall motivate curious students to explore real-world problems in operational environments. The book is hence appropriate for advanced undergraduate graduate level courses.

Through the book, readers can get familiar with the mainstream system architectures and integration methods that have been proposed in recent years. The book also discusses the open challenges and outlook from design, deployment, and operational aspects. Given the rapid pace of edge technology advancement, this book is not intended to be encyclopedic and likely to be an evolving "piece" for the foreseeable future. Overall, this book shall offer insights for active actors and stakeholders of mobile networking and cloud computing. By serving as a reference material and introductory guide on mobile edge computing and communications, this book shall appeal to a wide intended audience including academic researchers, service developers, computing professionals, curious university students, network operators and industrial experts.

Acknowledgments

To my wife Mia for her trust, encouragement and firm supports during those tough years. I express my sincere gratitude for the understanding and support from Sandra at Wiley. I appreciate the initial discussions with Lin Wang and Dirk Kutscher at the start of the journey for preparing this book.

Aaron Yi Ding

This book would not have been possible without the great help and supports of many. The concept of publishing this book on mobile edge computing and communications came into light during our research work in projects of EU H2020 SPATIAL project (Grant No. 101021808) and EU Marie Curie ITN APROPOS project (Grant No. 956090).

We would also like to acknowledge all the partners of those projects. Furthermore, we would like to thank our universities, Delft University of Technology, University of Bedfordshire and University College Dublin, for all the support extended towards the successful completion of this book. We would also like to thank chapter contributors, including Dewant Katare, Pasika Ranaweera, Yushan Siriwardhane, Anshuman Kalla for their invaluable contribution to complete this book. We also thank all the reviewers for helping us select suitable chapters for this book.

We appreciate the efforts of Professor for writing a nice foreword to the book. We are also grateful to Sandra Grayson, Teresa Netzler, Nandhini Karuppiah, and the whole John Wiley & Sons team for their support towards getting this book published.

Last but not least, we would offer our heartiest gratitude to our families, who gladly allowed us to share our time with them towards the completion of this book.

Aaron Yi Ding
Chamitha De Alwis
Madhusanka Liyanage

Acronyms

MEC^2	mobile edge computing and communications
BHCA	busy hour call attempts
BR	bandwidth reservation
b.u.	bandwidth unit(s)
CDN	content delivery network (CDN)
CDTM	connection dependent threshold model
CS	complete sharing
DiffServ	differentiated services
ETSI	European Telecommunications Standards Institute
FIFO	first in-first out
GoS	Grade of Service
ICT	Information and Communication Technology
IoT	Internet of Things
IP	Internet protocol
ITU-T	International Telecommunication Unit-Standardization sector
MMPP	Markov modulated poisson process
MPLS	multiple protocol labeling switching
PASTA	Poisson arrivals see time averages
PDF	probability distribution function
PFS	product form solution
QoS	Quality of Service
RED	random early detection
RLA	reduced load approximation
TCP	transport control protocol
UDP	user datagram protocol

Part I

Introduction and Edge Applications

'What exactly is an Edge App?' – this is probably one of the most common questions when you start the journey on edge computing and communication. Before we dive into the fundamental technologies, this part of the book will illustrate the focus of the book 1 and introduce the applications of MEC^2 for readers to comprehend the practical usage, including Multi-Access Edge Computing in Chapter 2, edge video analytics and augmented reality in Chapter 3, and smart healthcare in Chapter 4.

Mobile Edge Computing and Communications: Driving Forces, Technology Foundation, and Application Areas, First Edition. Aaron Yi Ding, Chamitha De Alwis, and Madhusanka Liyanage.
© 2025 John Wiley & Sons Ltd. Published 2025 by John Wiley & Sons Ltd.

1

Introduction

1.1 Focus of the Book

Besides the latest research on computing design and standardization efforts made by the European Telecommunications Standards Institute (ETSI) Industry Specification Group (ISG), the driving forces behind edge computing and its technology foundations deserve a thorough illustration and analysis. Along with the advancement of Internet-of-Things (IoT) and mobile communications, the rapid development on "Edge" is calling for a handbook that covers the service/application perspectives and how to integrate "Edge" in the future computing infrastructures.

Given the "Edge" domain is in a fast growing phase, in this book, we deliberately use the term **MEC^2** in a more generic manner to enclose the essential technologies for mobile edge computing and communications.

One of the key reasons for a seemingly old idea to take off after many decades is that other enabling technologies make it possible. This holds especially for MEC^2. For the rapid advancement of MEC^2, this book is not intended to be encyclopedic and likely to be an evolving "piece" for the foreseeable future. In this respect, readers' suggestions, feedback are more than welcome. By serving as a reference material and introductory guide on mobile edge computing and communications, this book shall appeal to a wide intended audience including academic researchers, service developers, computing professionals, curious university students, network operators and industrial experts.

In this book, we take the system and incremental deployment perspective, which is not to be mixed with incremental research. The MEC^2 research can be disruptive or ground-breaking (i.e., non-incremental) but still lend itself to incremental deployment and integration to existing mobile and cloud infrastructures. The research can hence go beyond the constraints of existing deployment context by envisioning what could be possible. Ultimately, it comes down to the deployment of MEC^2 that leads to long-term impact. Overall, this book shall offer practical deployment insights for actors and stakeholders in the future mobile and cloud infrastructures.

Mobile Edge Computing and Communications: Driving Forces, Technology Foundation, and Application Areas, First Edition. Aaron Yi Ding, Chamitha De Alwis, and Madhusanka Liyanage.
© 2025 John Wiley & Sons Ltd. Published 2025 by John Wiley & Sons Ltd.

1.2 The Vision of Edge

The vision for edge computing is brought out as early as 2005 [1] where the importance of network edge is highlighted as a new source of creative energy for system and applications, especially given the growing demand in mobile access networks. The initial concept is further formulated in recent years, along with a visible trend for data analytics moving toward this "edge." Gartner predicted that by 2023, over 50% of the primary responsibility of data analytics stakeholders will comprise data that is created, managed and analyzed in edge environments. The advantages may include greater data management flexibility, speed, governance, and resilience. In addition, the edge capabilities can support use cases ranging from real-time event analytics to autonomous driving services [2].

Albeit its great potential, the fundamental concept of edge computing, referred as MEC^2 in this book, is not new. By looking at the history of computing, it is not the first time for the network edge to receive tremendous attention. For instance, we have seen numerous computing design before the hype of MEC^2, including the peer-to-peer (P2P) networks, content delivery networks (CDN), and mobile cloud computing (MCC). It is hence appropriate to remind our readers that one of the key reasons for a seemingly "old" idea, such as edge computing, to take off after many years of undergoing is that other enabling technologies are making it possible. Those enablers and technology foundation are exactly what we intend to illustrate throughout this book.

In fact, the vision is largely built upon its promises to the rising demands for high bandwidth, low latency, better privacy and reliability. Those promises have been argued recently by experts as to whether it is merely an academic enthusiastic pursuit [3]. At the same time, as the Internet is transformed into a more ethical system, MEC^2 is gradually showing its potential in enhancing data privacy and sovereignty [4].

Regardless of the controversial opinions [3, 5, 6], MEC^2 is clearly far from reaching its "Hay Day." Many uncertainties and challenges are intertwined with the opportunities along its advancement. From a pragmatic perspective, this book conveys our vision to facilitate future development of MEC^2 towards a coherent edge-to-cloud continuum. This continuum path needs more collaborative efforts with experts from multiple disciplines even outside the technical sphere (e.g., law, ethics and public policies), for achieving the full potential of MEC^2.

Why Do We Need Edge?

Regarding a frequently asked question "Why do we need edge?," the general motives and potential benefits include:

- **Latency and Bandwidth:** MEC^2 brings the latency benefit by placing computation units closer to the data sources and end users. Given the bandwidth bottleneck to access a distant cloud via the Internet, MEC^2 can conserve bandwidth by processing data locally, hence avoiding transferring excessive amount of data to the Internet. This helps alleviate network congestion. The latency and bandwidth benefits are crucial for real-time services that also demand higher bandwidth, such as virtual reality (VR) gaming and autonomous driving.
- **Proliferation and Personalisation:** MEC^2 is envisioned to capture high quality data in a distributed manner from the massive deployment of IoT, augmented reality (AR)/VR

devices, and smart vehicles. *MEC*^{2} can achieve higher quality by filtering out the data "noise," labeling the data with more context and with better sampling. The high-quality data with locality context is demanded by analytic services for personalisation (e.g., for end users/clients).

- **Privacy and Sovereignty:** The edge-native computing and analytics (often referred as Edge AI) can keep the data ownership and control closer to the end users. This can when computation is managed and task distribution controlled from the user's own devices, and suitable security and privacy protection methods are in use. By utilizing local context, *MEC*^{2} can strike a balance between privacy and usability, while allowing ethical data management.
- **Energy Saving and Sustainability:** *MEC*^{2} avoids transmitting redundant data traffic to consume network resources. The energy required for transmitting data is proportional to the distance that data travels from the source (end user devices) to the Internet or centralized cloud sites. By cutting down the data volume to be transmitted and processed by the network units along the path, *MEC*^{2} can promote the energy saving of the Internet. In addition, the data flowing through the Internet is becoming one the primary drivers for CO_2 emissions. *MEC*^{2} can help reduce the carbon footprint.
- **Mobility and Collaboration:** *MEC*^{2} can support end user devices continue functioning even when they are disconnected from the cloud. This is particularly valuable for mobile scenarios where network connectivity is intermittent or unreliable. In addition, *MEC*^{2} is complementing the corporate data centers that facilitate computing, storage, networking, and data analytic functions at locations such as collaboratively training large-scale machine learning models.

What and Where is the Edge?

To approach the question **"What is the Edge?,"** it can often trigger heated discussions. Can it be a lightweight computing server, a layer of networking devices, or a set of IoT oriented services? For some of you who are studying *MEC*^{2} for the first time, this can be rather confusing.

One major reason for such blurring interpretation is the fact that *MEC*^{2} is in rapid development. The concept of "edge" is hence enriched and evolved along the way, sometimes mixed with different flavours. For example, a general expectation for edge is to extend the cloud computing capability to the access networks. For this prospect, *MEC*^{2} can take the form of lightweight computing servers that can offload the tasks from data centers (cloud) for performance considerations. Meanwhile, for mobile network operators that are enhancing their access infrastructure, *MEC*^{2} can also be regarded as a layer of network devices (e.g., routers) in the proximity to mobile devices and data sources. In addition, there is an increasing pressure to offer low-latency and high-bandwidth applications such as augmented reality (AR), virtual reality (VR), Internet-of-Things (IoT), and autonomous driving. From the service offering angle, *MEC*^{2} can be regarded as a "meta" service that enables numerous advanced applications demanded by the growing amount of connected things.

Closely related to the "What" question, **"Where is the Edge?"** also depends on which stakeholder we are asking. For an Internet service provider (ISP) that owns and operates the network access infrastructure, *MEC*^{2} can be a layer that consists of last hop gateways to connect end users to ISP's infrastructure. For instance, mobile operators can utilize *MEC*^{2}

to construct a computing layer on their base stations, so the MEC^2 enabled base station is regarded as where the edge is "physically" located. For cloud service providers such as Amazon and Google, MEC^2 is the extended computing and storage infrastructure closer to their customers for supporting real-time services, analytics, and content delivery. For instance, Content Delivery Networks (CDNs) can be regarded as where the edge is placed, for reducing the distance and latency of content delivery. In the future, MEC^2 may come in different form factors for different scenarios. The devices/entities with computational capability to perform computing near the data source could be considered as part of edge infrastructure This is in line with the motive of edge computing.

1.3 Stakeholders and Related Paradigms

Beyond the "What" and "Where" questions, it is important to understand the major stakeholders and related paradigms of MEC^2. The major stakeholders and actors involved in MEC^2 include:

- **Mobile Network Operators (MNOs):** MNOs are currently looking to adopt edge computing to offer responsive mobile services to their customers. For telecom's next generation mobile infrastructure, MNOs can harness the advantage of edge to create an integrated computing-communication hub in the "last-mile" network for efficient data processing, analytics, and communications.
- **Cloud Service Providers:** To meet the demands from a wide-range of latency-sensitive and data-intensive applications that will emerge in the near future, cloud service providers are adopting edge designs to enhance their service performance. This is in line with the new industry trend of computing continuum, which could lead to the convergence of cloud and edge.
- **Academic Researchers and Industrial Experts:** MEC^2 has created active research and practical solutions for academic researchers and industrial experts, respectively, to tackle challenges in computing sustainability and data sovereignty.
- **ICT Service/Application Developers:** As computing requirements and user demands can outpace the traditional tools, ICT service developers are constantly seeking new technologies such as MEC^2 that can facilitate them to create innovative ICT applications that cater to future demands from users, government regulations, and international standards.

Since MEC^2 is in its making phase, there are related computing designs that have been explored and inspired the trend of edge. The related computing concepts include P2P, CDN, MCC, Cloudlets, Fog and Mist:

- **P2P**: The concept of Peer-to-Peer (P2P) is driven by decentralization. It is intended to address the limitations of classic client–server design in terms of performance bottleneck and single point of failure on the server side. As a decentralized networking design, P2P can enable direct communication between distributed devices for better scalability, reliability and resource utilization. P2P has provided inspiration to MEC^2 from decentralization perspective.

- **CDN**: To optimize the latency in content-oriented ICT/web services, Content Delivery Network (CDN) forms a distributed proxy layer that can cache content and serve such content via CDN servers geographically closest to the end users. The latency benefit of CDN has inspired MEC^2 to achieve low latency, load balancing and service availability in a distributed environment.
- **MCC**: The design of Mobile Cloud Computing (MCC) integrates cloud computing with mobile computing to support efficient computational offloading. By harnessing the computing and storage capacity in the remote cloud, MCC can enable devices with limited processing power and storage to run resource-intensive applications. MCC has inspired MEC^2 to enhance end user experience in terms of resource offloading and battery saving on the end devices.
- **Cloudlets, Fog and Mist**: The idea of Cloudlets [7] is to deploy small-scale data processing units close to the edge of access network. Cloudlets aims to tackle latency and bandwidth challenges for applications that often need to exchange data with remote cloud servers. In line with the goal of improving latency and bandwidth, Fog computing [8] aims to transform the network edge to a distributed computing infrastructure for the rapid growth of IoT. Similar to Fog computing, the term Mist is geared towards embedded computing devices deployed in the access networks. In general, Cloudlets, Fog and Mist intend to extend the Cloud computing paradigm by allowing data to be processed and analyzed across various levels of the network hierarchy. The concepts of Cloudlets, Fog and Mist formed the groundwork for computing continuum that inspired MEC^2 to complement both centralized cloud and distributed end devices by offering an intermediate layer of edge resources to balance the benefits of local processing with the advantages of cloud services.

In the following chapters of Part 1 we will illustrate the applications of MEC^2 to help readers comprehend the practical usage of edge technologies.

2

Case for Multi-access Edge Computing*

2.1 Case for Multi-access Edge Computing

Multi-access Edge Computing (MEC) is a pioneer edge computing paradigm introduced by the European Telecommunications Standards Institute (ETSI) to overcome the intricacies of impending mobile communication networks. As the name derives, the focus of the MEC is to cater multi or diversified access to the existing mobile network for expanding its capability to meet the requirements of novel Internet of Things (IoT) based services. IoT, being the pinnacle of the Information Technology (IT) and electronics industries in the recent past, has raised the guaranteed intrinsic specifications of mobile and Internet technologies drastically. To have the "Things" such as physical devices/apparatus/components, sensors, actuators, or embedded software systems communicated through the Internet for IoT services, MEC provides the means to solve their scalability, interoperability, and compatibility challenges.

2.1.1 What Is MEC?

The main goal of MEC is to extend the Cloud Computing (CC) capabilities to the edge of the mobile network to overcome the constraints of prevailing cloud services [9]. Hitherto, CC was the solution that unburdened the service providers from the inconvenience of maintaining their own data centers. MEC creates a server/computing environment in mobile Base Stations (BSs) or 5G based gNodeBs (gNBs) integrated to the mobile network infrastructure [10]. As Figure 2.1 illustrates, MEC is complementing the corporate data centers that facilitate computing, storage, networking, and data analytic functions at locations in proximity to the data source [11]. In fact, these data centers localized in the mobile edge network are supported by the cloud servers, and 5G core network entities located at the mobile core network to cater enhanced service access to the radio access network (RAN).

2.1.2 Why We Need MEC?

Current cloud-based services require connectivity to the centralized cloud platforms to convey control information and authentication credentials for authorization mechanisms.

*This chapter is partly contributed by Pasika Ranaweera (University College Dublin, Ireland).

Mobile Edge Computing and Communications: Driving Forces, Technology Foundation, and Application Areas, First Edition. Aaron Yi Ding, Chamitha De Alwis, and Madhusanka Liyanage.
© 2025 John Wiley & Sons Ltd. Published 2025 by John Wiley & Sons Ltd.

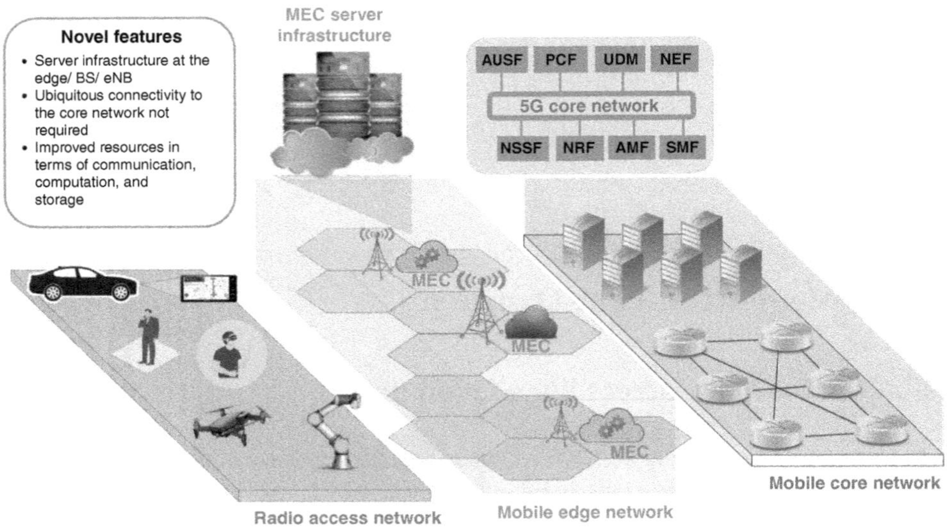

Figure 2.1 What Is MEC.

This connection, which is generally linked through the Internet with featured higher bandwidth consumption and embedded cumbersome cryptographic primitives; restricts the real-time service guarantees of forthcoming mobile technologies. Figure 2.2 illustrates the significance of the MEC paradigm concerning such impediments. Thus, the resource-enriched edge servers of MEC systems are promising real-time service guarantees for novel applications. Since the intrinsic processing and storage functions can be provided with the MEC edge platforms, cloud connectivity requirement becomes diminutive. It is obvious that Mobile Network Operators (MNOs) should invest more in forming the MEC edge server environment. The communication capacity of the MEC-enabled BSs should also be improved with the investment into the edge server environment. Thus, despite the investment put forward by the MNOs, the number of BSs can be reduced due to the improved communication range that serves better scalability [12].

The forthcoming fifth-generation (5G) mobile technology and its requirements serve as one of the rationales for the emergence of MEC. The performance metrics of data rates up to 10 Gb/s, service level latency below 1 ms, ultra-high reliability of 99.99999%, reduced energy consumption of 90%, and support for 300,000 devices within a single cell are guaranteed for 5G [9, 13]. To meet these requirements, service infrastructure (currently operating at clouds) should be migrated to a proximate location. Thus, the MEC paradigm is formed and designed with the above intentions.

As Figure 2.2 illustrates, storage and processing infrastructure facilitated with the networking platform in MEC deployment is ensuring the benefits of ultra-low latency, locational awareness, proximate data outsourcing, and improved capacity in the edge devices. These features enable higher bandwidth and real-time responsiveness to the subscriber applications. These features enable the launching of novel 5 G-based applications or use cases such as Ultra High Definition (UHD) video streaming, Augmented Reality (AR), Virtual Reality (VR), Mixed Reality (MR), Tactile Internet, Machine Type

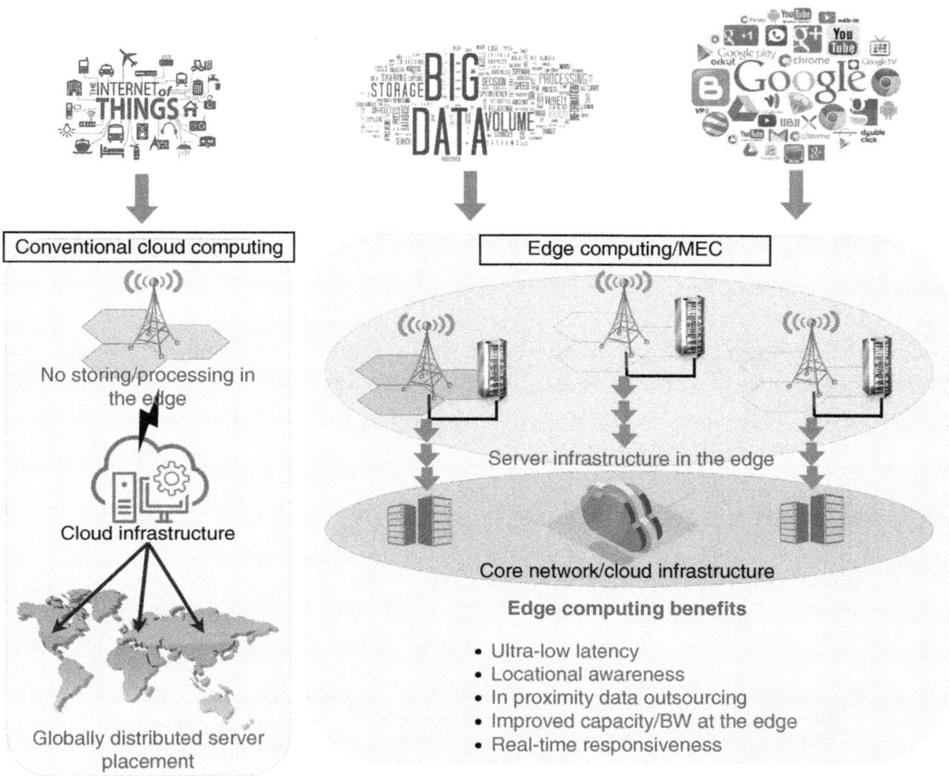

Figure 2.2 Benefits of MEC Compared to Cloud Computing. Source: Yeamake/Adobe Stock.

Communication (MTC), Machine-to-Machine (M2M), Industrial IoT (IIoT), Unmanned Aerial Vehicle (UAV), and Vehicle-to-Everything (V2E). These terms will be introduced later in the chapter.

2.1.3 Evolution of MEC

The data storage and processing services have evolved from mainframes (BITNET-1981) to dedicated servers, and to cloud computing for improving the convenience of data center maintenance. The growth in demand for Big data and data science over the years leading to IoT emergence has prompted the Internet Service Providers (ISPs) to invest more in their data centers; though faced issues in erecting dedicated vicinities for the purpose while expanding the capabilities of the network technologies to improve the network utilization. With cloud computing, the subscriber is facilitated with Infrastructure as a Service (IaaS), Platforms as a Service (PaaS), and Software as a Service (SaaS) directives by the Cloud Service Provider (CSP). Infact, the subscriber is free of data center maintenance and management concerns. Though, centralized and geographically dispersed placement of cloud infrastructure is raising concerns for the IoT paradigm and 5 G-related novel applications as specified in Section 2.1.2. Thus, MEC or any other edge computing

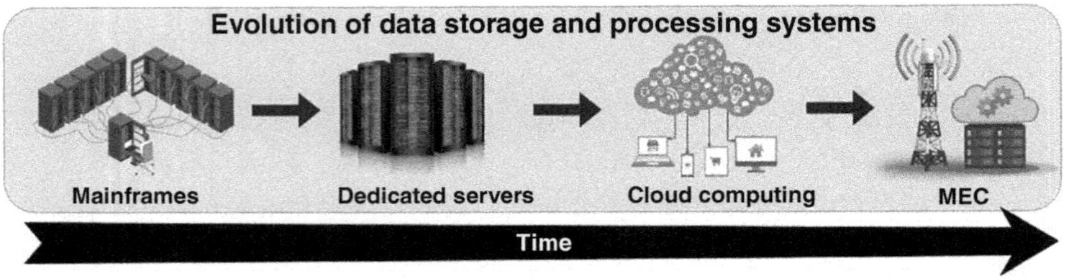

Figure 2.3 Evolution of MEC.

paradigm is envisaged to overcome these issues of cloud computing by shifting the data center infrastructure to a proximate location. Figure 2.3 depicts the discussed evolution of MEC in an illustrative context.

2.2 MEC Reference Architecture

ETSI has defined the reference architecture for MEC, which is illustrated in Figure 2.4 [14]. This architecture is formed with the idea of leveraging the deployed MEC infrastructure to

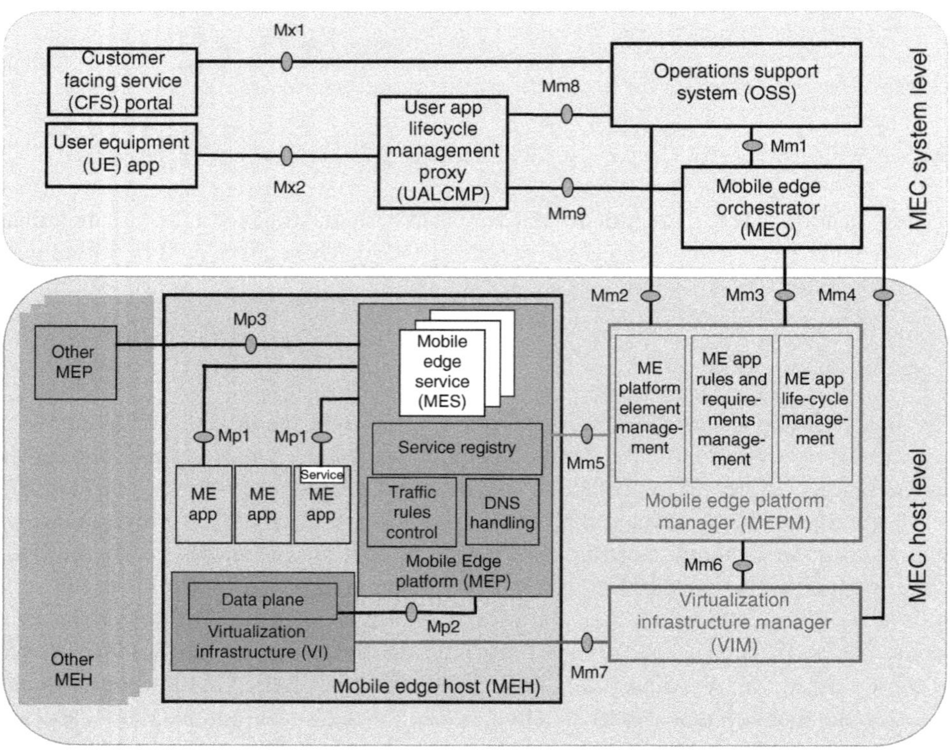

Figure 2.4 MEC Reference Architecture.

launch diverse services in an autonomous manner. The aspects of subscriber access through the mobile network, user access management, virtualization technologies for launching service instances, proper monitoring and isolation of virtual entities for convenient service handling and delivery, core and edge network segregation for better integration with prevailing mobile network layout, and orchestration of the entire MEC serviceable platform are considered in formulating this architecture. Though there is an updated version of the MEC architecture being published by pioneer research groups/institutions on integrating Network Function Virtualization (NFV) and 5G use cases, the version in Figure 2.4 depicts the fundamental formation of a MEC deployment. Conversely, NFV is a concept introduced for decoupling network functions and services from the hardware apparatus for moving them to the softwarized domain; where softwarized instances called Virtualized Network Functions (VNFs) are acting as programmable network functions that can execute independent of the hardware dependencies [13]. MEC reference architecture mainly has two levels: Mobile Edge System Level and Mobile Edge Host (MEH) Level. This segregation is allowing the MEC service providers to link the MEC system-level entities with the 5G core network entities to merge the two networks as a singular one.

2.2.1 Mobile Edge System Level

The MEC system level is the governing domain of the MEC system, where several MEC edge levels can be administered through a single system level. However, the governing capability of the MEC system level will depend on its resources and the scope and nature of the MNO domain. The 5G core network entities are tethered to the system level of the MEC (i.e., through mobile edge orchestrator [MEO]) for traffic steering and mobility management purposes [15].

2.2.1.1 User Equipment (UE) and Applications

User equipments (UEs) are any kind of device that provides an interface for MEC subscribers to interact with the MEC service framework. These UEs can be varied from a wearable IoT apparatus to UAVs, to an autonomous vehicle that might operate with human interaction or vice versa. The significance of the UEs are that they should be compatible with the communication access medium of the MEC system, either via 3GPP or non-3GPP means. 3GPP extends from Global System for Mobile Communication (GSM) or 2G to 5G New Radio (NR) Radio Access Technologies (RATs); where non-3GPP technologies are Worldwide Interoperability for Microwave Access (WiMAX), Code Division Multiple Access (CDMA), Wireless Local Area Networks (WLANs), or fixed networks such as Public Switched Telephone Networks (PSTNs). We can visualize that the UE Applications (UE Apps) running on the UE, interacts with the Mobile Edge Applications (ME Apps) launched at the MEC edge level. In fact, UE Apps are mapped with the ME Apps for entities operating at the access network and the mobile edge network, respectively. However, UE Apps are only capable of interacting with the mobile edge system through a user application life-cycle management proxy.

2.2.1.2 Customer Facing Service Portal (CFSP)

Customer Facing Services (CFSs) are a type of service category where the service and its specifications are exposed to the customer for perceiving a commercial view in an e-trading domain [16]. In contrast, Resource Facing Services (RFSs) are invincible to the customers because they act as internal and indirect processes even to the service providers. In the context of MEC and 5G, high-end service providers (i.e., automobile companies: autonomous vehicles, gaming companies: AR-based games, online trading and delivery companies: UAV based delivering) might leverage the resources in the MEC environment for launching their services. In such a scenario, the service is not offered by the MEC service provider, but by a third-party institution. Thus, for maintaining transparency in this third-party trading domain, exposing the service specifications and consensus regarding the service conduct (i.e., Service Level Agreements—SLAs) is a profound necessity from the customers' point of view. Hence, such services can be categorized as CFSs. In fact, customer facing service portal (CFSP) is an entity or an interface operated by the MEC operator for facilitating or outsourcing the MEC platform for third-party services, so that customers can directly negotiate the service delivery with the third-party service provider. CFSP is capable of recalling SLA information from the provisioned applications.

2.2.1.3 Operation Support System (OSS)

The operation support system (OSS) is the main access registry of the holistic MEC system. The service requests forwarded via the CFSP or UE application life-cycle management proxy are evaluated for approval of access. Generally, OSS entities are responsible for monitoring, controlling, analyzing, and managing network-type services. In the MEC context, OSS decides whether the user requests are granted access or not. The granted requests are directed to the MEC orchestrator for further processing. Though, OSS decides the instigation and termination states of any ME App registered under it. In addition, OSS is tasked with service or application discovery and relocation functions among the MEC system or any external clouds. Thus, this entity has a record of all the MEC applications registered under the MEC operator in addition to their statistics, while UE App to ME App mappings are envisaged to be conducted here. In fact, SLA negotiations can take place during the OSS process phases where a global perception of the SLA specifications, standardization (i.e., in terms of Quality of Service—QoS, or Quality of Experience—QoE), and measurement can be embedded into the OSS constructs due to its access to both UE App and ME App perspectives.

2.2.1.4 User Application Lifecycle Management Proxy (UALCMP)

This proxy is only accessible within the mobile network. The UE App requests are directed to this entity. The handling of multiple UEs for their application requests are pursued by this component. As the OSS is the access authority from the MEC perspective, user application lifecycle management proxy (UALCMP) is the access portal of the access network. In fact, UALCMP decides the instigation and termination states (i.e., life-cycle) of the UE Apps. Since it is a proxy, anonymization and user ID handling are probable functions. Further, authentication and authorization handling of the MEC system is pursued by this entity, which makes this ingress point a prominent location for cyber intruders.

2.2.1.5 Mobile Edge Orchestrator (MEO)

This component carries out the core functionality of the MEC architecture. Orchestration is the cardinal function of any virtualized deployment. It is in fact the function of automated configuration, coordination, and management of the entire softwarized system. Orchestration in cloud computing does not require a higher level of automation due to its centralized administrator involvement from the cloud service providers. On the contrary, automation is an utmost requirement for the edge computing paradigms, which eases the burden of the MEC operator. The emphasis on automation improves the responsiveness of the edge paradigms to meet the requirements of the emerging 5G and IoT applications. In the context of MEC, the orchestrator is responsible for delivering automated service provisioning, service mobility handling, interoperability among converged networks, and service optimization [13]. The granted requests from the OSS are forwarded for processing in the MEO. MEO maintains a holistic view of the MEC system focusing on: deployed MEC hosts at each edge level under its governance, access capacity of the system level, edge-level resource utilization, network utilization, available MEC services and applications, protocols, and topology. Being the main orchestrator, MEO performs the monitoring functions and constructs leading to system updates. MEO is responsible for the below functionalities;

- **Monitoring and Verifying**: Governing the virtualization infrastructure manager for application handling while checking the integrity, authenticity of the packages and validating application rules and requirements.
- **Service Mapping**: Selecting the appropriate MEHs for requested services considering service level requirements of latency, resources, and access capabilities.
- **Service Handling**: Triggering instantiation and termination of the applications.
- Application relocation.

2.2.2 Mobile Edge Host Level

The host level or the Edge level of the MEC system is the action domain of the MEC system. This dynamic environment, envisaged to be deployed via virtualization technologies is the actual EDGE part of the MEC concept. As there can be many edge levels governed under a system level, redundancy of the system-level entities is a challenge for the MEC concept. Further, the backhaul channels connecting the edge and system-level entities should be secured beyond the capabilities of resourceful cyber intruders. This level mainly consists of Mobile Edge Hosts (MEH), Mobile Edge Platform Managers (MEPM), and the Virtualization Infrastructure Manager (VIM).

2.2.2.1 Mobile Edge Host (MEH)

This is an entity that contains the corresponding ME Apps and the Mobile Edge Platform (MEP) on top of the underlying Virtualization Infrastructure (VI). In fact, MEH is operating on the *data plane* of the 5G mobile network, where it is a packet-based transmission platform that features a common encapsulation compatible with both fronthaul and backhaul traffic domains. Thus, access to the data plane guarantees direct connectivity for ME Apps towards the UE domain, in addition to the Internet. MEH mainly provides computing, storage, and networking resources to execute the ME Apps. In recent research advancements, MEHs are launched as Virtual Machines (VMs) that inherit formidable levels of resources and are specified for different service types [17].

Mobile Edge Applications (ME Apps) They are executed as software-only entities in the MEH. Different ME Apps are launched as service instances within the MEH to service UE Apps at the subscriber end. These ME Apps are to be pre-configured and hibernated within the MEH and dispense once the requirement arises with the UE requests. As the MEHs are envisaged to be launched as VMs, ME Apps can be deployed as lightweight virtualized entities such as docker containers [18]. The underlying virtualization infrastructure can be formed by a hybrid approach of integrating hypervisor-based and lightweight virtualization technologies. ME Apps should be optimized for resource utilization.

Mobile Edge Platform (MEP) Dispense functionality to run the ME Apps while facilitating the essential environment to discover, advertise, and consume mobile edge services. The traffic rules obtained from the MEPM are notified to the data plane by the MEP as the traffic rules controller, and the IP address translation is conducted for UE tokens. MEP Handles the Domain Name System (DNS) proxy/server according to the records conveyed from the MEPM. In a scenario that requires the interconnection of two or more MEHs (i.e., for a high-end mobile edge service that requires more than one MEH to implement the complete service function chain), the internal connections between the MEHs are established via the MEPs. In addition, the service registry element within the MEP maps the ME Apps to the mobile edge service IDs and triggers the internal resource allocations at the VI.

Virtualization Infrastructure (VI) Links are drawn towards the VIM and the MEP. This includes the *data plane* which executes the traffic rules set notified by the MEP and routes the traffic through ME Apps, DNS server/proxy, 3GPP network, local, and external networks. The most important task of the VI is to provision the intrinsic virtual resources to the ME Apps while monitoring their utilization. VI is notifying the VIM of all the states regarding virtual resources while corresponding resource allocations are conducted with the knowledge of the service registry in the MEP.

2.2.2.2 Mobile Edge Platform Manager (MEPM)

MEPM can be considered as the orchestrator at the edge level, where its main task is to notify the MEO regarding the status of the MEC edge platform. The function of the MEPM resembles the orchestration function in NFV (i.e., NFV Orchestrator—NFVO), where ME Apps or MEHs can be considered mapped entities to VNFs depending on their application deployment. The main functions of the MEPM can be listed below;

- Govern ME App life-cycle while informing the MEO of monitoring stats of each application.
- Element management provided for all the entities in the MEC edge level.
- Managing the application rules and requirements while performing the tasks of service authorization, traffic rules, DNS configuration, and resolving conflicts.
- Record the fault reports and performance measurements sent from the VIM.

2.2.2.3 Virtualization Infrastructure Manager (VIM)

VIM is responsible for allocating, managing, and releasing visualized resources (compute, storage, and networking) of the holistic virtualization infrastructure at the MEC edge level.

In fact, VIM is the hypervisor of the MEC edge platform, that is tasked with handling and managing the virtual resources. VIM is tasked to properly measure the resource consumption in terms of memory, processing, and networking; and to benchmark the general operation of the MEC edge platform for the sake of detecting any anomalous behavior in a compromised scenario. The functions of the VIM can be summarized as:

- Storing, configuring, and running the software images in the VI.
- Monitoring the virtualized resources in the VI for their performance and fault/ malfunction detection.

2.2.3 Mobile Edge Services and Interfaces

A service originated or facilitated by either the mobile edge platform or a mobile edge application is defined as a Mobile Edge Service (MES). All the ME Apps provisioned are registered under the mobile edge platforms' service registry through the Mp1 interface, where the ME Apps should subscribe to the relevant authorized MES. Sample MESs can be specified as Radio Network Information (radio network condition information, user plane statistics, UEs connected to the MEH), Location (location information of all UEs such as geo-location or cell ID connected to the radio nodes associated with a MEH), Bandwidth management which route, prioritize, and manage the traffic related to the ME Apps. The reference points in Figure 2.4 mentioned below are associated with:

- Mp: MEC Platform functionality
- Mm: Management reference points
- Mx: External entity connectivity

2.2.3.1 Mp–MEC Platform Interfaces

The three Mp interfaces are representing the connections inside a MEH. The Mp1 interface laid across ME Apps and MEP is facilitating service registration, service discovery, and communication support for registered MESs [19]. In addition, functions such as application availability, session state relocation support procedures, traffic rules, DNS rules activation, access to persistent storage, and time of day information are aggregated functions of this directive. Moreover, this reference point can be configured for provisioning service-specific functions deemed by the corresponding MES. MEP leverages the Mp2 connectivity for enforcing traffic rules, DNS rules to the data plane in the VI. The Mp3 interface connects MEPs of MEHs for control information exchange in case of a collaborative high-end MES.

2.2.3.2 Mm–Management Interfaces

The nine management interfaces distributed to both system and host levels are conveying statics and information related to the management aspect of the MEC system. In fact, MEO and OSS are the main entities relying on these interfaces. Mm1 between the MEO and the OSS triggers the instantiation and termination of MESs [19]. Mm2 tracks the configuration, fault, and performance management of the MEC platform through the MEPM and notifies the OSS. On behalf of the MEO, Mm3 monitors the application life-cycle, application rules, and requirements, and keeps track of available MEC services from MEPM; while Mm4 tracks the virtualized resources and records the stats of application images from VIM. Rules and platform configurations are enforced by the MEPM to the MEP in MEHs and are conveyed through the Mm5 interface. MEPM keeps track of the virtualized resource utilization through the Mm6 interface from the VIM. VIM extracts the utilization statistics from each MEH through the Mm7 interface. Both Mm8 and Mm9 interfaces are laid from UALCMP towards the OSS and MEO respectively. The UE requests once registered in the UALCMP, convey the instigation queries to the OSS via Mm8 and Mm9 acts as a feedback connection for the UALCMP to keep track of the launched ME Apps via the MEO and vice versa.

2.2.3.3 Mx—External Interfaces

Both Mx1 and Mx2 reference points in the MEC architecture are located in the mobile network, where the exposure is conspicuous to the adversaries due to their exclusive nature. Both these interfaces are conveying the unique credentials of the MEC subscribers for authentication or authorization purposes. Thus, the security of these ingress points is critical for the entire system. Mx1 interfaces the CFSP with the OSS, where access to third-party subscribers is granted through this connection. In contrast, Mx2 connects the MEC subscribers to the UALCMP, and the function of the interface can be formulated as an Application Programming Interface (API) [20]. Since CFSP has direct access to OSS, the tendency for OSS to be compromised is comparatively higher from the Mx1 interface. Though the UALCMP is at a higher risk considering the external access to the mobile edge system is maintained through it. However, the limited connectivity among entities at the mobile edge system level and the mobile edge host level would make access for malicious entities arduous.

2.3 MEC Standardization

Initially, IBM and Nokia Siemens network introduced Mobile Edge Computing in 2013 as an infrastructure that could run applications within a mobile base station; conversely did not consider the aspects of application migration and interoperability [21]. Later, European Telecommunications Standards Institute (ETSI) launched the Industry Specification Group (ISG) for standardizing Mobile Edge Computing (MEC) concept in 2014, which specified the operation of MEC as "Mobile edge computing provides an IT service environment and cloud computing capabilities at the edge of the mobile network, within the RAN and in close proximity to mobile subscribers" [22].

Nokia Networks, Intel, Vodafone, IBM, Huawei, and NTT DOCOMO as the pioneer mobile network solution providers, led the ISG representation while the European 5G Infrastructure Public Private Partnership (5G-PPP) conceded MEC as a prime emerging technology for 5G networks [23]. The main objective for the ETSI MEC ISG was to facilitate an open eco-system to multiple vendors for provisioning a variety of applications and services at the edge of RAN for overcoming the restrictions with cloud computing deployments [13]. In 2015, ETSI ISG published a Proof of Concept (PoC) framework for MEC highlighting the rationale, roles, responsibilities, and the activity process of the PoC framework [24]. Another white paper was released in the same year for evaluating the economic aspects of MEC service scenarios such as AR, intelligent video acceleration, connected cars, and IoT gateways to identify the market drives [25].

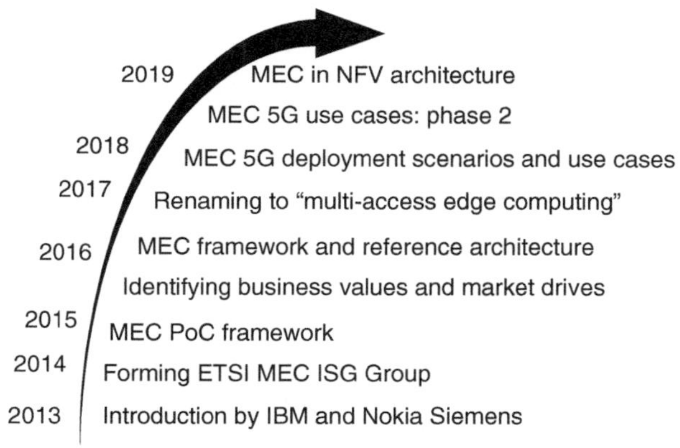

Figure 2.5 Standardization Timeline of MEC.

In 2016, the MEC framework and reference architecture was published by the ISG for designing the entities in the MEC system and to specify their intended function [14]. Realizing the potential of the MEC paradigm is reaching beyond mobile networks to Wi-Fi and fixed access technologies, the ETSI ISG group has renamed the concept as "Multi-access Edge Computing" that conveniently reinstate the acronym to MEC [9].

Moreover, a 2018 ISG release has accentuated the deployment scenarios and use case exemplifications of MEC with 5G integration [15]. This white paper investigates the deployment of MEC and 5G components in actual use cases with traffic steering, mobility, offloading, charging, and regulatory requirement considerations. An extension of the same directive was presented in [26] as the 2nd specification release that expands the scope of MEC use cases for Camera as a Service, video delivery, future factories, multi Radio Access Technology (multi-RAT), Internet Protocol Television (IPTV), In-vehicle systems and 5G use cases. In the latest release from ETSI in 2019 January [19], a variant architecture was released for conjugating Network Function Virtualization (NFV) and MEC while in-depth focus was drawn towards components in the architecture in contrast to the previous release of [14]. A concise timeline of the evolved standards on MEC is depicted in Figure 2.5.

2.4 MEC Use Cases and Applications

The features of MEC enable a wide range of applications. Here, we present some of the most interesting use cases and applications of MEC.

2.4.1 Critical Infrastructure

Critical Infrastructure based services such as energy, water distribution and sewage, Gas pipeline distribution, offshore oil drilling rigs, and emergency service applications have improved their scope by digitizing their controlling systems with IoT technology; which that leads to complete automation eventually. The controlling of most of such applications is achieved through Supervisory Control and Data Acquisition (SCADA) systems as

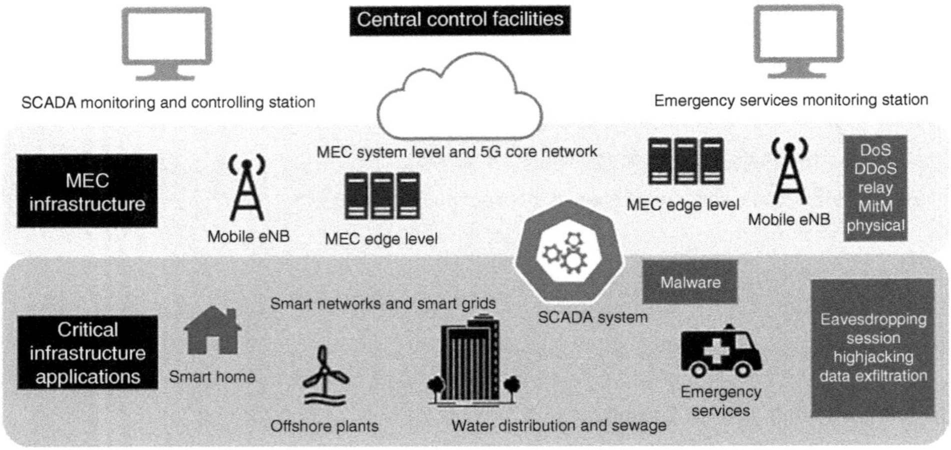

Figure 2.6 Critical Infrastructure Connectivity to MEC Platform.

indicated in Figure 2.6. Though the global expansion of these services to distributed global clusters is restricted by the existing centrally controlled data centers that are utilized for storage and processing. Thus, the MEC concept can be integrated into critical infrastructure services that improve the interfacing with the general consumers towards the services; in addition to rapid and real-time status updating of billing, consumer usage, and service interruption notifications.

The energy sector is leading the infrastructure services as it energizes all the other sectors and emerging smart grid technology is opening new avenues for power installations. IoT-based technologies are enabling the formation of Advanced Metering Infrastructure (AMI)/smart metering/net metering that overlay a monitoring framework for the smart energy solutions [27, 28]. In addition, amalgamating renewable energy sources demand a decentralized deployment of power coordinating activities (smart grids) that enforce bi-directional energy flow through the transmission grids [29]. Thus, employing an effective power utilization scheme is an intrinsic necessity and achievable through the system status analytics on consumer consumption, consistency of the generation, grid utilization, and performance of the operating devices.

In urbanized areas, water, and sewage treatment is a critical requirement to facilitate urban sanitation and improved quality of human life for the citizens; that eventually leads to sustainable development [30]. Current water and sewage governance techniques are characterizing complexities due to diversified cost structures formed by the origin of water sources and environmental externalities [31]. MEC can be adopted with such a governance system in a digital domain for overcoming such drawbacks. Resembling to the AMI concept, smart water meters can be deployed at household vicinities for automating the water and sewage billing. MEC edge servers or MEHs are responsible for facilitating a low latency communication platform between the consumer end, the water treatment plant, and the central monitoring station.

Petroleum extraction is a vital global industry that collects fossil fuels for the purpose of generating combustible energy, that energizes vehicular engines and electricity-generating plants. The continuous and rapid extraction of fossil fuels has depleted the petroleum reserves in the land areas; where the companies were forced to shift the drilling process to the offshore reservoirs [32]. Thus, offshore plants are critical for the petroleum industry despite the precarious conditions granted to the employees. Automation is an approach to be considered for entrusting the safety of employees at offshore plants. Access to communication or a mobile network is evidently restricted for offshore plants due to their remote positioning. Some of the existing offshore plants employ satellite communication due to their remote nature. Such a distant and unreliable communication medium is not ideal for an automation scenario that relies on a controlling system with plenty of insights extracted from data. Thus, a MEC-enabled BS can be placed within the offshore vicinity, which can be leveraged to achieve the communication/networking, and processing capabilities required for the automation.

The importance of emergency services is becoming evident to us during this era due to the increasing occurrences of natural disasters, terrorism, and pandemics (i.e., tsunamis, hurricanes, bombings, and COVID-19). This is an area where the existing infrastructure lacks formidably and emergency response should be as express as possible. In the current emergency service setups, response teams are notified or dispatched from a central location. In an emergency situation, with the already overloaded communication channels, conducting this emergency response is pragmatically arduous. With the decentralized storage and processing infrastructures that inherit enhanced communication capacity MEC BSs provide, emergency responses can be organized leveraging those MEC platforms. As the responding is locally managed and dispatching can be handled aligning to it, backhaul networks would not be congested by the emergency traffic. Figure 2.6 is depicting the central controlling, MEC infrastructure, and critical infrastructure application aspects in a three-tier manner. Probable security attacks that can compromise at each tier are mentioned there.

2.4.2 UHD Video Streaming and Analytics

UHD 4K resolution real-time streaming should inherit 3840×2160 pixels with the least bandwidth of 25 Mbps. Novel video streaming and video analytic applications require enhanced Mobile Broadband (eMBB) level guarantees in the streaming channels to ensure UHD resolution. eMBB is one of the 5G use cases that specify the minimum level of data rates, latency, and other parameters in broadband applications. According to [33], eMBB specifications can be summarized as peak data rate—Down Link: 20 Gbps/Up Link: 10 Gbps; peak spectral efficiency—Down Link: 30 bps/Hz/Up Link: 15 bps/Hz; user plane latency—4 ms; control plane latency—20 ms. The main reason for the eMBB level requirement is that current applications require more time spent on processing at encoding/decoding, compression/decompression, and UHD signal processing phases of the UHD level streams. In fact, video stream analysis-based applications such as vehicular license plate recognition, face recognition, and domestic/traffic surveillance require high computational complexity to execute their algorithms in addition to the processing associated with the UHD transmissions [23, 34].

Mobile gaming applications based on immersive VR and AR integration are probable deployments for high-level video streaming UHD channels that endanger the bandwidth provisioned for priority services. Conversely, the share of mobile video streaming is increasing rapidly, as predicted by Cisco; where the bandwidth-saving approaches to Over the Top (OTT) streaming channels are identified as intrinsic preliminaries to form the multimedia channels [35]. In addition, crowd-sourcing and social media-based media are uploaded into the servers through multimedia channels precipitately as in; 2.4 million pieces uploaded to Facebook, 72 hours of video content uploaded to YouTube, 347,000 and 216,000 images uploaded into WhatsApp and Instagram in a minute [36]. The perspective of subscribers regarding the bandwidth has extended from single view to multi-view, 2D to 3D, and single source stream representation to adaptive multi-bit-rate multi-resolution representation. Therefore, proper bandwidth utilization for OTT streams is a requisite, as well as a predicament. There are two scenarios where video streaming applications are deployed. Peer-to-peer (P2P) streaming traffic routed from a MEC-enabled gNodeB is conveyed to a UE directly; which preserves the backbone capacity and traffic of the network operator. In the case of a big event streaming, the streams are digested at a MEC host service subscribed by a local video production studio. This approach, however, could be subjected to amendments by the video editors. Figure 2.7 is representing various eMBB and video streaming applications that can benefit from MEC deployment.

2.4.3 Massive Machine Type Communication (mMTC) Applications

mMTC refers to the communication taking place among machines in a Machine-to-Machine (M2M) or Device-to-Device (D2D) manner that is independent of human involvement. As indicated in Figure 2.8, e-health wearables creating a Wireless Body Area Network (WBAN), IIoT, automated weather stations, and smart farm applications can be considered under this use case [28]. Majority of the IoT applications are composed of sensory devices and actuators that employ M2M communication for data transferring. M2M-enabled devices are called Machine Type Communication Devices (MTCDs) by

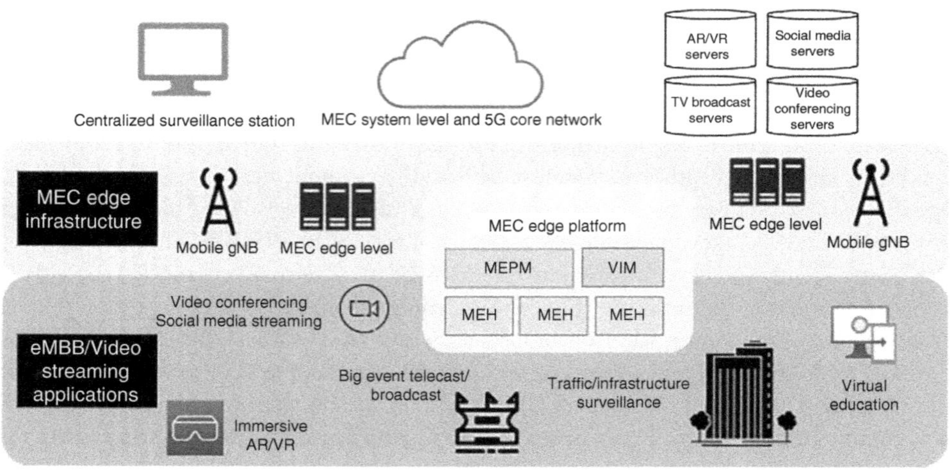

Figure 2.7 eMBB and Video Streaming Applications with MEC.

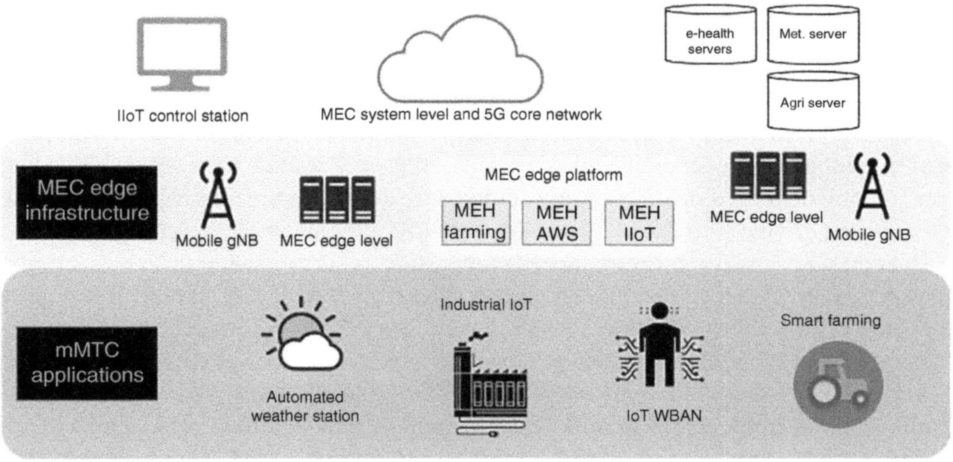

Figure 2.8 MTC Integration with MEC.

3GPP. The access networks formed for most MTCDs are non-3GPP technologies that are Ultra Wideband (UWB), WLAN, ZigBee, Bluetooth, Bluetooth Low Energy (BLE), Long Range (LoRa), Low Power Wide Area (LPWA), or Narrowband IoT (NB-IoT) [37–39]. For IoT applications, communication types are ranging from Human-to-Human (H2H), Human-to-Machine (H2M), or vice versa M2M [37]. A typical MTC architecture presents two communication scenarios: between MTCDs and MTC servers or inter-MTCD D2D type [40]. Healthcare applications are plausible with enabled remote patient monitoring through MEC MTC deployments that use WBANs for monitoring e-health statistics of the consumers [9]. The types of MTCDs employed in WBANs are Implantable Cardiac Defibrillators (ICD), pacemakers, neuro-stimulators, glucometers, oximeters, and vital sign monitors [38]. These WBAN bio-sensors are communicating to the gNB through Machine Type Communication Gateways (MTCGs) using non-3GPP network technologies. The General IoT concept can be applicable to automated weather stations and smart farming applications, where automated sensors are conveying atmospheric measurements to the MEC edge platform for further processing and forecasting. In IIoT scenarios, the MTC concept can be deployed for improving the efficiency in automation flows through error minimization with inherent optimization control techniques; that can be configured accurately with precision sensor fusion.

2.4.4 Intelligent Transportation System (ITS)

Intelligent Transportation Systems (ITS) is an initiative taken to form an advanced transportation system that extends to different transportation modes with more safer and coordinated traffic management. The communication aspect of the ITS is specified via the Vehicle-to-Everything (V2E) adaptation [41]. Even though this is a concept first introduced in the 1970s, the real advancement begin with the invention of sensors, microchips, and the IoT concept that enables intelligent capability with data extraction and analytics [42]. Vehicular Networks (VNs), being a formation of ITS deployments, have a significant standing in the 5G context [43]. Ultra-low latency and reliability factors of V2E applications urge MEC to be an evident requirement [9]. The connected vehicles technology defined by 3GPP is focusing on enhancing safety, reducing traffic congestion, sensing vehicle behavior, and servicing other vehicular value-added services. Such services are achieved by offloading computational and geo-distributed services to the roadside BSs or Infrastructure to enable autonomous driving with data connectivity that attributes the alleviated latency [34].

Though, with envisaged drastic development, transportation industries are becoming conspicuous cyber-targets for adversaries due to their rapidly evolving mobility structure as concluded by the report from IBM X-Force and Transport Systems Catapult [44]. The smart sensors deployed in vehicles enable the Advanced Driver Assistant Systems (ADAS), which is introduced as the preliminary stage of self-driving applications [45]. The embedded features that attribute to 100 million lines of program code and the processing ability of 25 GB data per hour improves the feasibility for deployment [46]. Vehicle automation approaches are entirely reliant on sensors. As sensors are electronic devices prone to be penetrated by adversaries, any successful malicious penetration could result in vehicle collisions, traffic congestion, or damage to properties or human lives.

The connectivity between the vehicles is different from the connectivity from a vehicle to the BS. The protocols and the communication technology employed for this connectivity depend highly on the manufacturer. Though, in the United States, the standard for V2V connectivity is Dedicated Short Range Communication (DSRC) technology which transmits the location, direction, and speed of the vehicle to the nearby vehicle [47]. The intention of this V2V deployment is to provide early warnings of imminent accidents detected through a smart system embedded in the vehicles. Figure 2.9 depicts the wide range of aspects in ITS deployments integrated with a MEC system. Further, possible attack vectors are indicated in an illustrative context.

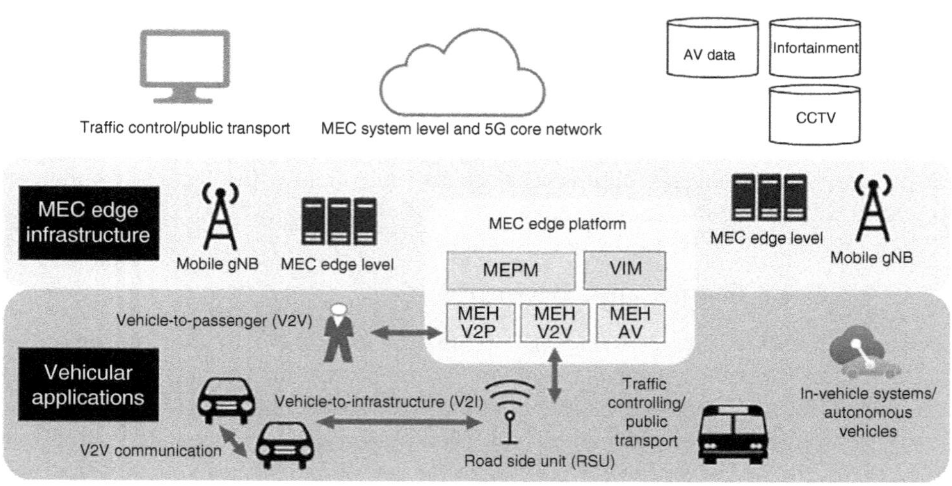

Figure 2.9 ITS Integration with MEC.

2.4.5 Augmented Reality (AR), Virtual Reality (VR), and Mixed Reality (MR)

VR, MR, and AR applications are relying on the 5G service categories of eMBB and Ultra-Reliable and Low-Latency Communication (URLLC) to be deployed with the corresponding communication channels [48]. 5G hotspots, in-vehicle infotainment systems, and gaming for educating/instructing are immersive experiences a consumer can endure under this directive [49]. The VR technology simulates the visualization 100% while AR and MR are differing by the extent of virtualization overlaid with digitization on visual perception [48]. A typical VR Head Mounted Display (HMD) occludes the user Field of View (FoV) and positions the virtualized elements through eye and head movement tracking. In the current trading, VR services are featured with mobile low-cost devices such as Samsung Gear VR and Google Cardboard while Oculus Rift, HTC Vive or PlayStation VR are high-quality streaming products that embed latency sensitivity. The Motion-to-Photone latency exceeding 15–20 ms for image rendering causes motion sickness for the users as a result of conflicted signals precipitated on the vestibular-ocular reflex.

The parameters of Latency < 10 ms, bandwidth > 1 Gbps, and cell capacity > 500 connections are the requirements for ensuring the AR services with performance factors of screen response $\approx$2 ms, sensory extractions $\approx$1 ms, refresh rate at 120 fps $\approx$8 ms, and network Round-Trip Time (RTT) processing $\approx$2 ms for 5G based deployments [49]. A typical AR mechanism combines the digital data generated through computed processing with the physical reality that intensifies the human experience. The mobile technologies such as Layar, Junaio, Google goggles, and Wikitude can be considered to enable AR integration towards MEC [23]. Error diagnosing in industries and fixing, remote life supporting by the Original Equipment Manufacturer (OEM), Human–Machine-Interface (HMI) functionality for machine operation and virtual training for operators are possible applications in the IIoT context [50]. In addition, AR can be utilized for e-trading, remote robotic surgeries, architectural/ real estate visualizations, education, and gaming applications as illustrated in Fig. 2.10. An AR process usually requires five critical components to operate: a video source

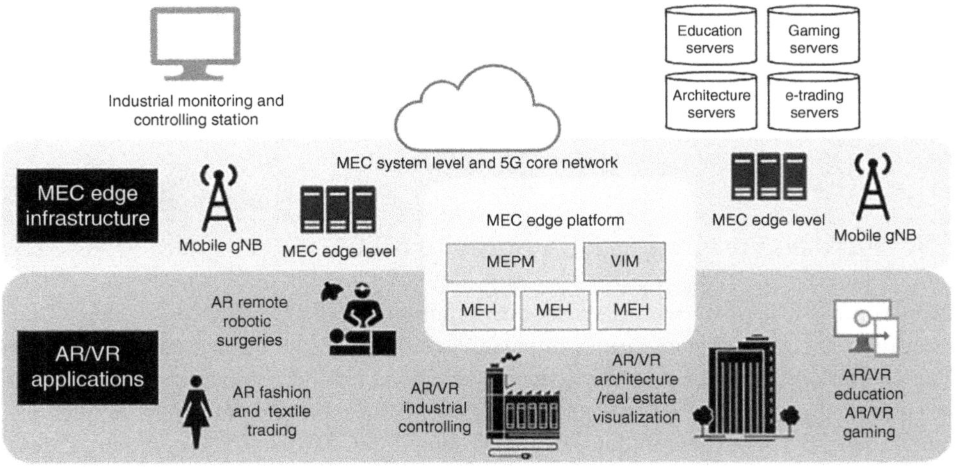

Figure 2.10 Augmented Reality Integration with MEC.

(mobile camera), a tracker (position tracker of the user), a mapper (modeling of the environment), an object recognizer (known object identifier), and a renderer (processing of the frames) where the components other than the locally deployable video source and the renderer could be hosted in the MEC server for computer-intensive offloading [34]. Findings in the research [51] where AR deployment leads to reduced latency and energy consumption by 88% and 93% respectively through computational offloading. This result encourages the adoption of MEC for AR applications. Moreover, web-based AR (web AR) is an approach that overcomes the cross-platform and extensive provisioning limitations that are inherent in device-based and app-based AR applications. MEC is a patent deployment option for web AR that is envisioned to achieve 1 ms latency with 5G integration [52].

2.4.6 Unmanned Aerial Vehicles (UAVs)

UAVs are quite popular these days to perform tasks that are deemed unreachable, especially in the case of a pandemic where contactless approaches are demanded. In addition, they can be employed for photography, disaster response, inspection, monitoring, precision agriculture, military, communication relaying, traffic control, and disaster relief services as illustrated in Figure 2.11 [34, 53]. Federal Aviation Administration (FAA) predicts that Unmanned Aircraft Systems (UASs) will triple its 2018 fleet size by 2023 with an annual growth rate of 24.7% [54]. UAV-based communication deployments attribute: the Line-of-Sight (LoS) transmission attained by hovering to targeted locations, dynamic deployment ability that features robustness to climatic effects and nullified costs for site installation in case of an acting BS, and UAV-based swarm networks that facilitate ubiquitous connectivity to ground users with high flexibility and various provisioning options [55]. UAV operations are categorized as Low Altitude Platforms (LAPs) and High Altitude Platforms (HAPs) which are distinguished by altitude, computation, coverage, power, capacity, and endurance capabilities. Moreover, the Size, Weight, and Power (SWAP) of UAVs are constraints for attaining desired performance metrics. The priority of the UAV is to conserve its battery life for flying while offloading the computational or storage content to the MEC servers for processing [9]. Thus, employing strong cryptographic primitives or prolonged security protocols would be infeasible. The controlling link to UAVs could be maintained from a ground station or by a remote station controlled through a MEC system. Figure 2.11 illustrates various UAV-enabled applications in MEC deployments.

2.5 Related Computing Paradigms to MEC

Apart from MEC, there are other Edge Computing Paradigms such as Mobile Cloud Computing (MCC), Fog computing, and Cloudlets. Though, the main intention of all the edge computing approaches is to extend the constricted cloud computing functionalities to reinforce the service access infrastructure. Despite the subtle differences in architecture and service provisioning means, most contexts about these flavors of edge computing are of resembling nature. The divergence in the architecture however is limiting the deployment options for certain applications and technologies.

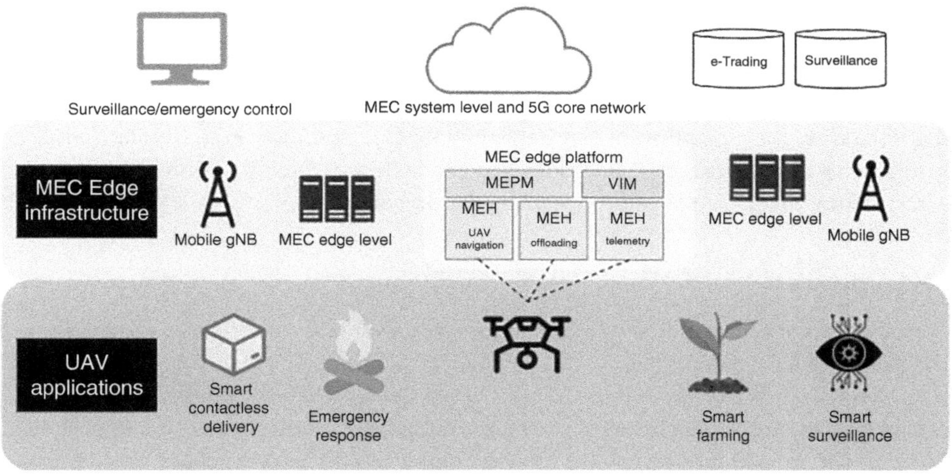

Figure 2.11 UAV-Based MEC Deployments.

2.5.1 Mobile Cloud Computing (MCC)

The MCC concept is formed with the intention of facilitating computing capabilities which were once offered in the cloud, to mobile devices for extending their battery life and storage capacity; while improving adaptability, mobility, scalability, self-awareness, and availability in mobile computing environments [56]. This initial principle was updated to delegate the mobile devices to be executed at the edge, resulting in the obstructions that transpired due to the centralized architecture [21]. In the MCC approach, cloud infrastructure is kept intact, while a clone or a partial content of the mobile process is offloaded to the edge. In addition, migration of program portions or threads of mobile processes in mobility scenarios are further deployments of MCC [57]. Novel applications of m-learning, m-gaming, m-healthcare, AR, and crowd-sourcing are constricted with cloud deployments due to the latency requirements attributed by the centralized nature, and affect the Quality of Experience (QoE) of the mobile subscribers [13]. Thus, 5G-based real-time applications are improbable with MCC.

2.5.2 Fog Computing

Introduced by Cisco Systems in 2012, Fog computing envisions an infrastructure formed by geographically dispersed edge nodes in collaboration with the cloud network to provision cloud services. The definition for Fog computing can be presented as, "a system-level horizontal architecture that distributes resources and services of computing, storage, control, and networking anywhere along the continuum from cloud to Things" [58]. This definition was declared by the OpenFog Consortium, which is the prime organization that standardizes and promotes fog computing. In this paradigm, IP or Multi-protocol Label Switching (MPLS) backbone is supporting the edge network for deploying cloud services for IoT devices in a proximate range. The strata of cloud, fog, and end-user stratum are forming the three-tier architecture for Fog computing [59]. Fog nodes in the fog stratum are placed in a virtualization infrastructure that links the end devices to the cloud data center while maintaining the fog-to-fog connections that expand the fog network. The applications of Brain–Computer Interface (BCI) using wireless electroencephalogram (EEG) headsets, AR, real-time video analytics, cyber-physical systems, and V2E are plausible with the flexibility of fog deployments [21]. The fog nodes which attribute comprehensive resources and functions are coalescing a holistic and dispersed service infrastructure that is deployable for realizing 5G use cases.

2.5.3 Cloudlets

The Cloudlet concept was first proposed by Satyanarayanan et al. in 2009 as a small cloud infrastructure located near mobile users, which could be deployed at small businesses or industries [7]. They are also referred as Micro Data Centers (MDCs), proposed by Microsoft Research in 2015 as an extension of traditional data centers used in cloud computing [60]. The cloudlet architecture resembles the fog computing three-tier model; where the end devices, cloudlet-based edge cloud platform, and centralized data center are forming the system structure [13]. Cloudlets are deployed as VMs that facilitate a transient environment for the users in a local vicinity. Offloading and caching tasks are probable with the small-scale cloudlets, where face recognition and video streaming applications are ideal deployments that emulate the scaled virtual environment [59]. In addition, approaches such as Hyrax, FemtoClouds, Superfluid clouds, Edge-Centric computing, Mist computing, Cloud of Things (CoT) and Edge Cloud (EC) are alternative edge technologies proposed for various deployments that operate at diverse scales [21, 60].

2.6 Comparison with Related Computing Paradigms

MEC, Mobile Cloud Computing (MCC), Fog computing, and Cloudlets tend to coexist in many technical contexts, hence the tendency for a misappropriation of these technologies given that they all have similar origins. However, these technologies are intrinsically different and each of them comes with its unique value proposition to both existing and future mobile networks. The efforts on developing an Open Edge Computing (OEC) platform are underway with the participation of several telecommunication companies (vendors and operators) and standardization entities. Their goal is to harmonize all the existing edge technologies, develop an open edge and reference platform for OEC and align it with the IT and telecom industries.

2.6.1 MEC and MCC

MCC mainly integrates cloud computing and mobile technologies optimally over wireless networks [61]. MCC uses computational augmentation approaches to deliver rich computational resources to end users and provide business opportunities for network operators and cloud service providers. Rapidly evolving mobile devices rely on several variable resources such as power, storage, and bandwidth; hence it becomes imperative to create a rich mobile computing platform that is capable of leveraging the elastic nature of these resources to create a more dynamic and optimized computing and networking platform that can accommodate the surge in computing devices for present and future networks. MCC provides a vast array of benefits; these include reduction in Operating Expense (OpEx) and Capital Expenditures (CapEx), flexible provisioning of a wider variety of features without resource redundancy, ability to extend developer services to a larger market, and creating a level playing ground for network providers with lesser means to compete favorably in the service market. Compared with MEC, MCC mainly aims at providing a robust platform for running a wide range of rich mobile applications on a vast array of mobile devices. This is accomplished by hosting all key resources that such mobile applications require as cloud services, hence relieving manufacturers the need for making high-configuration devices.

2.6.2 MEC and Fog Computing

MEC and Fog computing [8] are twin concepts that were proposed as a means of relieving IoT applications of dependence on local processing. Ultimately, the vast array of applications expected on IoT networks calls for a complementary combination of MEC and Fog computing to meet the requirements of efficiency and reliability for both industrial and utility IoT services. Such requirements include information security, latency, corporate policy, and data sovereignty. These concepts, though related still have some significant differences. First, in the context of tenancy, whereas fog computing is traditionally aimed at a single-tenant, enterprise-owned set of devices, MEC is a multi-tenant application that is aimed at multiple stakeholders. Second, fog computing finds its application mostly in gateway devices with computational and storage capabilities while MEC focuses on external applications. Third, in terms of point of data processing, both technologies are geared towards distributed analysis close to where the data originates, hence reducing the delay in the transmission of IoT data to the cloud. However, whilst fog processes data closer to the local area networks, MEC pushes this even closer to the source. Fourth, on the approach to connectivity, fog computing combines the functions of connectivity, whereas edge computing favors each item operating independently. In addition, the typical application areas for fog computing can be found in connected vehicles, smart grids, pipeline monitoring, smart traffic lights, connected rails, and machine-to-machine (M2M) communications while for MEC, typical application scenarios may range from augmented and virtual reality services, intelligent video acceleration services, connected vehicles and IoT gateway services, mobile Private Branch eXchang (PBX) for large enterprise services, and active location-aware application services [62].

Even though the formation fog and MEC architectures are visually different, there are similar functionalities to consider in terms of application processing in identifying threat vectors which could be common for both paradigms.

In both architectures, UE applications could be running on mobile devices such as a phone, iPad, and Personal Digital Assistant (PDA) or general-purpose hardware platforms such as Raspberry Pi, FPGAs, Arduino...etc. The connection of such devices to the fog nodes is authorized through the protocol abstraction layer of the hardware platform infrastructure while the network entity maintains the connectivity with the end devices. In contrast, UE app requests are forwarded to the OSS of the MEC architecture through the UALCM proxy, which possesses the capability to instantiate or terminate the UE apps. Thus, the applicability of the threat vectors at the node view of a fog node could resemble the Mx2 reference interface of the MEC.

Both paradigms employ virtualization infrastructures for processing the UE app requests. Fog nodes and MEC deploy both hardware and software virtualization functions. The threat models for virtualization infrastructure in both paradigms might be similar.

The entities MEH in MEC and the fog nodes execute similar functions in computing, storing, and controlling UE apps in a virtualized environment, except the access for UE apps is granted through the OSS in MEC, and fog nodes inherit that function in-built. Hence, the access control of the OSS and the fog nodes cannot be mapped. However, the access to UE apps in MEC is handled at reference points Mx2, Mm8, and Mm9. Thus, the access controlling of the fog node could be mapped to the Mm8, even though the OSS is not included in the MEH.

The connectivity between fog nodes is thoroughly investigated for their security in fog architecture-based research. The same threat domain would be applicable for links between the MEHs (Mp3) as they perform similar functions.

The traffic rules controlling function in the mobile edge host could be mapped to the network function and the Network level of the MEC architecture for forming a threat model.

However, MEPM and VIM are common entities for a set of MEHs in a particular trust domain. The fog nodes have those functions inside a single node, and a particular fog node creates its trust domain. These similarities between the fog computing architecture and MEC architecture would generate possibilities for extending the exploration of threats for MEC to fog computing architectures.

2.6.3 MEC and Cloudlets

Cloudlets are powerful computers strategically clustered as the data center in a box; the goal is to reduce the number of hops between the cloud and wireless devices to a single hop. This will optimize proximity and reduce latency and bandwidth requirements for such devices. For most developers and cloud service providers, the cloudlet is seen as the middle tier (i.e., between cloud and mobile device) of the three-tier hierarchy of the cloud architecture [7]. From the architectural layout of cloudlets, the difference between cloudlets and MEC becomes readily obvious, while MEC aims at extending cloud computing capabilities to the edge of the network, cloudlets aim at keeping the cloud services a hop away from edge devices. Typical cloudlet hosts run hypervisors to host multiple VMs: Kilo Virtual Machine (KVM) is a popular example of such a hypervisor. The hosted VMs then publish required information, such as OS and other properties to the network. The cloudlet hosts broadcast information via the cloudlet server to enable cloudlet clients like smartphones PDAs, and wearable devices to discover and connect to the cloudlet server using Wi-Fi or cellular data service of 4G LTE [7, 63]. Compared with MEC, cloudlets can provide resources for even more resource-intensive and interactive mobile applications. Typical application areas of cloudlets are in video streaming, speech processing, cognitive assistance applications, AR/VR, edge analytics in IoT, new automotive services, and drone technology [7].

2.7 Summary

As introduced by the European Telecommunications Standards Institute (ETSI), MEC represents a solid example for future applications to explore the mobile communication networks. Different from existing proposals of Cloudlets and MCC, the standardized approach adopted by MEC has advantage in terms of scalability, interoperability, and compatibility.

3

Edge-Based Video Analytics and Mobile Augmented Reality*

This chapter illustrates two data-oriented application domains enabled by MEC^2: edge-based video analytics (EVA) (Sections 3.1 and 3.2) and Mobile Augmented Reality (MAR) (Sections 3.3–3.6). Readers will gain an overview of the fundamental components, pipeline, and architectural options for these edge-based applications. In addition, practical use cases and deployment are discussed to provide readers with a pragmatic perspective on these emerging applications.

3.1 Video Analytics Overview

In this book, we refer to video analytics as a category of services that analyze the data collected from video systems for capturing and inferring temporal spatial events, activities, and/or objects [64–66]. Relying on computer vision and artificial intelligence, video analytics aims to extract meaningful insights from real-time video streams and/or recorded footage.

Together with the rapid deployment of cameras by governments and enterprises, video analytics has facilitated numerous applications such as security surveillance, traffic and emergency management.

Video Analytics Pipeline:

To perform video analytics, a processing pipeline needs to be constructed. The video analytics pipeline, commonly referred to as VAP, is a series of sequential stages through which the video data will be transformed, analyzed, and processed for specific applications [67]. The applications of VAP can be security surveillance in public and private venues, and digital assistants for autonomous driving.

Each stage in VAP represents a specific task or function that is applied to the video data. From data perspective, VAP forms a sequential chain through which the data flows from the initial stage to the final stage. By following the modular approach, the data output from one stage becomes the input for the next stage. Depending on the application category, a VAP can consist of several video processing modules, which may vary across applications. For example, a VAP for car counting will be different from a VAP for license plate recognition. The car counting VAP can include a video decoder, an object tracker, an object classifier, and

*This chapter is partly contributed by Yushan Siriwardhane (University of Oulu, Finland).

Mobile Edge Computing and Communications: Driving Forces, Technology Foundation, and Application Areas, First Edition. Aaron Yi Ding, Chamitha De Alwis, and Madhusanka Liyanage.
© 2025 John Wiley & Sons Ltd. Published 2025 by John Wiley & Sons Ltd.

a directional counter. The VAP for license plate recognition, on the other hand, can consist of a video decoder, a motion detector, a plate detector and a module for character/number recognition.

Video Analytics Components:

The key components of VAP for video analytics are highlighted as follows [67, 68]:

- **Video Encoding and Decoding**: Video encoding in VAP can reduce data communication volume by compressing consecutive frames into a compact segment rather than sending the raw frames one by one. Encoded data streams are transmitted via streaming protocols such as real-time transport protocol (RTP) [69], web real-time communication (WebRTC) [70], and dynamic adaptive streaming over HTTP (DASH) [71]. Video decoding is for reconstructing original frames by decompress the encoded data. Commonly used video encoding schemes in VAP include Advanced Video Coding (H.264), High-Efficiency Video Coding (H.265), and Versatile Video Coding (H.266) [72].
- **Pre-processing**: Pre-processing is to prepare the video data before the analysis process. It aims to enhance the input data quality and further benefit the subsequent procedures. Depending on the service needs, the pre-processing in VAP can include noise reduction, frame extraction, resolution adjustment, and/or color correction.
- **Image Segmentation**: Image segmentation is a process to partition an image to multiple segments such as motion, foreground, and background. The segments often represent distinct aspects of the image, which may provide useful insights for more comprehensive understanding of scenes.
- **Object Detection**: Object detection in VAP involves locating and recognizing specific objects in video frames. It often identifies the region of interest (RoI) and then classify RoI with object classifier.
- **Object Tracking**: Object tracking is to continuously monitor the objects as they move across frames in the video sequence. In VAP, the tracking process involves object locating (e.g., bounding box of the object) and estimating the moving trajectory.
- **Object Classification**: Object classification aims to identify the object and map it into a finite set of object classes. The process consists of extracting features from the image frame and feed the features to the classifiers (e.g., convolutional neural network based ResNet [73]).
- **Data Fusion**: Data fusion in VAP is to achieve consistent information by integrating raw data and analytics results from various sources. The data input can come from different devices such as surveillance cameras, drones, and mobile phones. One example is to identify the same object (e.g., a car) in multiple cameras located across the city during the day.

It should be noted that these VAP components need to be customized based on the requirements of targeted applications. In practice, we do not need to use all the components as a whole for every scenario. Depending on the application's specific needs, we often choose only the necessary components to construct a VAP.

Applications of Video Analytics:

Fueled by advancements in computer vision technologies and digital infrastructures, video analytics has been successfully utilized to facilitate surveillance, rescue and safety for numerous applications. Regarding domain-specific applications [68, 74, 75], we highlight four categories as follows:

- **Smart Cities**: Video analytics can facilitate urban management including road traffic monitoring, crowd surveillance, parking control, and stolen cars search. The insights from video analytics are valuable for improving the public safety and resource utilization (e.g., roads, buildings).
- **e-Health**: For healthcare systems, video analytics can be used to support patient monitoring, facility inspection, and remote care. With data-driven insights from video, emergency situations such as cardiac arrests or seizures can be efficiently detected, which allow healthcare providers to respond with timely interventions.
- **Agriculture**: For agriculture, video analytics can enable smart and sustainable farming by leveraging monitoring insights for efficient resource (e.g., crops, livestock) and equipment management. The video data from multiple sources such as cameras and drones can help automate decision making and hence reducing downtime and maintenance costs.
- **Education**: Video analytics can be used as an instructional resource for extracting valuable information for lecturers to improve their course offerings and understand student behaviors in the class [76]. The analytic insights such as usage patterns across students, lectures, and dates can be leveraged to adjust the course content according to reactions and study interactions.

3.2 Edge Video Analytics

Given the proliferation of cameras in urban environment, we are facing the challenges of real-time data processing at large scale. Although cloud computing has been successful to support numerous Internet services, it is difficult for cloud to strike a balance between transmitting large volume of data efficiently and generating on-demand actions in real time. For instance, it is often unnecessary to check every video stream manually by uploading video data from all the cameras to the cloud for the human inspectors. Instead of processing the video after an incident, it is useful to analyze the video streams in real-time and generate an on-demand action when an anomaly is detected.

In order to achieve a balance between responsiveness and data efficiency, edge-based video analytics has been regarded as a promising direction to exploit the available data and resources across the hierarchy of local/end, edge, and cloud devices for better video analytics performance.

3.2.1 Definition and Enabling Technologies

EVA Definition and Architecture:

In the context of MEC^2, Edge-based Video Analytics (EVA) [67, 68, 75, 77] is a video data processing paradigm that leverages the end-user devices, edge computing units, and cloud resources to maximize the performance of video analytics in terms of scalability, responsiveness and accuracy.

EVA architecture can take multiple forms, depending on the available hardware infrastructures. The hardware infrastructures include three types: IoT (end devices), edge, and cloud. To implement an EVA system for a specific solution and scenario, developers and engineers can choose one of these four architecture options as follows:

- **IoT-Only**: This architecture option includes only IoT and/or end-user devices, such as cameras, mobile handsets, and connected vehicles. The input data is collected by the sensing units of IoT devices. The collected data is processed locally on the devices. Given the resource limit of IoT, multiple devices can form a cluster and collaborate with each other to improve the performance.
- **IoT and Edge**: This option utilizes the edge resources that are close to IoT devices. To optimize performance (e.g., accuracy), computation workloads can be offloaded from IoT to the edge devices with better hardware. Since edge devices are often connected over wires, they can form a cluster to support more demanding tasks such as running deep neural networks.
- **IoT and Cloud**: This is a typical cloud based setting by using the abundant resources of the cloud. Since IoT devices often rely on wireless, communication overhead is an important concern in this architecture option. Meanwhile, owing to the scalability and elasticity of the cloud, this architecture can cope with dynamic workloads.
- **IoT, Edge and Cloud**: As a trending architectural option feasible to strike a balance across real-time requirement, resource consumption and scalability. In this architecture, video data can be captured by IoT (e.g., cameras) in the operation field. For better resource utilization, only part of the data is forwarded to the cloud for further analysis and storage. The aim is to pool heterogeneous resources across IoT, edge and cloud, by weighing their respective strengths.

To construct VAP for EVA applications, functional components serve as the basis to compose a suitable pipeline for optimal functionality and performance. The core functional components of EVA include:

- **VAP Optimizer**: The optimizer for EVA is for converting high-level queries into practical vision modules, which are used to compose the pipeline. Along the pipeline chain, each module uses pre-defined interfaces to receive the data and send the processed results to the next one in the pipeline. The VAP optimizer aims to estimate and optimize the resource consumption and execution performance of each module through profiling. The profiling process can capture execution information (e.g., accuracy, energy consumption) and share the aggregated insights to the resource manager, which handles the optimization decisions.
- **Resource Manager**: The resource manager takes the decision-making role to maintain service level agreement (SLA) and minimize resource usage. Given the SLA's service metrics and user-level expectations, resource manager shall determine the most suitable configurations in the VAP by considering the module profiles and available resources.
- **Executor**: The EVA executor is to execute the functions in the pipeline. Depending on the architecture setting (e.g., whether cloud or edge is involved), a group of executors can be distributed across the devices. The executor observes the working status and reports to resource manager in case of under-utilization or over-utilization.

- **Video Storage**: For EVA, raw video data can be optionally stored across various types of devices (e.g., IoT, edge, cloud). To facilitate future retrospective service queries, metadata can be used to summarize fundamental information about video data such as time, location, size, gender, age, color, etc. The metadata enables fast retrieval of video clips without re-processing the entire video data, which is essential for efficiency at scale.

EVA Enabling Technologies:

Enabling technologies are essential to develop and deploy EVA applications across domains. They form the basis to create a robust ecosystem for EVA. We highlight two major classes as follows:

- **Hardware**: To accelerate video analysis given the availability of computing resources, enabling hardware and accelerators are essential. The range of hardware for EVA span across the common central processing units (CPUs), graphics processing unit (GPUs), and more customized field programmable gate arrays (FPGAs) and application-specific integrated circuits (ASICs). CPUs can be considered when real-time processing not required, owing to their high availability on devices. For the parallel processing capabilities, GPUs are suitable for EVA tasks such as object detection and scene segmentation. With programmable support for reconfiguration, FPGAs can offer customizable hardware acceleration across diverse scenarios with low added development costs. Since FPGAs have the capacity to host multiple functions in parallel, it is suitable for complex EVA setups where a single FPGA can simultaneously handle video encoding/decoding, object detection, and data transmission. ASICs represent highly specialized hardware that can offer high performance and energy efficiency, making them suitable for resource-constraint devices. In practice, we can combine these hardware options for an efficient configuration where AI inference and processing workloads can be balanced and boost by dedicated hardware.
- **Frameworks**: The operational setting of EVA can be complex and demanding given the layer of heterogeneous devices. Orchestration and middleware frameworks are hence crucial to enable the deployment, management, and scaling of EVA solutions. For orchestration, the virtualization technique of containerization [78] is suitable for EVA as containers can simply and speed up the creation of isolated applications over heterogeneous hardware. To manage containers, Google initially proposed Kubernetes, which has become widely used for managing the lifecycle of containerized applications in a cluster. The KubeEdge [79] is an edge oriented framework, which considers IoT, edge and cloud layers from architectural perspective. For EVA specific AI workloads, KubeEdge provides its toolkit to facilitate the deployment of ML frameworks such as Tensorflow and Pytorch.

3.2.2 Edge Video Analytics in Practice

We discuss key considerations for EVA and two practical use cases in this section for readers to gain a pragmatic perspective on EVA applications.

Key Considerations for EVA Applications:

For developers of EVA applications, there are four criteria to consider, depending on the practical settings and the status of resources [67].

- **Single-input or Multi-input**: To collect video data, the input can come from one individual equipment (e.g., camera). The single-input approach is to detect, recognize, and track objects/events within its field of view (FoV). It is suitable if the area of interest can be covered by a single video capturing equipment. Meanwhile, the multi-input approach simultaneously analyzes the video input from multiple equipment. The video capturing equipment can be distributed in different locations without overlapping coverage, or concentrated in the same location with overlapping FoVs. The multi-input approach entails higher costs but offers a more comprehensive understanding of the environment. For instance, a target in a crowed scene can be missed by a single camera due to obstacles of the crowds. With multiple cameras, the lost target has a higher chance to be captured by other cameras since several cameras can form a holistic perspective of the area under surveillance.
- **Live or Retrospective**: EVA can analyze live video data on the go when it is captured. This live approach is useful for scenarios where timely responses are needed. To achieve the live target, all processes need to be finished within the deadline. The retrospective approach, on the other hand, is for analyzing video data after it has been archived. This "offline" style allows EVA to conduct in-depth examination of historical video data. The insights can help identify trends, behavior patterns, and event relationships, which are difficult to obtain directly from live (real-time) analysis.
- **Continuous or Event-Driven**: EVA can perform continuous analysis over video data without any event trigger. This continuous approach is often used in live surveillance, where a scene or environment can be monitored continuously for identifying potential threats or anomalies as they occur. Meanwhile, the event-driven approach is for processing video input in response to pre-defined event triggers. To extract insights, this approach can associate other sensors (e.g., sound, temperature) with video analysis to detect events. The event-driven approach can be used for smart home automation where appropriate reactions can be performed when specific events are detected. For instance, when doorbell camera detects someone ringing the doorbell (e.g., sound input), the EVA procedure can be triggered to analyze the video feeds and identify the person/object at the door.
- **Stationary or Mobile**: Stationary equipment such as cameras for traffic monitoring are fixed in specific location. Given the location context, the video capturing configuration on stationary equipment is often fixed (e.g., brightness, contrast). The stationary approach can harness the resources of edge and cloud to offload heavy duty tasks. To support mobile scenarios, equipment can be carried by drones or body-worn. Due to the changes of surroundings, EVA needs to apply pre-processing techniques such as noise reduction, rotation correction, and motion blur elimination. These techniques can improve the quality of video frames for VAP. Given the challenging conditions of mobility, real-time processing is performed mainly in an on-device manner, supported by dedicated hardware and lightweight pipeline designs.

To implement EVA solution, we need to make tradeoff depending the practical requirements, since all the criteria cannot be satisfied without compromising one or the other.

Because video analytics relies on data that may contain private and sensitive personal information, the applications of video analytics need to handle the data privacy matter with care, besides the performance considerations (e.g., latency, accuracy, throughput). With the help of edge devices, the privacy-preserving operations of EVA can utilize the edge layer, hence avoiding the exposure of private information directly to the remote cloud. We highlight three privacy preservation considerations for EVA along the data-flow stages, data collection, data analytics, and data storage [67, 75], as follows:

- **Privacy Preserving Video Collection**: When EVA applications are collecting video data, the risk can come from unreliable application owners and unreliable network conditions. To avoid illegal usage, the sensitive information such as bio-metric facial data shall not be collected by untrustworthy parties. Since the transmission of raw data can be eavesdropped, lightweight encryption schemes suitable for IoT devices need to be enforced.
- **Privacy Preserving Video Analytics**: To protect privacy while still extracting valuable insights from video data, EVA can utilize blurring technique to anonymize individuals and retain the overall context. In addition, differential privacy techniques can add statistical noise to the video raw data or the analysis results, while still allowing us to perform aggregate analysis.
- **Privacy Preserving Video Storage**: Since raw video storage on the remote servers (cloud) may lead to leakage of data, EVA need to balance the raw data storage across the IoT, edge and cloud. Selectively storing relevant video segments or frames with critical information can help reduce the volume of sensitive data stored in EVA.

Use Cases:

To illustrate practical use cases of EVA, we discuss two recent solutions that are dedicated to smart car collaborative perception and ride sharing services.

First, the Edge-Assisted Multi-Vehicle Perception (EMP) framework [80] is designed to enhance the perception capabilities of multiple vehicles through the assistance of MEC^2. Such collaborative perception design of car and edge can extend the field of view for each vehicle, enabling better detection of objects and obstacles outside the range of any individual vehicle's sensors coverage. Given the dynamic network conditions, EMP provides robust algorithms for resource-efficient view merging and data sharing under potentially fluctuating network access. Owing to its offloading design, in addition to ground transportation, the core concept of EMP can be introduced to other domains such as UAVs (drones).

Second, from safety and security perspectives, SafeShareRide is a solution for ride-sharing services [81]. The focus is to protect both drivers and passengers from various security threats, and hence ensuring safer ride-sharing experience. By using edge devices that are near the vehicles, SafeShareRide can support real-time monitoring of ride-sharing activities. SafeShareRide analyzes multiple input data (e.g., GPS, accelerometers, on-board cameras) locally at the edge, for rapid detection of suspicious behavior and potential attacks. The core feature SafeShareRide is its 3-stage detection technique: the first stage is by using speech recognition from audio sources to detect keywords such as "help" or loud arguing statement. The second stage utilizes external sensors for driving behavior monitoring, for instance, by collection insights of speed, acceleration and angular rate. The third stage is video driven analysis for emergency detection. The combination of these three stages can enhance detection rate with low bandwidth demand for video uploading.

3.3 Edge-Enabled Augmented Reality (AR)

3.3.1 AR Overview

Augmented Reality (AR) technology provides an enhanced perception of the real world, by including additional information over the real-world view. In general, an AR system projects computer-generated augmentations on top of a real-world view, combining both real and virtual objects so that they function in relation to each other [82]. This augmentation enables real and virtual objects to act with synchronicity and provide proper depth of perception. In addition to providing visual augmentations, the recent advancements of AR technology is capable of providing audio outputs and haptic feedback to make the experience more sophisticated and informative. Thus, AR technology analyzes the environment, derives important details, and augments it with additional knowledge provided from databases [83]. Generally, an AR system has the following three properties [84]:

- Combines real and virtual objects in a real environment;
- Runs interactively, and in real-time;
- Registers (aligns) real and virtual objects with each other.

AR applications running on mobile devices which enable users to experience AR while supporting device mobility are often referred to as Mobile Augmented Reality (MAR).

The distinction between AR and Virtual Reality (VR) and their placement in the continuum of real to virtual environment is defined in [85]. AR is defined as part of Mixed Reality (MR) where the surrounding real environment is enriched. Conversely, in both augmented virtuality and virtual environment/VR, the surrounding is virtual. This is depicted in Miligram's continuum of real to virtual environment in Figure 3.1.

In general, a MAR application has its inputs, processing functions and the outputs [86] as depicted in Figure 3.2. The inputs are captured by cameras, microphones, gyroscopes, GPS or any other companion devices including wearable devices having multiple inbuilt sensors [86]. Processing is a collection of functions that generates one or several forms of outputs. Tracking, rendering, interaction, calibration and registration are some of the processing techniques [87, 88]. Output is responsible for presenting the processed virtual content to the user in different modes such as visual (via AR displays), voice and haptic feedback [89].

3.3.2 Recent Advancement Towards MAR

Even though the concept of AR has been there for over two decades, recent technological advancements made the concept a reality. Those technologies include computing

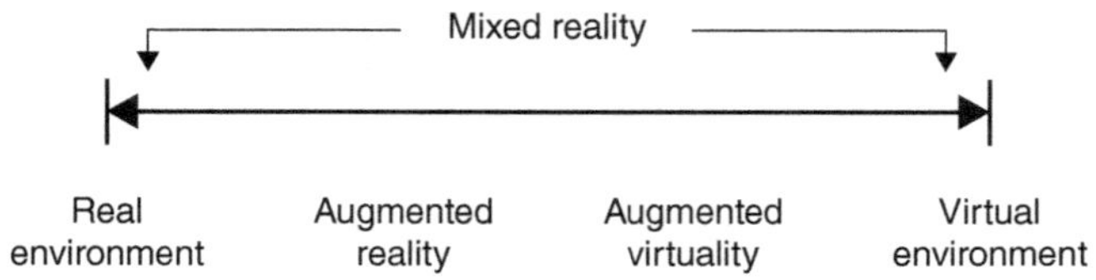

Figure 3.1 Miligram's Reality–Virtuality Continuum.

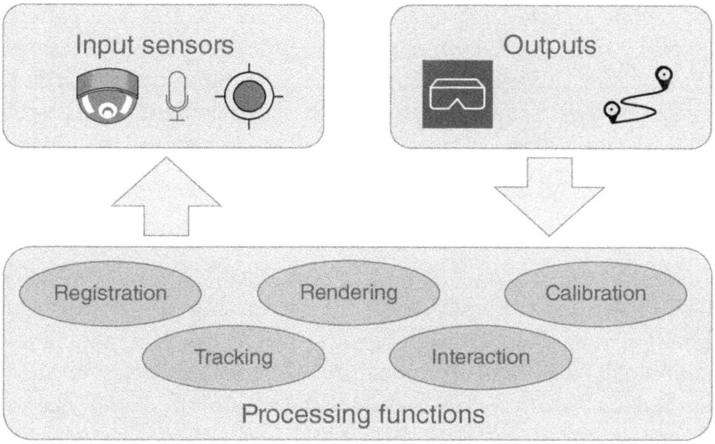

Figure 3.2 Components of a Generic AR System.

hardware and software, wireless communications and networking, displays, and wearable devices [90]. Recently, AR technology has become more popular in number of application areas such as industry, healthcare, entertainment, and education and training [91]. Extending the capabilities of AR to mobile devices such as general mobile phones and wearable devices realizes more benefits from the technology [92–95]. Using the navigational capabilities and situational awareness of mobile devices, features of MAR applications can be enriched to serve the device user in a more sophisticated manner. MAR is part of the AR that users can take with them wherever they go [96]. MAR does not only mean that the devices are portable, but also means that the users can access the AR capabilities conveniently using mobile devices anytime anywhere. Characteristics of MAR against AR and the difference between AR and MAR is illustrated in Figure 3.3. Google Glass [97], which was first commercially available in 2014, was one such product that demonstrated the enhanced user experience in the future [98, 99]. Since then, the technology has been evolving and available to widespread users in mobile phones via many applications such as Pokemon Go [100], Google Lens [101], and in recent wearable devices such as Microsoft HoloLens [102].

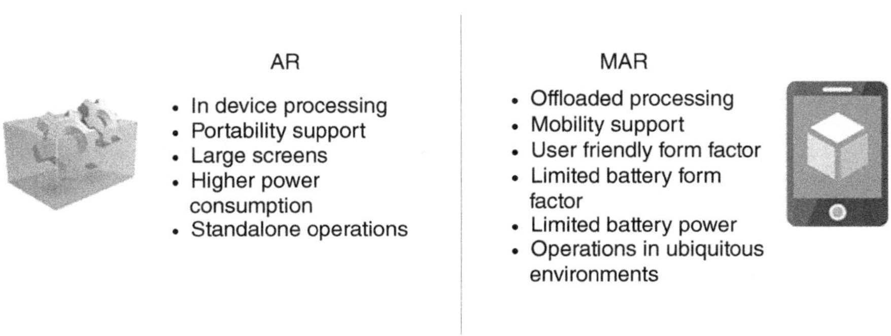

Figure 3.3 Characteristics of Augmented Reality vs. Mobile Augmented Reality.

Resource constraints of mobile devices such as limited computing power and battery power, hindered the emergence of MAR. Even though cloud offloading provides a solution [103–106], it is not suitable for latency critical applications. In the context of *MEC*², since the proposed Multi-access Edge Computing (MEC) brings the computations closer to the end devices [13, 25], the processing of latency critical MAR applications can be offloaded to MEC via the radio network [52].

3.3.3 MAR in 5G and Beyond

MAR places strict demands on wireless communication and networking and the practical adoption of MAR is still limited due to current technological constraints. The capabilities of 4G Long Term Evolution (LTE) networks do not satisfy the strict requirements such as ultra-low latency required by MAR. As an example, the minimum Radio Access Network (RAN) of LTE networks is 10 ms [107, 108] while scenarios associated with MAR assisted surgery use case requires End-to-End (E2E) latency less than 10 ms. Certain extreme remote surgery use cases demand E2E latency of 1 ms [109].

5G will be a paradigm shift from present wireless communications technologies, and it will also be highly integrative to provide universal high-rate coverage and a seamless user experience [110]. The Mobile and wireless communication Enablers for the Twenty-twenty Information Society (METIS) project has derived 5G requirements into the five technical objectives [111]. These objectives include 10–100 times higher user data rate, 10 times longer battery life for low power massive machine communication, and 5 times reduced E2E latency. As MAR systems demand higher data rates, extended battery life of terminal devices, and ultra-low E2E latency, 5G systems likely be providing solutions to the technological constraints associated with wireless communication aspects, which would otherwise be highly challenging/not possible with the current 4G systems. A stereoscopic High Dynamic Range (HDR) 360 video at 8 K 90 fps requires a data rate higher than 200 Mbps. A 6 Degree of Freedom (6DoF) video that allows translational movement requires a data rate from 200 Mbps to 1 Gbps per user [112]. The key focus of 5G is on enhanced Mobile BroadBand (eMBB), Ultra Reliable and Low Latency Communication (URLLC) and massive Machine Type Communication (mMTC) [113, 114]. eMBB focuses on high data rates while URLLC aims to provide extremely low delays, both are mandatory to realize MAR. AR and VR applications require very fast request-response cycles [115], which will drive the design of 5G network architectures.

3.3.4 MEC for MAR

As MAR requires high data rates and ultra-low latency, the edge offloading is preferred over cloud offloading. Along with 5G capabilities such as higher data rates and ultra-low latency [116], edge processing at mobile networks likely to guarantee the requirements of MAR applications. For a general AR application, particular environment and the details about the surroundings are highly localized and often irrelevant beyond the point of interest [25]. Few details or a summary of the processed data may be required beyond the MEC server. Hence, by processing the data close to the MAR devices, bandwidth of infrastructure beyond the MEC server can be saved. This concept is illustrated in Figure 3.4, where the

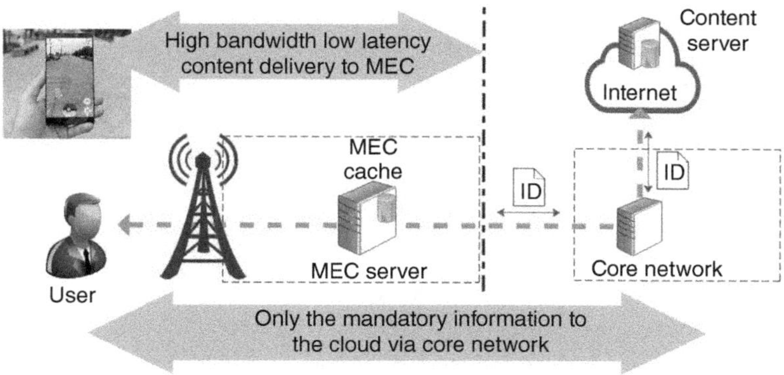

Figure 3.4 A Mobile Augmented Reality Service Scenario with MEC processing.
(a) Source: PaulShlykov/Adobe Stock.

local information is processed at the MEC server and only the mandatory information is passed to the cloud via the Mobile Network Operator (MNO) core network.

In general, MAR users in a given environment tend to seek similar type of data at different times. The environments they capture using the MAR device cameras might also be similar. This requires the transfer of redundant data over the network causing the network resources to be utilized inefficiently. Caching techniques overcome this problem by keeping the popular content at MEC server, which significantly reduces the processing at the MEC server in applications like AR [117], ultimately resulting in reduced E2E latency.

3.4 Architecture Options for Mobile Augmented Reality

This section presents cloud based, MEC based, localized and hybrid architectures for MAR systems. The categorization is based on the location of execution of the AR processing functionality. Different architectures are needed for use cases with diverse requirements and the architectures are based on the fact that whether the AR function is hosted at a cloud server, MEC server, local server, within the client AR device, or in a mix of the above.

3.4.1 Cloud Based Architecture

MAR system operates in a client server model in the cloud based architecture. MAR devices that perform a defined task act as the clients. AR server is located at the cloud and accessible from anywhere via the internet. The client's task is to capture the images, carry out minimal mandatory processing, transmit the images to the server and display the augmentations received from the server. AR server performs computationally intensive tasks related to image processing [118]. The generic cloud based system architecture for a MAR system is illustrated in Figure 3.5.

A MAR system that identifies books using the captured images [119] has a cloud based implementation. Once the device camera is directed to a known identifier in the book cover, the application identifies the book and provides details about the book on the device screen.

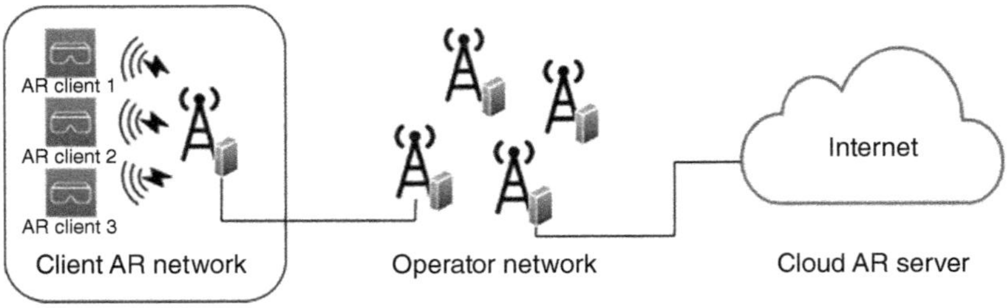

Figure 3.5 Cloud Based System Architecture for a MAR System.

As this application is not latency critical, cloud offloading is a viable option. An analysis of different applications that can be hosted in a cloud based MAR system includes indoor and outdoor navigation, collaborative urban design, and multi user interactive motion learning systems [120]. The possibility of providing AR enhanced education using cloud computing architecture is explored in [121]. A cloud based low latency MAR system which performs most of the operations locally to achieve low latency [122], offloads the images to the cloud only when necessary. The purpose of the cloud offload is calibrating its internal localization results. This system is designed to achieve minimal offloading overhead. A seamless and energy efficient AR navigation system for indoor environments [123], utilizes a cloud offloading architecture. Rather than executing complex tasks in resource constrained mobile devices, offloading improves performance and reduces device power consumption. A MAR system enriches and enables educational content transmission, favoring medical training [124] hosts the image recognition function in the cloud. Students can visualize the enriched content on the screen of the mobile device, based on the images scanned by the device camera. A mechanism to substantially reduce the cloud data transfer in a cloud based MAR system [125], proposes a solution to the network latency issue in MAR applications.

3.4.1.1 Advantages and Disadvantages of Cloud Based Architecture

A cloud based MAR system offloads the processing functionalities to the cloud server. MAR clients perform only the mandatory pre-processing tasks such as image compression/coding. Therefore, these devices are called "thin clients" [126]. Apart from achieving a longer battery life for the MAR devices, offloading helps the client to be designed with minimum hardware components. This leads lightweight and user friendly MAR clients which are easy to handle and wear as head mounted devices or smart glasses. Due to the minimal processing at the terminal MAR devices, administration of the system is easier. For example, feature upgrades of the core image processing software is easy to perform only at the server, rather than upgrading all the terminal devices. The same applies to software troubleshooting as well.

The main disadvantage is the increased transport latency due to the cloud setup. The radio delay of the 5G system and the delay of the backhaul links are aggregated for increased one way E2E latency. The possibility of a single point of failure is another disadvantage, where a failure at the AR server affects the entire MAR system. The cloud based MAR architecture highly depends on the condition of the infrastructure. If the infrastructure network is congested, the E2E latency increases. This causes difficulties in satisfying the application service level requirements. A failure of any infrastructure link may also affect the performance of the AR system. Additional security and privacy threats arise in a cloud based architecture because the data is transferred outside [18]. If the AR data stream contains sensitive information, then the information is more vulnerable to attacks than a standalone system.

3.4.2 MEC-Based Architecture

As highlighted in Section 3.3.4, certain AR use cases require ultra-low latency which the cloud based architecture can not guarantee. To serve the ultra-low latency AR use cases, MEC is the most suitable computing paradigm as it performs the computations at the edge of the network [25, 127]. MEC based architecture also operates in a client server model in which the AR server is located at the edge of the radio network. MAR clients are equipped with minimal functionality and the server performs all the computationally intensive tasks such as pre-processing of the images, performing AR tracking, rendering the augmentations. The generic MEC based architecture for an AR system is depicted in Figure 3.6. The internet connection establishes the connectivity with the outside for actions such as upgrading device software, but does not involve in core AR processing. Similar to cloud based architecture explained in Section 3.4.1, MEC based architecture also comprised of lightweight, energy efficient MAR clients due to computational offloading. MEC based architecture can serve latency critical applications which would otherwise not be possible with cloud based architecture.

An edge based architecture for driver assisting AR systems [128] is defined around three layers called device layer, access network layer and the core network layer. The device layer consists of AR clients, edge servers located at both access network level and core network level. The access network level edge servers serve the latency sensitive applications and core network level edge servers provide a large scale of services. The optional cloud is used as data backup and does not actively involve in any AR communications. Caching of popular content and performing computation at the edge in 5G to reduce the E2E latency, and dynamic adaptability of AR/VR applications have been discussed in [117]. This will remove the bottleneck of bachhaul links and enable ultra-low latency services with 5G networks. A latency minimization resource allocation for a multi-user AR system [129] proposes a MEC based data sharing model for AR task processing. An edge based AR architecture that provides low latency communications in vehicular networks is presented in [130].

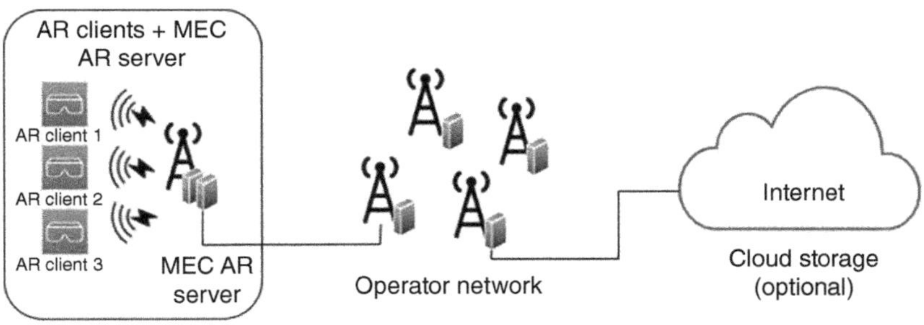

Figure 3.6 MEC Based System Architecture for a MAR System.

3.4.2.1 Advantages of MEC-Based Architecture

The main advantage of MEC based architecture is the reduced E2E latency compared to cloud based architecture. The AR function depends less on the infrastructure links compared to the cloud based architecture as the server is located closer to the clients. This ensures a more reliable communication than the cloud based architecture. As most of the information is localized, it is possible to perform caching at the edge servers as highlighted in Section 3.3.4. This also reduces the E2E latency and congestion at the infrastructure network beyond the edge server. The user data does not transfer over a public network, ensuring more secure communication. MEC based architecture also allows lightweight and energy efficient MAR devices due to computational offloading, making the devices more user friendly.

3.4.2.2 Disadvantages of MEC-Based Architecture

As in cloud based architecture, the single point of failure still exists because the processing occurs in the MEC server. When the MEC AR server fails, it affects the functionality of all AR clients leading to a system failure. MEC based architecture needs more server resources in certain use cases when the application spreads geographically. As an example, consider two distant branch locations of a factory that uses the same AR functionality. In a cloud based architecture, this requirement could be served from a centralized server but the MEC based architecture needs two servers at respective network edges. The higher operational overhead compared to cloud based architecture is also unfavorable. Activities such as server upgrades are centralized in cloud based architecture while MEC based architecture needs upgrades in all the MEC servers.

3.4.3 Localized Architecture

This book presents two options for the localized architecture. The first option is a client server architecture having AR server very close to the MAR devices as depicted in Figure 3.7, preferably in the same environment as the MAR devices. The MAR devices and the AR server are connected to the nearest next generation NodeB (gNB) via wireless links. The server performs the computationally intensive processing tasks and the MAR clients perform only a limited set of mandatory processing.

Alternatively, the function of the AR server can be eliminated as depicted in Figure 3.8. In this case, the MAR device is responsible for performing all the functions including

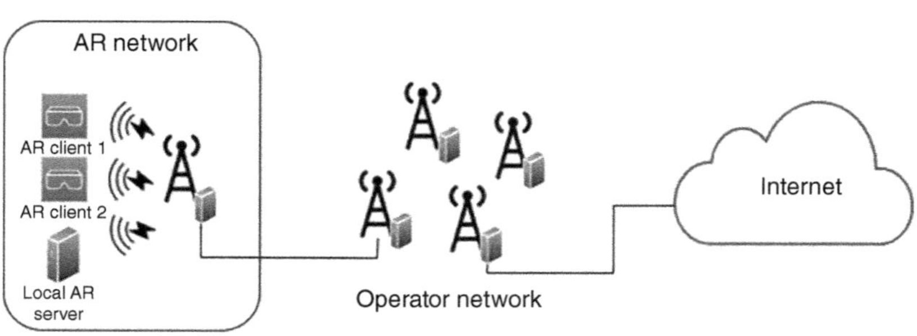

Figure 3.7 Localized System Architecture for a MAR System Using a Dedicated AR Server.

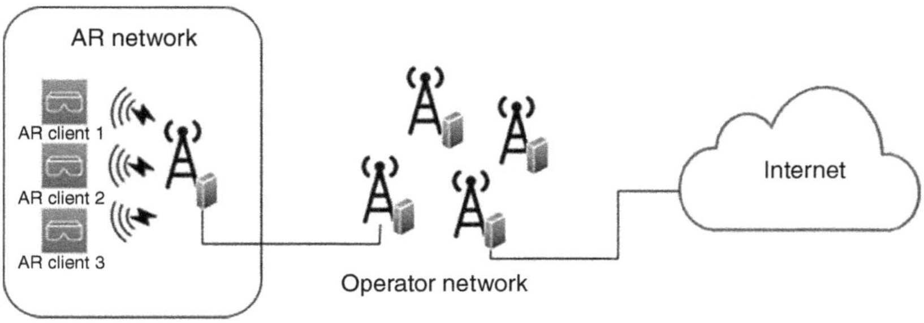

Figure 3.8 Localized System Architecture for a MAR System Using only MAR Devices.

capturing the images, performing AR tracking, rendering augmented image and displaying it on its own screen. This is also said as a standalone MAR system.

An android based AR application that provides support for interior decoration [131] utilizes the inbuilt sensors of the smartphone to acquire the inputs and perform the processing within mobile device to produce the output. A survey on different types of AR applications [132] presents different types of handheld device based AR applications used in five areas called sports, games and edutainment, cultural heritage, medicine, and education and training, marketing/advertising. An analysis of handheld device based AR [133] explains how the evolution of mobile devices can be used for AR and the current limitations of such devices. An AR based indoor navigation application for mobile devices [134] uses handheld devices, providing visual content and activity based instructions as the augmentations.

3.4.3.1 Advantages of Localized Architectures

The MAR system having a local server can achieve significantly lower latency compared to the cloud based approach. Moreover, all the data related to AR communication stays within the premises, making them less vulnerable to security attacks and privacy issues. The day-to-day administration activities such as server upgrades, client setups are easier because the whole system is deployed locally. The system no longer depends on the condition of infrastructure links, making it more reliable [12].

Ultra-low latency is the main advantage of the second approach where the entire AR operation is performed within the MAR device. The system in totally independent of the infrastructure as the data stays within the device. Security and privacy protection is very high in comparison with all other architectures as the system is standalone.

3.4.3.2 Disadvantages of Localized Architectures

Even though the AR server is hosted locally in the first approach, the latency will be higher than the MEC based architecture due to the fact that it is connected to the base station wirelessly. A single point of failure still exists as the server hosts the AR processing function.

Shorter battery life is the main disadvantage of the second approach. As the device performs all the processing, it consumes more power than all other MAR architectures leading the batteries to drain fast. As the terminal MAR devices possess less computing power than the cloud, MEC, or local servers, it is not possible to run complex MAR applications.

This limits the number of applications that can be run with the standalone MAR architecture. If any additional hardware is used to increase the battery power or computing power, the MAR device becomes bulky and heavy making it operationally difficult for continuous use. Operational activities such as software upgrades are not centralized. This needs to perform upgrades at each device. Another disadvantage is the absence of a centralized control in a use case that needs multiple devices.

3.4.4 Hybrid Architecture

A hybrid architecture for a MAR system is an architecture having a cloud node as well as an edge node along with the client AR devices. Since the overall network latency has a direct relationship with the length of a signal path and the number of intermediary routing/switching elements, the latency critical interactive MAR functions should be executed at the edge node. Conversely, the applications which can tolerate a certain delay are hosted at the cloud node. The control and management plane functions of the network may be performed at the cloud node or at the edge node. These functions include control plane signalling, service flow processing, policy scheduling and capability opening, and network operation and maintenance, node management and slice management [126]. These functions have no impact on the interactive nature of the MAR experience. The edge node is responsible for user plane functions that assure the ultra-low latency requirements of the required MAR applications. The hybrid architecture is depicted in Figure 3.9.

A hierarchical architecture to serve the AR applications [135] comprised of an edge layer between the conventional user layer and the cloud layer. A fog computing and cloudlet based AR system for the Industry 4.0 shipyard [136, 137] has an architecture that consists of three layers called the node layer, edge layer and the cloud. The node layer consists of Industrial Augmented Reality (IAR) devices. The edge layer consists of a local gateway that is used for data caching for sending content to IAR devices and a cloudlet capable of carrying out computations requiring low latency. The cloud hosts the services that require more processing power. A practical evaluation of commercial IAR systems in an Industry 4.0 shipyard is discussed in [138]. The evaluation was conducted on an IAR system that is based on a fog computing architecture which comprises three layers called node layer, fog layer and cloud layer. A MEC-based collaborative web AR solution that has complimentary

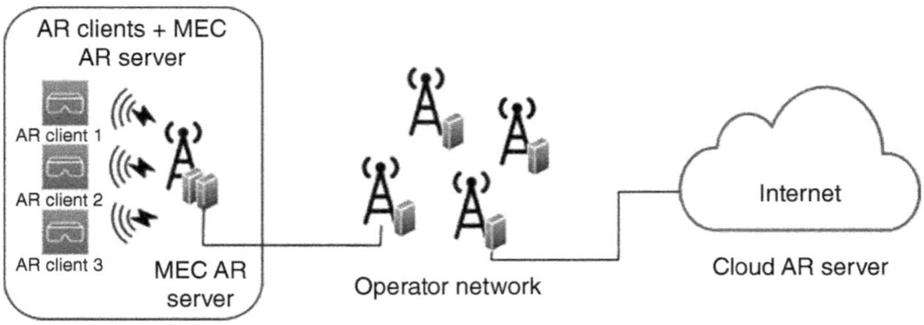

Figure 3.9 Hybrid System Architecture for a MAR System.

edge and cloud servers [139] reduces the network latency while decreasing the bandwidth usage of core networks. The edge server consists of MEC application platform, Web AR runtime environment and Web AR applications.

3.4.4.1 Advantages of Hybrid Architecture

The advantages of hybrid architecture over cloud based architecture are as follows. Reduced latency is the main advantage over the cloud based architecture due to the edge based processing. Edge processing reduces the application dependency on the infrastructure links leading to more reliable communication. In a case where most of the information is highly localized, caching at the edge servers reduce the E2E latency while decreasing the congestion at the infrastructure network beyond the edge server. Hybrid approach has better security and privacy than the cloud based architecture because the data stays close to the terminal devices. The advantages of hybrid architecture over the MEC based architecture are as follows. MEC server has less overhead because it does not need to process the control messages, but the cloud server does it. Cloud based control offers a more centralized approach for controlling in comparison with MEC based architecture. Energy consumption of the terminal devices is low compared to the localized architecture due to the fact that only the mandatory operations are performed at the device level.

3.4.4.2 Disadvantages of Hybrid Architecture

Since the control functions are handled by the cloud and the interactive AR functions are handled by the edge servers, the single point of failure still exists. The system does not work properly if either the edge server, or cloud server or the infrastructure network fails. Hybrid architecture is generally more complex than the other architectures. Therefore, the effort for the initial design, operational overhead and administrative overhead is higher.

3.4.5 Summary of Architecture Options

Four possible architecture options are discussed in Section 3.4 i.e. cloud based, MEC based, localized and hybrid. The selection of the most suitable architecture for a MAR system depends on the requirements of the applications it serves. Even though there are few uses of the localized architecture discussed in the literature, it is not the preferred option for many use cases. The reasons are twofold. To produce a better experience with an AR system, the underlying processing functions should be sophisticated and this requires more computing power causing the batteries of the mobile devices to drain fast. MAR devices have limited battery power and this leads to frequent charging of the MAR devices, making it operationally difficult. If MAR devices are equipped with larger batteries, then they will be bulky and heavy making them less user friendly. Secondly, the network level limitations are currently diminishing and paving the way to achieve ultra-low latencies. This is possible due to the availability of high speed connectivity and it makes computational offloading a viable option. With these network characteristics, cloud based and especially MEC based architectures are promising to provide the required ultra-low latencies for MAR applications. A mapping of different AR applications against the possible architecture options is depicted in Figure 3.10.

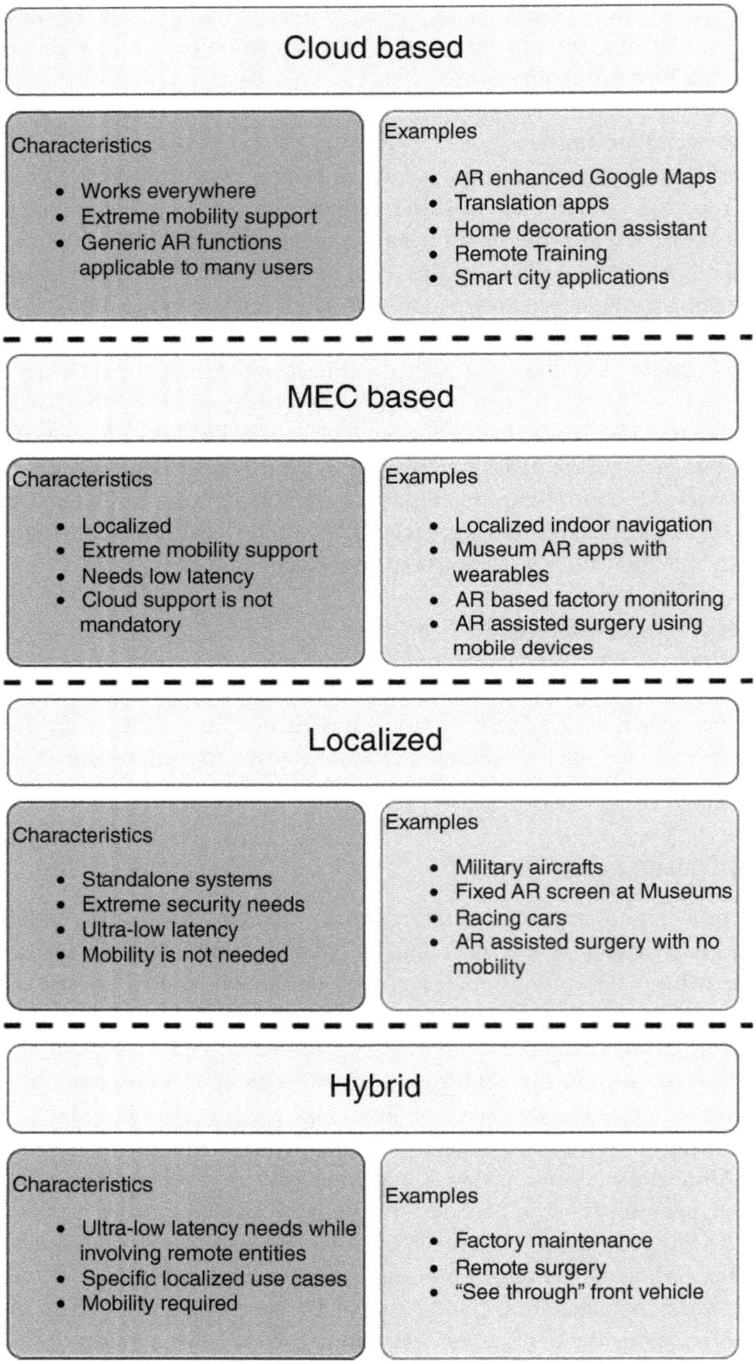

Figure 3.10 Applicability of Different Architectures Based on the Characteristics of MAR Applications.

3.5 Applications of MEC-Based MAR

The use of MAR has been a limited number of specific applications in the past. Even though technological limitations still remain, current MAR applications are spread across multiple domains. With the introduction of 5G systems, the realization of MAR technology in different application areas will be more practical since MEC plays a crucial role in the realization. This section discusses the potential MEC based MAR application areas and the use cases. This section considers application areas of the Industrial Internet of Things (IIoT), healthcare, travel and tourism, smart city and smart home, and automotive.

3.5.1 IIoT

Use Cases:

MAR is an essential part of IIoT which serves multiple use cases. For example, remote maintenance in IIoT allows a skilled remote operator to connect with an unskilled operator working on site, and guide the unskilled operator through a maintenance task. The unskilled operator receives the maintenance instructions via a MAR device and the use of paper material is obsolete. The expert does not need to be on site, thus bringing cost benefits to the industry. A similar approach is followed in MAR based remote training in industries. Both the trainer and trainees use MAR devices, share a synchronized view and the trainer guides the trainee(s) through the material while sitting at a remote location. In productivity improvement of assembly processes, workers can monitor the assembly processes and identify any defects using MAR technology. MAR technology in IIoT is used to provide safety in day-to-day activities. If an operator does not perform a given activity in an accurate and safer way, the MAR system can block the operator and stop progressing to the next step to assure safety.

How MEC Improves MAR in IIoT:

Operators generally use MAR devices as head mounted, wearable or handheld devices in IIoT environments. This requires the devices to be smaller and easy to handle. IIoT use cases are typically of the same nature and capture similar type of environment, therefore a dedicated server is preferred over performing the processing at the individual client. To achieve these, the AR related computations should be offloaded. By having the processing at a MEC server rather than a cloud server, ultra-low latencies required by the use cases can be achieved [9].

3.5.2 Healthcare

Use Cases:

MAR assisted surgery is a widely discussed use case in the healthcare sector. The healthcare officers take the aid from wearable devices to conduct the surgery. In surgeries, these devices should be worn for a considerable amount of time, therefore they should be easy to handle. This makes the wearable devices to be equipped with limited processing power and battery life, ultimately leading to offloading computations. The same use case may be supported by remote entities with a fast and reliable communication infrastructure, paving the way to remote surgeries. The remote professional can either provide guidance for the surgery or in a more sophisticated case, the remote professional can actually conduct the surgery if he is having the necessary equipment to capture his movements and transmit them real-time to a robot at the theater who actually performs the surgery of the patient. Similar to IIoT remote maintenance, remote medical training is another use case for MAR in healthcare. It could be either remote training or self-training using wearable devices. Personalized medicine enhanced by wearable MAR devices, can be used for eye tracking, speech recognition, and motion tracking. It reveals the patients' health related information via the wearable MAR device and is used to provide personalized medication for patients.

How MEC Improves MAR in Healthcare:

Similar to IIoT use cases, the provision of ultra-low latency is the main advantage of MEC. The devices support greater mobility when they are lightweight and easy to handle. The MEC enables the process offloading to realize this objective. Moreover, the healthcare related data are sensitive and includes personal data, which needs greater security [140]. By having a MEC server for processing the data stays close to the client devices assuring better privacy and security.

3.5.3 Travel and Tourism

Use Cases:

The most familiar MAR application in the travel and tourism sector is the AR enhanced Google Maps application. In addition to the directions, the AR enhanced Google Maps provides more information on the screen such as virtual signs. Google Maps application uses the camera of the mobile phone to capture the surroundings and determines the position of the user. Then the application augments the screen with virtual signs and provides audio notifications appropriately. The same concept applies in number of MAR based guidance applications that are available at specific locations such as parks, museums, libraries, hotels, shopping malls. Enhancing user experience of museum visits with the help of MAR enables the visitors to visualize additional information about the artifacts. Automatic translators such as Google Translate, MAR based restaurant menus, hotel brochures are other applications.

How MEC Improves MAR in Travel:

In general, travel and tourism MAR applications utilize the general mobile devices rather than specialized devices, such as navigational MAR applications. Rather than operating with a cloud based architecture, MEC can be used to perform the processing closer to the MAR clients to reduce latency. This reduces the network congestion beyond the MEC server.

3.5.4 Smart City and Home

Use Cases:

MAR based applications to assist elderly and disabled people at home provide the necessary support in carrying out day-to-day activities such as switching lights on and off, operate electrical appliances, opening the doors, etc. The same concept can be applied in smart cities to increase accessibility to the disabled people. Navigation in smart cities is also a popular application which can be improved with MAR technology. MAR based remote home control allows the user towards a more convenient and sophisticated way to monitoring the home from a remote location. MAR enabled building/home management systems are also a popular MAR application. A MAR application can automatically recognize the dimensions and geometry of the building interior and visualize those values on top of the application screen for the user to decide the size of the furniture that fits in.

How Edge Improves MAR:

Smart home applications mostly operate on personal data. MEC can be used to impose additional security compared to a cloud based architecture. This makes the application less dependent on the network infrastructure.

3.5.5 Automotive

Use Cases:

AR enhanced windshield allows the driver to see vital information on the windshield, allowing the driver to keep constant focus on the road. Vehicle speed, present speed limit, navigation information, and the best possible driving path for in a racing event, are some of the possible information that can be shown on the windshield. More advanced AR systems allow the vehicles to see through the front vehicle, i.e., the driver at the rear vehicle sees the road as if there is no front vehicle blocking the road. This AR application artificially removes the obstructing view of the front vehicle making the road more visible to the driver. Collaboration between the front and rear vehicles is mandatory for the success of this use case. Driver training is also an application that can be facilitated by AR technology. Both the trainer and the trainee can wear HMDs and see the same view. AR can be used in the automotive industry to compare the real part of a vehicle against the prototype design. If the real part does not match the prototype perfectly, then the part will not be assembled in a car.

How MEC Improves MAR:

Automotive applications require fast responses, hence ultra-low latency support is mandatory for MAR systems. MEC architecture aims to reduce the communication latency by implementing the servers closer to the end devices guaranteeing the ultra-low latency.

3.6 Technical Aspects of MEC-Based MAR

We discuss technical aspects in relation to future MAR systems, namely security, privacy, communication, energy, offloading, service migration, and mobility.

Security:

Security is crucial in 5G MAR systems to assure the service levels required by its use cases. The data flow of the MAR system should be available accurately to its users without any tampering or modification. Because the harm that a malicious party can incur may lead to incorrect decisions and subsequently undesirable outcomes. The data should be available to the users within the defined latency requirements. Attacks like overload attacks slow down MAR applications in a way that the required ultra-low latency requirements cannot be fulfilled. As an example, sensory overload attacks such as playing loud sounds, flashing bright lights on the display can cause malfunctions in the system. In MAR context, cross-app sharing is a type of threat where one application shares the sensor data to another unauthorized application of the MAR device. Clickjacking is another type of attack that tricks users into clicking on sensitive user interface elements that originated from malicious applications [141]. Deception attacks on AR populate the screen with distracting content such as a modified road sign [142]. Therefore, during the design and development of future 5G MAR systems, proper security mechanisms should be established to prevent from such activities of malicious parties.

Implementing security algorithms is mandatory to protect the MAR applications so that the required service times are realized. However, with the resource constrained MAR devices, the complex security algorithms should be placed outside the terminal devices and the most preferred option is the MEC server. This way the security algorithms can be optimized to suit the local needs and support low latency communication.

Privacy:

Privacy concerns of AR technology are mainly twofold, the privacy of the user of AR system and the privacy of the people in the environment in which the AR system captures. The user's privacy concern is that the AR system captures and stores information on what user sees and then the system knows information about user actions. Other people in the environment are also captured by AR devices, sometimes even without their knowledge. Necessary protection in MAR systems must be imposed on ensuring confidentiality, anonymity, undetectability, unlinkability [143].

Similar to the security algorithms, privacy protection approaches should also be implemented at the MEC server to assure the service times and to utilize the resource constrained terminal devices more efficiently.

Communication:

The most important factor for MAR systems in terms of communication requirements is latency. A study in communication for automation in vertical domains [144] reveals that the E2E latency between capturing a new image and displaying the augmented image should be smaller than 50 ms to avoid cyber-sickness. Other studies [145] suggest that the real time applications would need one-way latency of 100 ms and online gaming needs 75 ms. These strict requirements demand ultra-low latency support from future 5G systems. Moreover, the same study suggests that the communication service availability should be higher than 99.9%, which means that 99.9% of all the video frames should be delivered successfully within the given latency constraints. Required bandwidth for the communication depends on the resolution of the video and the frame rate, which basically arises from the application. For a general industrial applications, video streams with a frame rate greater than 60 Hz and 1280 × 720 High Definition (HD) resolution is desirable. It is also required that the video stream between the MAR device and the image processing server should be encrypted and authenticated by the 5G system.

Ultra-low latency, higher reliability and higher bandwidth can be supported with a MEC based architecture. As the base stations are located close to the clients, the latency in the transport network is minimal and less depending on the infrastructure. Increasing the capacity does not affect the infrastructure beyond the MEC AR server hence, the system can adapt easily in changing requirements in terms of bandwidth.

Energy Management:

A MAR device can maximize its energy efficiency in two ways. The first one is internal process optimization. It is done by adjusting the power settings of the various processors and controller chips to allocate optimal resources to maximize power saving based on the running application. This can be achieved using Dynamic Voltage and Frequency Scaling (DVFS) technology. The second approach is to design MAR devices such that they offload most of their processing tasks to a MEC server, thereby helping the device batteries to last long. This leads to another problem of transmitting a significant amount of data wirelessly, which also causes battery drain. The transmission power is also affected by the characteristics of wireless networks such as available bandwidth. Therefore, a MAR device typically carries out some mandatory pre-processing to reduce the amount of external data transfer to the MEC AR server, but the MEC server performs the intensive task of image processing.

Service Offloading:

MAR applications require high computing power due to underlying image processing. When 5G becomes a reality, a whole set of novel MAR applications will emerge which require more intensive processing at the device level. MAR devices themselves are not equipped with such powerful processors to perform most of those computations. Computational offloading to a MEC server is the most viable solution. On the other hand, if there are a number of MAR devices used for similar types of tasks with similar types of processing, performing the tasks within the device is highly inefficient due to the redundant use of resources. As an example, consider a MAR based training session where all the MAR devices perform similar type of processing. In this case, it is beneficial to offload the image processing tasks to a MEC AR server and use the terminal devices to display the augmented data received from the MEC AR server.

Service Migration:

It should be possible for a user of a MEC based MAR system to move away from the coverage area of one base station to another. Since MEC based MAR operates on an offloading model, assuring service continuity via service migration is mandatory. Service migration helps to effectively transfer or migrate a service from a given server to another server. To support a seamless migration of services while the user is moving, the destination MEC AR server should possess similar capabilities as the source MEC AR server. These capabilities include processing power, operating system compatibility, applications, and other supporting services. The migrations should happen within a defined time to provide a seamless experience for MAR application users. The state data must be passed to the destination MEC server which contains information about the present service instance.

Mobility Management:

With the increased capabilities of future 5G systems, it is expected that the MAR devices will be directly connected to 5G [144] to maximally utilize the benefits and deliver the expectations of the users. Since 5G coverage will be provided mostly with the small cells, frequent handovers will also be seen in future MEC based MAR systems. To provide a seamless experience for the users while they are moving, mobility support is one of the major design considerations in MAR technology. The ultra-low latency and higher bandwidth requirements make mobility support challenging in MAR systems. The 5G system supporting the MAR communication should provide seamless mobility so that a handover from one base station to another base station does not have any observable impact on the application [144]. Moreover, the horizontal indoor positioning accuracy should be better than 1 m with 99% availability for a moving MAR device with speed up to 10 km/h [144].

3.7 Summary and Outlook

The adoption of EVA and MAR in different applications are expected to rapidly increase with the introduction of 5G and beyond services, given the provision of edge supports, enhanced Mobile BroadBand (eMBB) and Ultra-Reliable and Low Latency Communication (URLLC). The extreme and versatile demands such as ultra-low latency and high bandwidth by MAR lead to the design of different network architectures to serve those with the aid of edge and cloud computing. The outlook for EVA and MAR depends on the advancements on hardware, AI integration, and user demands. For these two data-oriented application domains, MEC^2 will be a key enabler by providing the necessary infrastructure to support new intelligent and interactive services. As the technologies behind EVA and MAR continue to evolve, more sectors will benefit from the analytical insights and interactive experience.

4

MEC^2 for eHealthcare: Empowering Next-Generation Healthcare Systems*

4.1 Introduction

The novel requisites for e-Health type applications align with IoT service requirements of ubiquitous connectivity to cater to the data aggregation needs of wearable devices, Augmented Reality (AR) based surgery, and Virtual Reality (VR) based surgical training applications [146]. In eHealthcare, wearable devices are pivotal in monitoring patient health, collecting vital signs, and tracking various physiological parameters. These devices generate a vast amount of real-time data that needs to be seamlessly aggregated and analyzed to provide timely insights for healthcare professionals. In this context, Mobile Edge Computing (MEC) plays a significant role in healthcare by addressing modern healthcare systems' unique challenges and requirements. Enables low-latency data processing, edge-based analytics, enhanced connectivity, security, scalability, and supporting the integration of emerging technologies. By leveraging MEC, healthcare systems can achieve improved efficiency, enhanced patient care, and transformative advancements in healthcare delivery.

In 2020, the world faced one of the most changing health issues, which is the pandemic of COVID-19. World Health Organization (WHO) has declared COVID-19 outburst a pandemic considering its rapid dispersion and the level of contagiousness. The amount of infected personnel and resources allocated for treating them create issues for governments. In addition, establishing and maintaining quarantine facilities withers the resources in terms of monetary and manpower perspectives. More importantly, the lack of knowledge on the cause of the pathogen and its transmission exposes the health care officials while treating the infected. This is a vital conundrum for most nations where hospital medical care and treatment protocols are designed and practiced via human-to-human interaction. Thus, COVID-19 is leading the future of healthcare towards contactless treatment strategies with technological engagement.

In addition, epidemic or pandemic threats of COVID-19, Ebola-2015, MERS-CoV-2012, H1N1-2009, and SARS-CoV-2003 have urged for robotic influence in the medical care and daily hospital activities; that can contribute in the areas of clinical care, logistics, and reconnaissance. To cater to all these novel technologies, Ultra-Reliable Low-Latency Communication (URLLC) capabilities are pertinent for controlling robotic appliances,

*This chapter is partly contributed by Pasika Ranaweera (University College Dublin, Ireland).

Mobile Edge Computing and Communications: Driving Forces, Technology Foundation, and Application Areas, First Edition. Aaron Yi Ding, Chamitha De Alwis, and Madhusanka Liyanage.
© 2025 John Wiley & Sons Ltd. Published 2025 by John Wiley & Sons Ltd.

enhanced Mobile Broadband (eMBB) based audio/video patient monitoring, aggregating medical data, and notifying health officials on their schedules and emergencies. However, the prevailing cloud computing-based service infrastructure is inadequate to launch these envisioned directives. The geo-distributed placement of cloud servers and weaker access capacity inherited with standalone Base Stations (BSs) create bottlenecks in traffic flows at various points in the mobile network [147]. These limitations are enforcing an unintended latency and jitter on network nodes. Thus, URLLC and eMBB specifications demanded by robotics and AR applications are infeasible with the prevailing networks. The concept of edge computing, which offers communication, storage, processing, and networking capabilities within a proximate operating range, is an ideal solution for such circumstances [13]. On the contrary, edge paradigms are promisingly adopted for healthcare directives with prolific success over cloud-based systems [148, 149].

Multi-Access Edge Computing (MEC) is a leading-edge computing paradigm introduced by the European Telecommunications Standards Institute (ETSI) to provision cloud computing capabilities at the mobile BS in proximity to the accessing User Equipment (UE) [18]. This emerging concept is one of the pillars in pragmatically launching the fifth generation (5G) mobile technology; that reforms the prevailing mobile network architecture towards a softwarized core and a backhaul network.

One of the main intentions of 5G is to launch micro or macro-scale BSs for delivering specialized services with featured URLLC capability. Considering health services in a hospital vicinity as a specialized service, a MEC-enabled In-building Base Station (IBS) can be placed on the premises. MEC's compatibility and interoperability features ensure the interfacing of myriads of medical, sensory devices and their protocols towards a centralized data center. The highly dynamic and virtualized service provisioning platform of the MEC IBS will allow servicing of diverse healthcare applications with real-time performance. The orchestrator functionality at the MEC IBS, Mobile Edge Platform Manager (MEPM); guarantees the proper distinguishing of various services, scalable access to all IoT medical devices with improved link capacity, and maneuvering the storage and processing resources to mediate the seamless operation.

Here are some critical roles of MEC in healthcare:

- **Low-Latency Data Processing**: MEC brings computational capabilities closer to the network edge, reducing data processing and analysis latency. In healthcare, where real-time decision-making is critical, MEC enables healthcare applications to process data quickly and respond promptly. MEC-based services are vital in scenarios such as remote surgeries, remote patient monitoring, emergency response systems, and telemedicine consultations.
- **Edge-based Analytics**: MEC allows healthcare organizations to perform data analytics and extract valuable insights at the network edge. Edge-based analytics improves response times and reduces the burden on centralized cloud infrastructure. It will also reduce the analytic-related communication cost between the edge and the cloud. By processing data closer to the data source, MEC enables real-time analysis, vital for applications like real-time patient monitoring, early detection of anomalies, and predictive healthcare analytics.
- **Enhanced Connectivity**: MEC provides improved connectivity solutions in healthcare settings. By leveraging edge computing resources, MEC enables local data offloading,

reducing network congestion and ensuring efficient use of network bandwidth. MEC is particularly beneficial in environments where tons of IoT-based devices, such as wearable health trackers, medical IoT devices, and imaging systems, generate massive amounts of data that must be transmitted and processed reliably.

- **Security and Privacy**: MEC helps address security and privacy concerns in healthcare. By processing data at the edge, sensitive patient information can be kept within the local network, reducing the risk of data breaches and unauthorized access. MEC also enables the implementation of privacy-enhancing techniques, such as data anonymization and encryption, to ensure patient data confidentiality. Moreover, MEC enables the deployment of security-as-a-service type secure service delivery models.
- **Scalability and Resource Optimization**: MEC allows healthcare systems to scale their computational resources based on demand. MEC enables flexible resource allocation and optimized utilization by distributing computing capabilities at the edge. MEC is particularly valuable in scenarios where healthcare systems experience fluctuations in workload, such as during peak hours or sudden influxes of patients.
- **Support for Emerging Technologies**: MEC facilitates integrating and deploying emerging technologies in healthcare. For instance, MEC can support implementing Augmented Reality (AR) and Virtual Reality (VR) applications for remote surgeries, surgical training, remote consultations, and medical education. MEC's low-latency and high-bandwidth capabilities enable seamless transmission of rich multimedia content, enhancing the user experience and opening doors to innovative healthcare solutions.

Therefore, this chapter discusses the edge computing approach to satisfy the intrinsic technological requisites of the novel MEC-based contactless and remote medical procedures within a medical facility while improving the feasibility of smart health solutions proposed in [146, 150–153].

4.2 MEC Enabled Use Cases

MEC-enabled IBS deployed at medical institutions contrives an IT ecosystem for digitizing the holistic medical infrastructure. In this section, three plausible use cases are discussed with reasoning. As illustrated in Figure 4.1, our proposed use cases share the MEC IBS's digital infrastructure. These use cases are attributed to contactless capabilities that enhance the accuracy and management of healthcare operations for optimal performance. Further, various communication and networking parameters required for realizing the proposed use cases on their primary goals/applications are introduced in Table 4.1.

4.2.1 Use Case 1: Augmented Reality (AR) Based Remote Robotic Monitoring

AR-based remote robotic monitoring is essential in pandemic situations where the hospital must treat highly contagious patients. During Covid-19 pandemic, the most observed fact during the first phase was health officials, including the physicians, were infected rapidly while treating the patients regardless of the extremely secure protective gear being worn by them [154]. This is impacting the health workforce directly. Thus, it is evident that

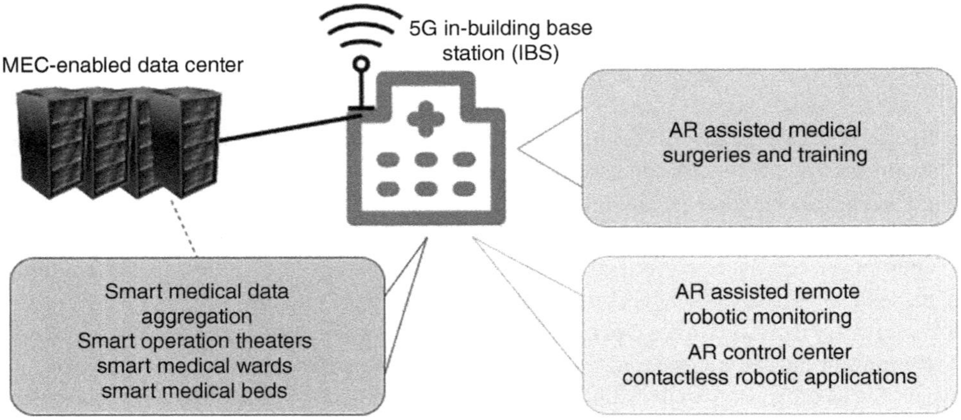

Figure 4.1 MEC Based Health Services.

measures should be taken to remotely monitor and treat the patients in a controlled environment to mitigate the infections toward health personnel. Various remote monitoring approaches are followed at present through robotics and sensory acquisitions aggregated via sensors placed in medical wards. Augmented Reality (AR) offers a novel approach to reaching patients with Remotely Operated Vehicles (ROVs) or robots that interfaces robotic entanglements with improved precision than typical remote operations [155]. As illustrated in Figure 4.1, AR-operated robots can be used for surface decontamination through ultraviolet (UV) emission, delivery and handling of toxic or bio-hazardous waste, collection and testing of nasopharyngeal and oropharyngeal swabs, and patient monitoring with the delivery of minor treatments. These functions cover the majority of the daily routine work that medical staff personnel are engaging in.

4.2.2 Use Case 2: Smart Medical Data Aggregation/Patient Monitoring and Treatment

Monitoring patients' vitals rapidly leads to the automated and accurate commandeering of medical treatment devices, which is critical for the fast recovery and survival of critically ill patients. Further, medical apparatus such as Polymerase Chain Reaction (PCR) testers, pulse oximetry measuring scope (that leads to National Early Warning Score (NEWS2) and the National Paediatric Early Warning System (PEWS) early warning scores), ventilator, and chest imaging (radiograph, CT, lung ultrasound) are intrinsic within a COVID-19 treatment facility according to the WHO treatment protocol [156]. Therefore, smart and sophisticated medical wards embedded with remotely operable medical apparatus are required for early detection and treatment of COVID-19 and to protect the medical staff from exposure. If the medical bed comprises the intrinsic measuring and treating devices (as seen from Figure 4.1), medical officers can operate the bed remotely while contacting the patient via audio and video. With the multitude of electronic devices, this digitized infrastructure is forming an mMTC environment that demands communication links with a formidable capacity and continuous connectivity. In addition, aggregation and visualization of medical

Table 4.1 Requirements of MEC Use Cases.

Use Cases/ Primary Applications	Requirements: End-to-End Latency	Jitter	Packet Loss Rate	Bandwidth/ Bit Rate	Availability	Max. # of UEs	Serving Area (m^2)	Security Level
Use Case 1: AR Controlled Robotic Monitoring								
Stereoscopic 4 K (3840 × 2160 pixels) 120 fps real-time video stream with lossless compression	<2 ms	< 10 μs	10^{-3}	>24 Gbps	>99.99999%	1	100	MEDIUM
4 K 120 fps real-time video stream with lossless compression	<50 ms	<2 ms	10^{-3}	>12 Gbps	>99.99999%	10	100	MEDIUM
Robot telemetry/ motion control data stream	<2 ms	<2 ms	10^{-4}	>16 Mbps	>99.999999%	10	100	MEDIUM
Haptic feedback data stream	<2 ms	<2 ms	10^{-4}	>16 Mbps	>99.999999%	1	100	MEDIUM
Use Case 2: Smart Medical Data Aggregation and Monitoring								
Physical vital signs monitoring data stream	<250 ms	<30 ms	10^{-4}	>1 Mbps	>99.999%	20	10^8	HIGH
CT/MRI scan data	<1 s	<30 ms	10^{-4}	>240 Mbps	>99.99%	20	10^8	HIGH
Uncompressed 512 × 512 pixels 20 fps video stream from ultra-sound probe	<50 ms	<30 ms	10^{-4}	>160 Mbps	>99.99%	20	10^8	HIGH
High quality audio stream	<100 ms	<30 ms	10^{-2}	>128 Kbps	>99.99%	20	10^8	MEDIUM
Stereoscopic 4 K 60 fps color coded real-time video monitoring	<250 ms	<30 ms	10^{-3}	>2 Gbps	>99.99%	20	10^8	MEDIUM
Use Case 3: AR-based Remote Surgery								
Uncompressed 4 K 120 fps real-time video stream	< 750μs	< 10μs	10^{-4}	>30 Gbps	>99.99999%	1	100	MEDIUM
3D 256 × 256 × 256 voxels 10 fps ultrasound unicast data stream	<10 ms	< 10μs	10^{-3}	>4 Gbps	>99.9999%	1	100	MEDIUM

Legend: Capabilities of Pre-5G; Capabilities of General 5G; Capabilities of MEC enabled 5G data from a remote and a centralized hub is vital for early diagnosis of COVID-19, where the treatment plan for the patient is determined accordingly.

4.2.3 Use Case 3: AR Assisted Remote Surgery

There are tendencies to adapt AR for remotely operated surgeries in addition to visualizing internal organs externally to improve invasive medical procedures with enhanced precision. Further, Haptic feedback approaches can be induced in AR applications; where kinaesthetic communication is practiced with touch or motion-based perception is conveyed to the user for enhancing the feedback retrieval through additional human-sensory adaptions. The requirement for a Robot-Assisted Surgery (RAS) arises in emergencies where a qualified surgeon is unavailable within the premises or the possibility for surgical staff to be exposed to a pathogen while performing the surgery constitutes a COVID-19-like circumstance. American College of Surgeons identified the surgeries of rupture tubal-ovarian abscess, tubal-ovarian abscess not responding to conservative therapy, emergency cerclage, Cancer or suspected cancer, and Cerclage of the cervix as plausible RASs [157]. These AR-based RASs demand eMBB, mMTC, and URLLC requirements with higher processing and communication resources.

4.3 Realizing Use Cases with MEC

To facilitate the requirements of the use cases proposed in Section 4.2, an IBS with MEC capability is ideal, considering its dynamic service provisioning capability. Since the MEC platform is locally deployed, the medical services can be classified as local and remote operations. As in Use Case 1, the local services operate without the core network connectivity. This section presents a MEC-based edge platform architecture implemented in a hospital environment to achieve the requirements demanded from the use cases mentioned above. Figure 4.2 presents the high-level architecture of the MEC edge platform that is formed according to the ETSI standards.

4.3.1 AR-Based Remote Robotic Monitoring with MEC

As indicated in Figure 4.2, AR Robots or AR bots proposed under this use case are controlled via the MEC edge platform that is deployed with a 5G radio BS within the hospital vicinity. This AR robot controlling platform is interfaced by a control station where users operate the AR bots. The remote connectivity between the robots and the control station is established through the MEPM. MEPM, in accordance with the requirements of the AR bot to be launched (depending on the requirements of services out of decontamination, logistics, or interacting with the patients), will request the Virtualization Infrastructure Manager (VIM) to form a Virtual Machine (VM) with intrinsic resources to handle a single AR bot. These AR bots will be embedded with 5G radio transceivers to enable wireless and rapid communication with the IBS.

The Mobile Edge Host (MEH) formed as a VM, launches the required functions as Mobile Edge Applications (ME Apps), that are governed by the Mobile Edge Platform (MEP). The

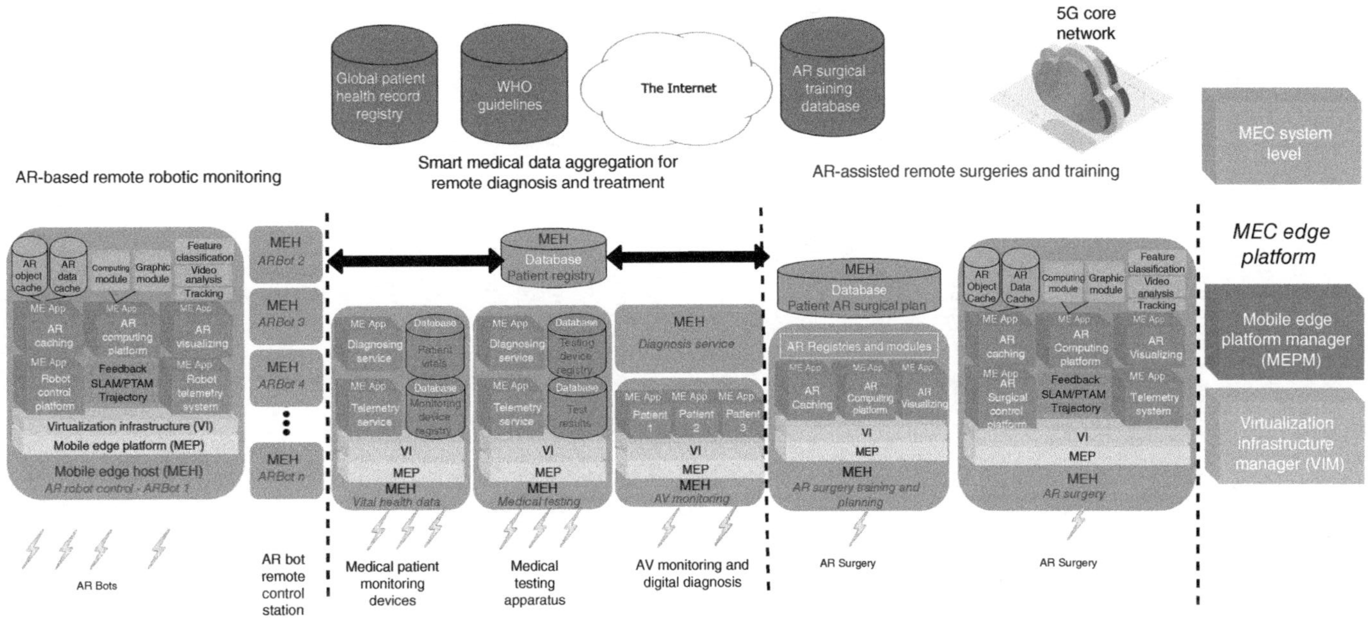

Figure 4.2 Realization of Use Cases with MEC.

AR-related function of caching, computing, and visualizing are handled by separate ME Apps [135], where the necessary sensory inputs are conveyed through the Robot telemetry service. The key significance of this service in contrast to a typical AR system, is its integration of the robotic control system. The robot control station is directly linked to the robot control platform inside the MEH. The design and formation of the robot-controlling platform launched as a ME App rely on feedback, trajectory planning, control interfacing (brain–computer interface, touch-based, or haptic), and telemetry methods. Thus, the resource specifications of each MEH might vary from each other. As this service is localized, MEHs controlling the AR bots do not require Internet connectivity. However, internal networks and protocols operating within the MEH should feature dynamism. Therefore, we propose lightweight virtualization or containerization for implementing the ME Apps, and the MEP [17].

4.3.2 MEC Assisted Smart Medical Data Aggregation/Patient Monitoring and Treatment

The main objective of this use case is to aggregate vital medical information related to contiguous patients remotely while retrieving them for diagnosis and treatment following the WHO protocols and guidelines. Thus, the MEC edge platform should incorporate storage facilities for medical data. The patient registry, proposed as a MEH with a formidable capacity, is linked with the entities of two other use cases and with a centralized global Patient Health Record (PHR) server. For diagnosis, both patient vitals and testing results are required. Therefore, a MEH is assigned to store patient vitals via a telemetry service connected to a pulse oximeter, ventilator, and smart medical beds. Similarly, records of medical tests conducted under each patient are stored in another Mobile Edge Service (MES) where the data are aggregated from the medical testing apparatus of CT, ultrasound, X-ray, and PCR. Further, an Audio Video (AV) monitoring service runs on a MEH while it is interfacing with the medical officials through handheld devices. Under the AV monitoring MEH, each patient is monitored with a ME App. Access to the PHR level depends on the health official's designation. The diagnosing MES can access all the health records and registries hosted at different MEHs and the global PHR database. Thus, once a patient has been diagnosed and a treatment plan is finalized, treatments are carried out in the reconnaissance mode with AR bots, explicated in Use Case 1.

4.3.3 AR Assisted Remote Surgery with MEC

With AR, Virtual Reality (VR), and Mixed Reality (MR) technologies, surgeons can train and plan all kinds of surgeries in a virtual environment that emulates pragmatic circumstances. Thus, use case 3 includes a training and planning function launched as a MEH and saves the surgical plan to a database, where the surgical attendants can revisit the strategies before the surgery. Further, surgical trainees could use this module for training in various surgeries available in the AR surgical training database. This function is implemented as a typical AR streaming application in the MES. However, access to the system should be controlled by an internal function. For each AR-assisted remote surgery, a MEH is launching the required facilities. The content of the MEH resembles use case 1, except for the AR surgical controlling platform that links the surgical bed and the remote AR console platform. To improve the accuracy and perception of the surgical environment, a Haptic feedback system (such as PHANToM [158]) should be implemented with the surgical controlling function.

4.4 Deployment Challenges and Solutions

In this section, the challenges and limitations related to the proposed MEC implementation are discussed and possible solutions are introduced for improving their feasibility.

4.4.1 Security and Privacy Issues

Healthcare is an area that requires a formidable level of security and privacy in its IT infrastructure. The sensitivity of PHRs—maintaining doctor-patient confidentiality and protecting the MEC edge data center from physical and cyber intrusions that intend for manipulations- are the main objectives for ensuring a secure IT ecosystem in hospital premises. Further, in the proposed context, due to the higher reliance on time, service-impeding attacks that are intended to delay the services have a higher impact. Thus, securing the communication protocols with the proper level of cryptographic primitives is a major requirement.

Possible Solutions: Blockchain is one of the main approaches that researchers have employed to ensure privacy and security in healthcare systems. Using Blockchain technology adds a higher complexity to crypto-analysis and alleviates the cryptographic overhead on the information. Blockchain-based enhanced authentication mechanisms and securing methods are promising ways to secure the healthcare environment [159].

4.4.2 Communication Issues

The future medical wards and operating theatres are going to be embedded with a multitude of sensors and actuators that are controlled via wireless interfaces. This IoT environment which is commandeered by a MEC edge infrastructure as proposed, requires the level of 5G radio communication to satisfy URLLC and eMBB standards. However, IoT devices might not have the capability to embed and serve 5G transceivers due to their high power consumption. In addition, on AR and VR perspective, the motion-to-photon latency that exceeds 20 ms for image rendering causes motion sickness for the users as a result of conflicted signals precipitated on the vestibular-ocular reflex. This is harmful to AR users in UC 1 and UC 3. Further, interoperability and compatibility are challenges that should be addressed to manage scalability.

Possible Solutions: In order to address the limitations on embedding 5G TRXs on IoT devices, Intermediary Access Points (IAPs) that interface 5G radio and other low power communication methods such as Bluetooth low energy and Zigbee; that are common with IoT devices can be deployed. These IAPs would act as a bridge between the IoT domain and the edge platform. Launching IAPs provides the ability to scale the miniature sensory devices, while the compatibility issues would be solved from a communication perspective.

4.4.3 Scalability Limitation

Though MEC is effective for provisioning services with ultra-low latency to a limited coverage area, the available resources in an edge platform are not as rich as in a cloud data center. The MEC concept was introduced to offer specialized services that do not scale up to typical services with higher scalability. Therefore, for applications that consume and demand higher resources, as in UC 1 and UC 3, MEC is limited to serving the 10th and 11th application instances as specified in Table 4.1. Further, in a pragmatic situation where all three proposed use cases are operating simultaneously, load-balancing the resources of the MEC platform is an obvious challenge. Thus, a dynamic service provisioning control framework is a prime requirement.

Possible Solutions: Allocation of excessive resources are the only means for mitigating scalability issues. For MEC deployments -if the existing resources are inadequate to launch the services- the most recommended method is to append more resources into the system. However, if there is another MEC-enabled BS in close proximity, such services can be either migrated to the novel MEC BS, or the content to be processed can be offloaded to it. Further, in an overwhelming situation, MEC BS can offload the service to a cloud platform while balancing the load of its own platform.

4.4.4 Legal Adaptation and Ethics

"Robot ethics," a branch of machine ethics, is a clear application for healthcare robotics that is used in both use cases 1 and 3. Even though the robots are human-controlled, an intentional or unintentional malfunction would cause harm to human beings which implies ethical and legal repercussions. Thus, robot ethics deal with such dilemmas in a legal context, for the purpose of prioritizing human safety, ensuring superior human command, and protecting the robots' own existence without inflicting harm to human beings [160]. In the proposed applications, such standards on ethical conduct should be clearly defined and practiced as these human-to-machine interactions might create unforeseen conundrums. User consent is imperative in such a circumstance. Further, aggregation of healthcare data should be pursued in accordance with the EU General Data Protection Regulation (GDPR) standards [161] to ensure the privacy of patients and healthcare officials. When a hospital or a medical institute is digitizing its holistic system, a legal framework should be established to govern all the entities within the institution's jurisdiction. All the unspecified legalities regarding the cyber-domain should be contextualized, prior to practicing these novel medical applications since the causes might be quite lethal.

4.5 Conclusion

The main intention of this article was to emphasize the significance of MEC resembling an edge computing platform for futuristic applications in health care; that would be beneficial in mediating a pandemic situation similar to COVID-19. The proposed use cases are requiring an IT eco-system that optimizes the network and computational utilization to novel extents that reach beyond prevailing limitations. Proposed use cases are demanding a higher level of specifications. Therefore, it is obvious that the use cases cannot be achieved in an absence of a storage and processing infrastructure at a proximate distance. Thus, the proposed use cases are realizable with MEC platforms deployed within medical institutions. The findings of this research are beneficial for state-of-the-art medical companies to form an IT infrastructure that launches their applications with featured synergism.

Part II

Technology Foundation

The core functionalities of MEC^2 are the foundation to meet diverse demands from end users, device vendors, network and service providers. This part of the book will take a close look into the technology foundation of MEC^2, including edge intelligence in Chapter 5, edge offloading in Chapter 6, edge caching in Chapter 7, edge communication, hardware and enabling frameworks in Chapter 8.

Mobile Edge Computing and Communications: Driving Forces, Technology Foundation, and Application Areas, First Edition. Aaron Yi Ding, Chamitha De Alwis, and Madhusanka Liyanage.
© 2025 John Wiley & Sons Ltd. Published 2025 by John Wiley & Sons Ltd.

5

Edge AI*

This chapter introduces one of the most exciting development in MEC^2 by integrating Artificial Intelligence (AI) to edge computing, referred in this book as Edge AI [162]. Readers can obtain an overview in Section 5.1 and learn the core concepts of Edge AI training and inference in Section 5.2. Section 5.3 illustrates the practical applications of Edge AI, and Section 5.4 further discusses the open challenges and future directions for Edge AI.

For clarity, we use the term Edge AI and Edge Intelligence interchangeably in this book, as these two notions share essential commonalities [162]. To stay focus on edge technologies, we take the perspective of "AI on the edge," which enables the execution of machine learning/deep learning tasks [163] directly on the interconnected edge devices [164–166].

5.1 Overview of Edge AI

Motivation for Edge AI: Compared with Cloud AI, which refers to the deployment of AI algorithms and models on cloud or datacenter servers, Edge AI brings the "intelligence" close to the end devices and date sources, as illustrated in Figure 5.1. Edge AI can enable the data to be processed directly on the network edge, regardless of the Internet connection between end devices and the cloud. With Edge AI, the data-driven analytic services become more independent with less reliance on cloud infrastructure. In addition, decentralized data analytics as enabled by Edge AI can effectively reduce data transmission to central cloud location.

Definition: Edge AI is considered as an evolving concept especially given the rapid advancement of AI and computing technologies. In this book, Edge AI refers to the set of enabling technologies that facilitate the execution of AI/machine learning tasks directly on the interconnected edge devices and infrastructures. Besides this definition, there is a related track focusing on utilizing AI for managing edge computing infrastructure from topology and orchestration perspectives. Readers interested in this line of development can read further from recent publications [167, 168].

Architectural Scope of Edge AI: Given the advancement of computing and AI technologies, Edge AI is expected to bridge the cloud, cellular (mobile) systems and IoT towards the envisioned future computing continuum [169]. As depicted in Figure 5.2, the scope of

*This chapter is partly contributed by Dewant Katare (TU Delft, Netherlands).

Mobile Edge Computing and Communications: Driving Forces, Technology Foundation, and Application Areas, First Edition. Aaron Yi Ding, Chamitha De Alwis, and Madhusanka Liyanage.
© 2025 John Wiley & Sons Ltd. Published 2025 by John Wiley & Sons Ltd.

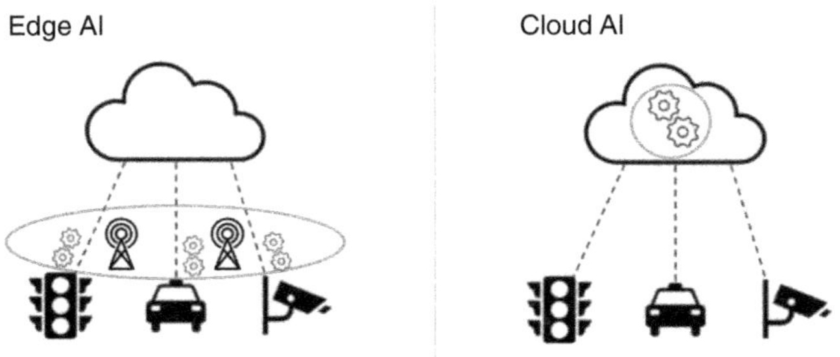

Figure 5.1 Overview of Edge AI and Cloud AI Architectures.

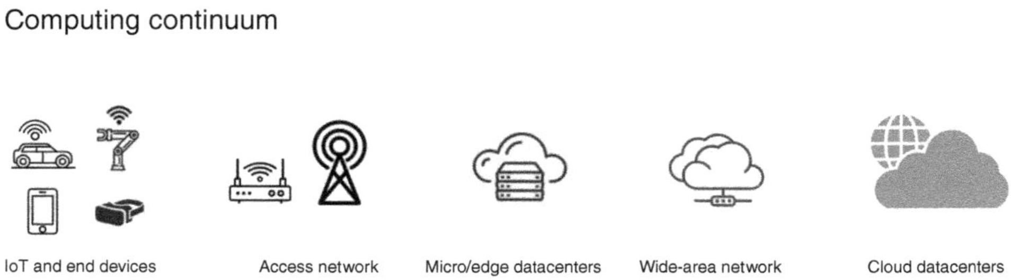

Figure 5.2 Computing Continuum.

Edge AI is not restricted to the execution of AI only on edge devices. From the convergence perspective, Edge AI shall fully exploit the resources across different layers of computing and networks, including end (IoT) devices, access networks, micro datacenters, wide-area networks, and cloud datacenters.

In the context of computing continuum, Edge AI can be explored through seven levels in terms of AI training and inference [164]. As illustrated in Figure 5.3, the scope of seven levels reflects the degree of decentralization and data distribution in the synergy of cloud, edge, and end (IoT) devices.

- **Level 0**: This is the baseline level of conventional fully-cloud AI, where model training and inference take place in the cloud. In this level, the involvement of edge and end devices are minimum.
- **Level 1**: In this level, model training will take place in the cloud while the edge and cloud will jointly carry out the model inference. The co-inference of edge and cloud indicates that data can be partially offloaded to the cloud.
- **Level 2**: This level shares the similarity with Level 1 in terms of cloud training, while the edge will carry out the inference tasks.
- **Level 3**: In this level, model training takes place in the cloud. Compared with Level 2, the inference will take place locally on the end (IoT) devices.
- **Level 4**: In this level, cloud and edge will jointly collaborate on model training and inference.
- **Level 5**: In this level, both model training and inference will be carried out by the edge.
- **Level 6**: This level represents the highest degree of decentralization as both model training and inference will take place on the end (IoT) devices.

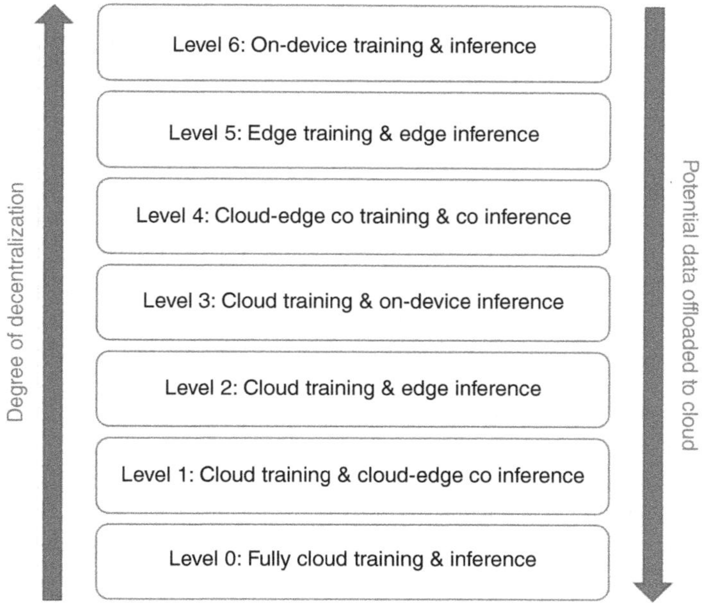

Figure 5.3 Scope of Seven Levels for Edge AI. Source: Adapted from [164].

The seven-level scope in Figure 5.3 provides a structural view on Edge AI from decentralization and data offloading (details in Chapter 6) perspectives. Going from Level 0 up to Level 6, the degree of decentralization increases as both training and inference are shifting from central cloud to the edge and end (IoT) devices. Meanwhile, going from Level 6 to Level 0, the amount of data offloaded from end devices to the cloud increases, as needed for model training and inference that involve the cloud.

5.2 Training and Inference in Edge AI

From the AI perspective, bringing intelligence and analytics to the edge encompasses two fundamental elements: training and inference. This section will dive into the training (Section 5.2.1) and inference (Section 5.2.2) of Edge AI.

5.2.1 Edge Training

Edge training refers to a distributed AI training procedure that learns the optimal values for the model weights, bias, and/or the hidden patterns from the training set cached at the edge [170]. For AI developers and operators, it is essential to understand the key considerations and enabling techniques for edge training, so the design and deployment of AI models can be optimized in heterogeneous edge environments.

Key Considerations for Edge Training:

The fast growing amount of multi-modal data generated from network edge and end (IoT) devices is calling for Edge AI to fully unlock the potential of data and manage its overhead.

Different from centralized AI training, edge training needs to tackle specific considerations [166, 170, 171]. Training AI models at the edge can benefit from a careful balance over the key considerations, highlighted as follows:

- **Communication**: For edge devices involved in training, communication is a key concern. Since training ML models such as deep neural network (DNN) is data intensive, where raw data or intermediate data needs to be exchanged across the devices. The overhead of communication depends on input data size, bandwidth, and transmission methods (wired or wireless). Data aggregation and compression methods can be used at the edge to reduce the volume of data and save bandwidth. For instance, data acquired from end devices can be sent to and used for training/re-training directly on the edge devices. This helps reduce the communication overhead of transmitting training data to remote cloud servers.
- **Training Latency**: For training in a distributed environment involving IoT and edge devices, latency can directly influence when the trained model is ready for use. The edge training latency includes both computation and communication latency. The computation latency depends on capability of the training equipment (e.g., CPU, memory size). The communication latency can be affected by the size of training data and the conditions of network connection. For instance, poor connectivity can lead to prolonged communication latency, which can hold up the training process.
- **Convergence**: Since edge training is often organized in a decentralized manner, the training processes occurring on various distributed devices need to converge to a consensus. The consideration of convergence is to measure how fast and whether a consensus is made in a decentralized setting. For instance, the convergence of DNN training depends on how the gradient is updated and synchronized across participating nodes.
- **Privacy**: Edge training involves data coming from heterogeneous end user devices, which may contain sensitive and private personal information (e.g., voice and facial bio-metrics). It is hence essential to preserve privacy by transmitting less privacy-sensitive data out of the end devices. For instance, when cloud participates in the training, we shall avoid sending private data to remote servers to prevent potential data leakage.
- **Energy Consumption**: In edge training, both computation and communication can consume energy, which is often limited on end and edge devices. The energy consumption is also affected by the type of the training models and the hardware capability of training devices. Energy efficient mechanisms and hardware are hence desirable. For instance, end devices can offload heavy computing tasks to edge or cloud servers so to reduce the computational energy consumption on the end devices.

Edge Training Architectures:

Training architectures of Edge AI can be conducted in three modes, in order to match specific considerations (as discussed above):

- **Centralized Mode**: In this mode, cloud servers are responsible for training the AI models by collecting data generated from distributed resources such as cameras, mobile phones and cars. This mode corresponds to Levels 0–3 as indicted in Figure 5.3.
- **Decentralized Mode**: In this mode, end devices conduct local training of AI models using their local data. The decentralized mode has a nature advantage to preserve private information. This mode corresponds to Level 6 as indicted in Figure 5.3.

- **Hybrid Mode**: This mode is a combination of centralized and decentralized modes. In this mode, the edge layer is serving as a hub between end devices and cloud. The edge devices can train the models in a decentralized manner by exchange updates with other edge devices. Meanwhile, the edge layer can also take the centralized training mode by sending data to the cloud. This mode corresponds to Levels 4 and 5 as indicted in Figure 5.3.

Enabling Techniques:

To support edge training, academic scholars and industrial experts have proposed several enabling techniques [164, 170–172]. For readers looking for a systematic overview, we highlight three categories of enabling techniques of edge training:

- **Training Acceleration**: Given the resource constraints and need for real-time experience, this category of acceleration techniques is crucial for edge training. The core techniques include hardware acceleration, aggregation frequency control, and transfer learning. The hardware acceleration technique exploits GPUs and specialized AI accelerators such as FPGAs and ASICs for high parallel processing and power efficiency. Instead of relying only on CPUs for training operations (e.g., matrix multiplications), edge training can collaboratively utilize a selection of available GPUs and accelerators to meet dedicated training requirements. The aggregation frequency control technique is to accelerate the communication, which is essential for edge training in distributed environments. Since the updates from local training need to be aggregated, this technique controls the aggregation procedures, including the aggregation content as well as the aggregation frequency. The goal is to explore optimal tradeoff between the local update and global parameter aggregation under the available resource constraints. The transfer learning technique helps reduce training time by using pre-trained models (e.g., from base datasets) and then fine-tuning process on target edge devices for rapid deployment. This technique is useful for edge training environments due to the limitations of data and computation on edge/IoT devices. Besides the transition from generality to specificity, transfer learning also reduces resource demand by fine-tuning a pre-trained model, rather than training models from scratch under limited datasets.
- **Privacy-Preserving Training**: The privacy-oriented techniques for edge training include federated learning and DNN splitting. The federated learning [173] enables multiple edge/IoT devices to collaboratively train shared global models without sharing their local data. Rather than uploading all raw data to remote cloud, this technique helps edge training comply with policy protection regulations since the local data remains on the edge/IoT devices. To minimize the risk of privacy breaches, only model updates (gradients or weights) are shared. For bandwidth saving, federated learning can use sparse update or compression to further reduce the amount of data that needs to be transferred. With the hybrid IoT-Edge-Cloud mode, federated learning can take a hierarchical approach in which the edge layer can harness IoT devices within their coverage to train AI models. Upon receiving the updates from IoT devices, the edge layer will upload aggregated updates to a remote (central) cloud server for further aggregation. Once the aggregation is completed, the cloud server will share the updated model parameters to the edge layer, which can further forward the parameters to target IoT devices. The DNN splitting technique [174] is dedicated for ML training on edge devices

by partitioning a DNN model into two successive layers and deploying the partitioned layers across IoT and edge devices. To protect privacy, only partially processed data is exchanged between IoT and edge, rather than transmitting raw data. The key matter of DNN splitting is to select the splitting point so that latency requirement can still be met when distributed training is completed. By balancing data parallelism and model parallelism, it is necessary to strike a balance between communication overhead and utilization of computation resources.
- **Decentralized Information Sharing**: For edge training in a decentralized setting, the decentralized information sharing techniques include gradient compression and gossip training. The gradient compression technique targets at a critical edge training scenario in which the last-mile communication channel (e.g., wireless) can be constrained in terms of bandwidth. To reduce the data size of updates, gradient compression can utilize gradient quantization and sparsification. The gradient quantization approximates the floating-point gradients with low-bit width numbers (e.g., 32-bit can be approximated by an 8-bit number, with size reduction by factor of four). Meanwhile, the gradient sparsification selectively discards unimportant/less relevant gradient updates by transmitting only the updates that exceed a pre-defined threshold. The gossip training is to enable information exchange in a peer-to-peer way. It take advantages of asynchronization with less dependency on centralized computing devices. In fact, gradient compression and gossip training can be combine to reduce update size and improve resilience against malfunction of centralized devices.

5.2.2 Edge Inference

Similar to edge training, it is essential for AI developers, operators and end users to comprehend the key considerations and enabling techniques for edge inference. From the computing continuum perspective, efficient and effective utilization of edge devices can help tackle the resource bottleneck on an individual computing layer.

Key Considerations for Edge Inference:

Edge inference is performed locally on edge devices, which can benefit from a careful balance over the key considerations [164, 170], highlighted as follows:

- **Inference Latency**: The latency of edge inference refers to the time spent in the entire inference procedure. Under the edge AI setting, the inference procedure can include pre-processing of raw data, model inference, data transmission, and post-processing. For edge and end devices with resource limitation, the inference latency can also depend on the capability of the hardware and how the system executes the AI models.
- **Accuracy**: The inference accuracy is reflected by the ratio of correction predictions on input samples from inference to the total number of input samples. As a performance factor, the accuracy of edge inference can be influenced by model selection, data quality, and deployment configuration.
- **Resource Footprint**: Given the resource constraint on edge and end devices, the footprint of computing such as memory and CPU/GPU utilization must be considered. It is often a balance between complexity and expected performance (e.g., accuracy, latency), given the high-precision AI models may contain large amount of parameters, which consume memory and prolong inference latency.

- **Privacy**: To protect potential privacy leakage (e.g., by eavesdroppers) for edge inference, we need to preserve privacy by enforcing techniques such as obfuscation and/or adding noise to the data uploaded by end (IoT) devices to the edge. Meanwhile, we shall also consider secure computation schemes that can ensure end devices only obtain inference results without learning the model details, and edge devices only obtain model updates without learning the raw data from end devices. One example in secure computing is homomorphic encryption [175], which allows computation to be performed on encrypted data.
- **Energy Consumption**: Since end and edge devices are often limited by battery power, the energy consumption from both computation and communication for inference needs to be considered in system design. The energy consumption of edge inference can be affected by the model in use, and also the computing resources on the edge devices.

Edge Inference Architectures:

To support performance requirements and considerations as stated above, we highlight four architectural modes for edge inference as follows:

- **Edge Mode**: In this mode, the model inference is carried out on the edge devices and the input data is contributed by end (IoT) devices. The performance of inference in this mode depends on the network conditions (e.g., bandwidth) between the end and edge devices.
- **Device Mode**: In this mode, end devices perform inference locally using the model sent by the edge/cloud. Although the inference has privacy advantage under local inference, the performance of device mode depends on the capability (e.g., CPU, GPU, memory) on the local devices.
- **Edge-Device Mode**: With collaboration between end and edge devices, AI models are often partitioned and distributed across end and edge devices. The partition depends on multiple factors such as bandwidth, workload, and device capability.
- **Edge-Cloud Mode**: This synergy mode is similar to the Edge-Device mode. It is suitable for handling heavy workload where edge devices still need additional supports from the cloud.

Besides these four modes, we can also combine them to simultaneously carry out complex inference tasks across the end-edge-cloud layers, by pooing heterogeneous resources.

Enabling Techniques:

Edge inference needs to bridge the growing gap between AI computational requirements and capacity across end, edge and cloud. For readers to comprehend the latest advancements [164, 170–173, 176, 177], we highlight the major enabling techniques of edge inference as follows:

- **Compression**: Given the constant increase of AI model demand on storage, memory, and computing powers, edge inference needs compression supports to reduce the model complexity, which helps improve energy efficiency and inference performance on resource-constrained edge devices. The aim is to enhance latency, energy saving, privacy, and memory footprint. The core techniques for edge compression include network pruning, parameter quantization, knowledge distillation, and low-rank approximation. The network pruning involves removing less significant parameters (e.g., weights or neurons from a neural network) to reduce its size and computational requirements. The parameter quantization approach aims to eliminate redundant parameters since highly precise parameters in a model are not necessarily leading to high performance. For a neural network, quantization can reduce the precision of the weights and activations from high precision to low precision (e.g., from 32-bit to 8-bit precision), which reduces the resource usage after the quantization process. The knowledge distillation approach involves training compact models (often referred to as "student") to replicate the behavior of more accurate models ("teacher") by learning from the teachers' soft outputs. As a matrix could be decomposed into multiplication of multiple matrices of smaller size, the low-rank approximation aims to reduce the size of weight matrices by approximating them with lower-rank matrices, hence decreasing the number of parameters and computational load on end/edge devices.

- **Partition**: Model partition in edge inference uses the edge and cloud to offload the computation-intensive part of the partitioned model. Similar idea of split computing (SC) was first proposed in the Neurosurgeon work [174], which searches for the best partitioning layer in ML model for minimizing total (end-to-end) latency or energy consumption. The model partition can be categorized into two types: partition between different layers (e.g., edge to cloud, IoT to edge) and partition between devices in the same layer. The key for edge partition is to seek an optimal partition point that can make the collaborative inference meet the latency and/or energy requirements. By carefully partitioning models and distributing them across end, edge and cloud, effective model partition can strike a balance between performance and resource utilization.
- **Early-Exit**: This technique aims to reduce the latency and computation of edge inference by engineering the target AI models with specific exit points. Instead of going through all the layers of a model, these exit points allow intermediate results to be produced before reaching the final layer. It essentially enables us to selectively terminate the inference under service level objectives, which is suitable for real-time applications.
- **Caching**: The caching technique for edge inference is to cache the inference related data (e.g., prediction of image classification) for potential re-usage. The cached results help reduce the querying latency. In a distributed environment, a key concern is to locate where to cache and what to cache. For instance, frequently accessed data, models, and inference results can be cached at an aggregated edge device to reduce latency and bandwidth usage for more responsive services. The cached inference results also avoid redundant processing, which lead to energy saving for edge devices.
- **Model Search**: For edge devices, the model search technique aims to seek and optimizing AI architectures that are tailored for deployment under various constraints and requirements. Given the edge devices often possess limited computational power and resources, the key is to find the model architecture that can operate efficiently within these constraints. The considerations for model search include both task requirements and hardware specifications. For instance, Neural Architecture Search (NAS) [178] is proposed to automate the design of neural network architectures by using dedicated algorithms to explore a wide range of possible architectures and identify the most efficient ones for the target ML tasks.
- **Acceleration**: The acceleration techniques for edge inference come from both hardware and software. Owing to the growing capability of chip design, hardware acceleration can benefit edge inference through the parallelization of model execution on available hardware (e.g., GPUs, FPGAs, ASICs). Meanwhile, software acceleration focuses on the

optimization of resource management and pipeline design. In general, the hardware approach can help speed up edge inference through higher computing power, but it may not necessarily affect the accuracy. On the other hand, software acceleration approach tends to maximize the performance under the given resource budget, which may affect the accuracy for certain cases. From energy perspective, both hardware and software acceleration can lead to energy saving by selecting the right combination of accelerators and resource management schemes.

5.3 Edge AI in Practice

In this section, we discuss new development (Section 5.3.1) and practical use case (Section 5.3.2) of Edge AI.

5.3.1 Energy-Efficient Edge AI Using Approximate Computing

For sustainability goals, approximate computing represents a new paradigm shift towards energy-efficient AI by intentionally allowing for controlled imprecision in calculations, trading off computational accuracy for significant gains in energy efficiency and performance. This approach is particularly relevant for Edge AI applications, where power and processing resources are constrained, yet the demand for real-time analytics and decision-making is high:

- **AI Models**: The suitability of AI models for approximate computing varies. Convolutional Neural Networks (CNNs), given their inherent robustness to minor inaccuracies, are particularly well-suited for techniques like quantization. In contrast, the sequential dependency of Recurrent Neural Networks (RNNs) might limit their compatibility with aggressive approximation strategies.
- **Data**: The effectiveness of approximation techniques is also contingent on the volume and variety of training data. Larger datasets can help fine-tune models to maintain performance despite reduced precision, enabling more aggressive approximation without compromising essential accuracy.
- **HW Platform**: The design and capabilities of the target edge hardware dictate the feasible scope of approximation. Devices with limited computational power or memory might only support simpler forms of approximation, such as reducing floating-point precision to integer operations. This necessitates a careful balance between hardware limitations and the computational demands of deployed AI models.

5.3.2 Applications at the Edge

As supported by advanced hardware (detailed in Chapter 8), several application domains can benefit from Edge AI.

- **Smart City**: In smart city applications, such as real-time traffic flow optimization and environmental sensing, approximate computing enables the deployment of AI models on edge devices like the NVIDIA Jetson Xavier NX. These models can perform tasks like vehicle detection and pollution level monitoring with reduced precision to save energy while delivering timely insights.

- **Wearable and Smart Devices**: Energy efficiency is paramount for AI-powered wearables and smart home devices. Approximate computing allows these devices to execute voice recognition or activity tracking with minimal power consumption. With its low-power profile, the Google Coral Edge TPU exemplifies the integration of approximate computing in compact, consumer-facing technology.
- **Healthcare**: In healthcare applications, approximate computing facilitates near-patient analytics, such as real-time image analysis for diagnostic purposes. Devices like the Intel Movidius Neural Compute Stick can process imaging data directly at the point of care, balancing computational efficiency with the precision required for medical analysis.
- **Mobility as Service**: The mobility sector, encompassing autonomous vehicles and ride-sharing services, demands both real-time processing capabilities and stringent energy efficiency. Approximate computing can optimize on-board AI functions, from navigation to passenger entertainment, on platforms like NVIDIA Drive AGX Orin, ensuring operational efficiency without sacrificing critical decision-making accuracy.

5.4 Summary and Outlook

Overall, Edge AI has potential to enable mission-critical and personalized AI [179] for future mobile and IoT services that are intensive on computation and energy consumption, and sensitive to latency and data privacy. From data perspective, the growing amount of data generated at the network edge can benefit from Edge AI to fully unlock the potential of data while respecting its privacy. From application perspective, Edge AI can facilitate more innovation by harnessing the advantage of edge computing in terms of latency and bandwidth. From infrastructure perspective, Edge AI is becoming a core enabler to democratize AI in the future computing continuum as to make AI ubiquitous and accessible by more users and organizations in our society. It paves the way towards a digital continuum infrastructure, which can prosper AI with meaningful data and ensure societal values such as privacy, inclusiveness and sustainability are also preserved.

Given the open challenges in edge AI [162, 170, 171, 180], there are two future directions that deserve our attentions:

Data Scarcity and Consistency: Data is crucial for developing AI solutions. Because of data privacy regulations and difficulties to access multi-modal data from heterogeneous sources, research directions such as synthetic data generation, data augmentation, zero-shot learning and automated validation shall be looked into. Besides data sharing strategies such as open data marketplaces, we also need robust and scalable system solutions that allow edge AI to perform effectively in data-scarce and consistency-sensitive environments.

Trustworthy and Sustainable Edge AI: The success of edge AI depends on its trustworthiness for wide adoption and how sustainable it can incentivize key stakeholders through viable business models. One critical direction for trustworthy edge AI is accountability design, which allows us to maintain audit trails of AI decisions made at the edge. For social and economic sustainability, industry standards are crucial for stakeholders to maintain compatibility and interoperability. From energy angle, low-power computing chips for edge devices is a viable direction for reducing the overall energy footprint.

Since the pace of AI innovation remains high, we expect new technical and societal challenges for edge AI will still emerge in the future. On the positive front, those challenges help create opportunities for scientific and engineering development in AI and MEC^2.

6

Edge Offloading

This chapter introduces the fundamental concepts and applications of edge offloading. To provide a comprehensive view on the subject, Section 6.1 illustrates the design concepts for data offloading and computation offloading. We then dive into the latest development of edge offloading in Section 6.2 to learn its design and enabling technologies. We further study the practical applications in Section 6.3 to showcase how edge offloading is used to tackle new demands and join up with artificial intelligence (AI) technologies. We summarize in Section 6.4 to pinpoint potential directions for future investigations.

6.1 Motivation and Fundamental Concept

This section covers the fundamental elements for readers to discern the rationale and design behind edge offloading in Section 6.1.1, including both data and computation offloading, covered in Sections 6.1.2 and 6.1.3, respectively.

The term of offloading in the context of mobile computing refers to a general mechanism to balance the data traffic and computation complexity, for instance, by shifting data traffic from one network access to the other available network accesses, and shifting the function executions from one device to other available computing resources.

In the context of MEC^2, the general goal of offloading is to balance the data traffic and computational workloads across multiple entities such as IoT/end-user devices, edge and cloud servers. From the system architecture perspective, there are two major categories: cloud-oriented offloading and edge-oriented offloading. The cloud-oriented model is to shift the functional complexity from resource-constraint systems (e.g., low-power sensors) to the resource-abundant systems such as cloud or data centers. Meanwhile, the edge-oriented model, referred to as ***edge offloading*** in this book, is to take advantage of the data locality by selectively bringing back the data processing and computational functions from the cloud to the systems nearby where data is generated.

6.1.1 Rationale for Edge Offloading

The main drive for edge offloading comes from the fast growing data volume and demands for low-latency and versatile functionalities. Compared with the conventional

Mobile Edge Computing and Communications: Driving Forces, Technology Foundation, and Application Areas, First Edition. Aaron Yi Ding, Chamitha De Alwis, and Madhusanka Liyanage.
© 2025 John Wiley & Sons Ltd. Published 2025 by John Wiley & Sons Ltd.

cloud-oriented offloading, one particular motivation for edge offloading origins from the need for decentralization, for instance, by the advanced IoT devices such as augmented reality (AR)/virtual reality (VR) and connected vehicles. The development of those devices have changed their role from simple data consumers to "rich" data providers. The consequences of this new trend include: (i) an unprecedented increase of data traffic from bottom-up, which stresses the network operators and hence calling for traffic reduction; (ii) on top the data volume increase, the emergence of real-time applications such as gaming and self-driving services can make the traditional cloud-oriented solutions undesirable in terms of latency; (iii) along with data volume and latency considerations, there is a growing desire to keep the sensitive data local at a protected region, especially for highly regulated healthcare and critical industry scenarios.

By allowing the cloud to offload lightweight and fine-grained functionality/services to run on constrained devices on the edge, the rationale for edge offloading includes:

- To utilize proximate context, computing, and storage resources.
- To ease the burden on network infrastructure.
- To alleviate wireless interference on mobile access networks.
- To enhance performance and user experience for data and computation intensive applications.
- To relieve the workload burden and energy (battery life) scarcity on an individual device.
- To offer more control to data sources and users to manage the generated data.

The discussion below focusses on data and computation offloading towards providing an understanding on the internal mechanisms of offloading in general.

6.1.2 Data Offloading

Definition of Data Offloading: In this book, data offloading refers to the set of mechanisms that shift data traffic from one network access to the other available network accesses for load balancing purpose.

The enabler for data offloading is the fast development of wireless communication technologies. The modern mobile devices such as smartphones and tablets are supported by a rich set of communication technologies including LTE/4G, 5G, WiFi, and Bluetooth. New standard such as WiFi-direct [181] further provides the feasibility and flexibility for communications among peer devices. The mobile network operators have been upgrading the network infrastructures for better coverage and performance. WiFi hotspots are gaining popularity in urban areas, which complement cellular networks to offer convenient wireless connectivity.

There are two general types of data offloading [182, 183]: infrastructure-driven data offloading and device-driven data offloading.

As illustrated in Figure 6.1, infrastructure-driven offloading is typically initiated by operators to dynamically manage and balance the traffic load. It requires dedicated signaling protocols such as router advertisement [184]. Meanwhile, device-driven offloading is often triggered by applications that need to access the Internet or nearby resources, based on the demand of the end users.

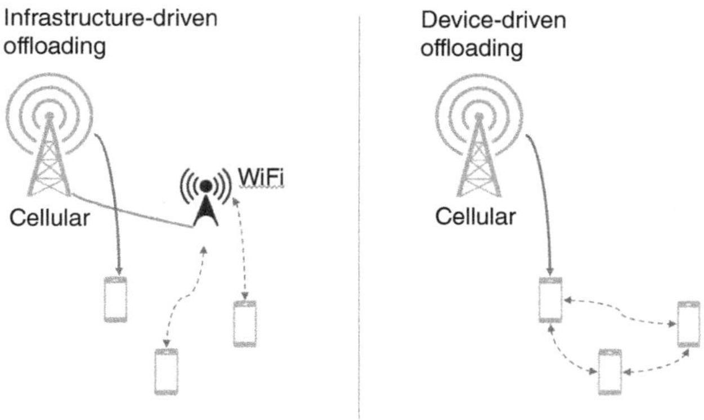

Figure 6.1 Infrastructure-driven and device-driven offloading.

The infrastructure-driven offloading can offer timely and optimized offloading guidance based on the comprehensive knowledge from the network side such as network layout and connectivity conditions. However, such approach introduces overhead in terms of extra signaling and potential energy cost. In addition, infrastructure-driven approach cannot utilize the local context, which is only visible to the end devices such as wireless signal strength of nearby access.

While the device-driven offloading avoids the extra signaling costs, it lacks the network context, hence making it less efficient for demanding scenarios such as users with high moving speed in vehicles.

Data Offloading Procedures: There are six general steps in data offloading [182, 183]. The details of each step are illustrated as follows:

- **Initiation**: The initiation of data offloading can be triggered by the network infrastructure (infrastructure-driven) or by the end user device (device-driven). The infrastructure-driven data offloading requires dedicated signaling protocols while the device-driven data offloading is triggered by user applications, which reflect the user demand.
- **Context Collection**: This phase is to enable context collection from the operator network or from the nearby access environments. The key context to collect includes user location, conditions of network connectivity and connection details such as signal-to-noise ratio of WiFi access points, MAC addresses, and extended service set identification (ESSID), etc. The collected information is often sent to remote or local management units for decision making.
- **Target Selection and Decision**: The target selection and decision typically require algorithmic or pre-defined logic operations. By taking the context as input, the decision can be made either by the remote network unit or directly on the local device. If decision is handled by the network unit, the powerful hardware resources can improve the efficiency. However, this approach requires stable network connectivity and depends on the infrastructure support. Meanwhile, the local decision approach can be robust to network conditions but it entails extra energy cost for the local devices.

- **Network Association**: According to the offloading decision, network association will take place to enable data offloading. This association process consists of network/peer discovery through pre-defined protocols or configurations to establish network connectivity with the offloading target, which can be an alternative access network or proximate computing devices. For mobile devices that are moving at high speed, association must be efficient to avoid decreasing the actual data transmission time since the connectivity period can be short.
- **Data Transmission**: The core part of data offloading is to transmit data to alternative links, which determines how much is shifted away from the original access network. Depending on the type of data traffic such as real-time streaming, web surfing or delay-tolerant traffic, the characteristics of the traffic type can be harnessed to optimize the offloading design.
- **Offloading Completion**: A successful offloading session is completed by terminating the temporary offloading connection and smoothly switching to other feasible access networks. To achieve a smooth termination, guidance can be obtained from the network side or from local heuristic prediction. The available handoff techniques [185, 186] can facilitate the seamless migration.

Metrics to Evaluate Data Offloading: The assess the outcome of data offloading, there are a set of metrics [183, 187] to facilitate the evaluation including: *offloading ratio*, *offloading overhead*, *energy consumption*, and *Quality of Experience (QoE)*.

As highlight in Table 6.1, the *offloading ratio* is an important metric to quantify the capacity achieved by data offloading, especially from the operator perspective. The metric is measured as the ratio of bytes offloaded to other channels against the total bytes generated from the data source. The *offloading overhead* is to evaluate the amount of extra control signaling data that is required to complete the data offloading, such as in the context collection and network association phases. The overhead typically depends on the additional traffic that is introduced to overall offloading procedure. The *energy consumption* is a metric associated with the power saving potentially attained by offloading data to more efficient channels. The *Quality of Experience (QoE)* is another user oriented metric that is directly linked to the overall service satisfaction. The throughput, packet loss, jitter are representative components in the QoE metric.

Representative Solutions of Data Offloading: To illustrate the concepts of data offloading, we look into two representative design concepts: the infrastructure-based MADNet [188] and the device-based Opp-Off [189].

Table 6.1 Metric for Assessing Data Offloading.

Metric for Data Offloading	Remarks
Offloading ratio	Ratio of bytes offloaded against the total bytes
Offloading overhead	Extra amount of signaling and control data
Energy consumption	Power saving or loss from offloading
Quality of Experience (QoE)	User perceived service quality

The Metropolitan Advanced Delivery Network (MADNet) is a collaborative WiFi-based data offloading design by utilizing infrastructure supports. MADNet offers three modules to enable collaborative infrastructure-based offloading, including Proxy-Cellular, Proxy-WiFi, and Smart-Client. The design principles of MADNet include (i) to make the data offloading transparent to end users; (ii) to connect cellular, WiFi and end users together for sharing context important for data offloading; (iii) to reduce energy consumption. MADNet offers three modules to enable collaborative infrastructure-based offloading: Proxy-Cellular, Proxy-WiFi, and Smart-Client. The Proxy-Cellular is the decision making module for collecting context from both infrastructure and end devices, and coordinating the offloading based on the MADNet algorithm and signaling channels. The Proxy-WiFi can prefetch content and exchange with Proxy-Cellular the local offloading events. For collaborative offloading, the Smart-Client module shares the essential context such as location and content preference with the Proxy-Cellular to make the data offloading decision.

As a device-based data offloading design, Opp-Off [189] utilizes the social connections among nearby users and arrange data offloading via Bluetooth in an opportunistic manner. Opp-Off proposed three target-set selection modes: greedy, heuristic, and random. Those three modes provide flexibility to set up opportunistic communication depending on the requirements. For example, in the random mode, Opp-Off uses a simple target-set selection in terms of function complexity, which is still effective. Besides active social participation, Opp-Off also explores the feature from delay-tolerant data traffic to further optimize the data offloading ratio.

Readers who are interested in data offloading can further check the proposed solutions [125, 129].

Overall, the infrastructure-based MADNet and device-based Opp-Off bring theoretic design into practice. Data offloading is an important building block to enable core functions of MEC^2.

6.1.3 Computation Offloading

Definition of Computation Offloading: In this book, computation offloading refers to the set of mechanisms that transfer computational tasks from the original device to external computing devices [51, 190–192]. The external computing devices can take the form of processors, hardware accelerators, or computing clusters.

The general goal of computation offloading is to relieve the workload burden and resource scarcity on an individual device, for better performance and user experience. From the service perspective, a main target is to reduce the response delay of the service and improve the service quality.

Computation offloading is motivated by the device restrictions (e.g., limited battery and memory) and computing intensive applications. By shifting the function executions to other computing resources, computation offloading can balance the computational complexity of applications on a single resource-constraint device.

From computational perspective, there are two general styles for computation offloading: full offloading and partial offloading [51, 192]. Full offloading typically requires each task to be offloaded to the target as a whole, which is often referred as binary offloading.

Meanwhile, partial offloading allows the task to be partitioned and executed both locally and at remote targets such as a data center.

Computation Offloading Procedures: The computation offloading include five general steps as follows:

- **Offloading Preparation**: The preparation of computation offloading often entails application/logic partition, which can be partial or full.
- **Target Selection**: The target for computation offloading can be either the cloud servers or peer/edge devices that have computing resources to fulfill the service requirements.
- **Execution**: The execution phase includes the transmission of functions and data (e.g., model parameters) to the offloading target. By processing the request, the offloading target will send back the expected results through the communication link, often according to pre-defined protocols.
- **Completion**: In the final step, the end device shall process the response received from the offloading target. Depending on the service logic, the end device can terminate the offloading process by releasing the resources such as communication link and storage.

Metrics to Assess Computation Offloading: As indicated in Table 6.2, the effectiveness of computation offloading can be assessed by a set of metrics [193], including the *Latency*, *Cost*, *Energy consumption*, *Quality of Experience (QoE)* and *Fairness*.

The *Latency* metric is important for delay-sensitive applications. It is typically measured by overall execution time to complete the computation offloading, including the time spent in communication and computation.

The *Cost* metric considers computing related costs needed for offloading, including storage usage, migration management, security operations and deployment overhead in terms of materials and monetary expenses. Related to the cost, incentive is a general challenge for computation offloading.

The *Energy Consumption* metric is to assess the potential energy saving of offloading, often for resource-constraint devices such as battery powered mobile phones. The metric shall consider both network level and device level consumption.

The *Quality of Experience (QoE)* metric concerns users' perceived performance, service reliability, and responsiveness that can be affected by computation offloading.

The *Fairness* metric is a system level measure that can reflect the distribution of resources without favoring any particular group. This metric is critical especially for peer to peer scenario where an unfair design can result in resource depletion such as battery drain on a few individual devices.

Table 6.2 Metric for Assessing Computation Offloading.

Metric for Computation Offloading	Remarks
Latency	Overall execution time to complete the offloading
Cost	Equipment, materials, money and time from offloading
Energy consumption	Power saving or loss from offloading
Quality of Experience (QoE)	User perceived service quality after offloading
Fairness	Fairness for resource distribution

Representative Solutions: To illustrate the concepts, we discuss the classic design of full offloading in terms of live migration [194], and the partial offloading design of MAUI [190].

The full computation offloading design [194] explored the live migration of operating system instances to facilitate load balancing and service continuity. With its "pre-copy" technique, in-memory state can be transferred between computing nodes in a consistent manner, including application state and kernel-internal state such as TCP control block for active connections. This can enable the live migration of real-time service (e.g., on-line game server) without requiring clients to reconnect, which creates values for both system availability and service experience in terms of downtime minimization. Without dividing the process or task, this design migrates the OS instance as a whole to enable full-scale computation offloading.

The partial computation offloading design of MAUI [190] focuses on the energy and performance of mobile applications. With its fine-grained code partition technique, MAUI creates two versions of mobile application so that one of which can run locally and the other one runs remotely (e.g., cloud infrastructure). To guide the partition and be efficient on resource usage, MAUI utilizes continuous network and device level profiling and extracts only the program state needed by the remote methods. As one of the earliest solutions to use cloud infrastructure for computation offloading, MAUI explores a data-driven optimization approach by using the data collected from its profiler to determine which methods shall be executed locally and which shall be executed remotely.

Readers who are interested to acquire more insights on computation offloading can further check those proposed solutions [104, 187, 195–199].

Overall, both data offloading and computation offloading aim to utilize resources in the best possible manner. The major difference is that the first one focuses on data flows while the latter one focuses on computing functions.

6.2 Edge Offloading

This section illustrates the edge offloading from the data (Section 6.2.1) and computation (Section 6.2.2) perspectives. Since edge offloading is an architectural approach to harness appropriate distributed computing resources, we further discuss the enabling technologies (Section 6.2.3).

One particular note on the incentive of edge offloading is that it is often not driven by the convenience to offload to the edge, but more towards compelling reasons not to offload to the cloud (e.g., for privacy and regulation concerns).

6.2.1 Edge Data Offloading

The critical question why data needs to be offloaded to edge can come down to three major factors: data volume, bandwidth, and privacy.

The first one concerns the data volume, especially given the recent rapid increase of data generated from end devices such as smart vehicles [200]. The data volume factor directly results in energy and monetary costs for users and Internet Service Providers (ISP). Since not all data (e.g., noise, repetitive input) is necessary to transfer, a smart "gatekeeper" such as edge, which sits between the data generator and receiver, can perform intelligent filtering and sampling of the data inputs. In this regard, edge data offloading is expected to improve data utilization and reduce redundant data transmission. For mobile access networks, less data also alleviates the potential wireless interference.

The second factor is the network bandwidth, which is a limited resource especially when data transmission is conducted via wireless. In addition, comparing with data transmission over the Internet, there is a potential throughput advantage when data can be accessed nearby such as over one-hop access link.

The third factor is the growing concern of data privacy from end users and regulatory requirements (e.g., GDPR [201], California Consumer Privacy Act [202], PIPEDA of Canada [203]). By offloading data to the local edge computing nodes, especially under the same administrative policy, data privacy can be better enforced by respecting the regulatory requirements.

Definition: In this book, edge data offloading can be illustrated as an advanced process to utilize the edge layer to manage the data. In the typical setting, the edge layer sits between the data generator (e.g., IoT devices) and data receiver (e.g., cloud). In this regard, data can be offloaded to the edge from either IoT devices or cloud, in order to fulfill performance and regulatory requirements. Within the same edge layer, different computing nodes can also offload/exchange data with each other to fulfill specific service requirements such as latency and mobility [200].

How to Offload: Regarding how edge data offloading is conducted in practice, one common form of edge data offloading is data caching on the edge nodes. This concept is closely related to the content distribution network (CDN) [204, 205]. Regarding the data offloaded to the edge, there are two general types: raw data from end devices, and application data (e.g., results after processing the raw data).

Owing to the location proximity and contextual awareness of edge (close to the data source), the relevant "raw data" can be cached temporarily on the edge to serve multiple users/devices. For example, the road condition data for a specific region can be stored on an edge device by the road side unit (e.g., lamp-post, traffic light system, base station). When other cars approach this region, the data cached (offloaded) on the edge device can be shared with the corresponding cars that need such data [206]. Since edge applications can process similar contextual raw data that lead to approximately same outcome, such "application data" outcome can be cached and reused to avoid redundant computation [207].

For mobile scenarios such as offloading vehicular data, selecting proper wireless communication channels and protocols [183, 208] is crucial to strike a balance between the data offloading metrics as discussed in Section 6.1.2. Regarding to caching, data offloading on the edge needs to tackle data freshness, i.e., how long the data is still relevant to the target users.

6.2.2 Edge Computation Offloading

The motivation for computation to be offloaded to the "edge" comes down to three factors: performance, resilience, and energy.

The first motivating factor is the performance, especially given the growing demands from various real-time and vehicular services in terms of service response time and latency.

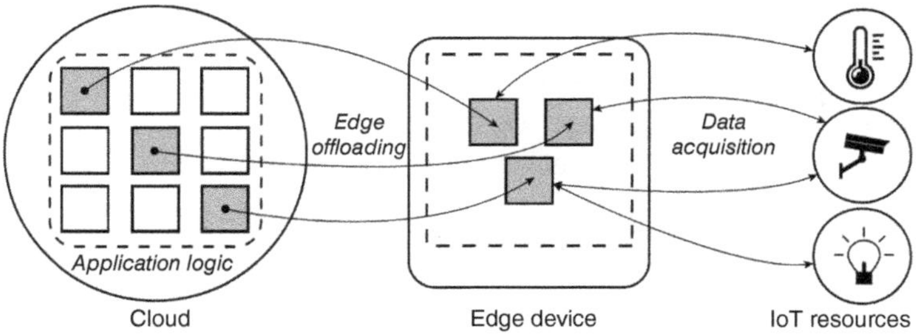

Figure 6.2 Edge Offloading: Cloud-to-Edge Paradigm.

For data-driven AI applications, model training and inference can be offloaded from resource-constraint devices to the edge, which has more powerful computing capacity. Offloading this type of computing (ML training and inference) can speed up the overall service performance.

The second factor concerns the resilience. For safety critical services, edge can provide an extra "buffer" in case cloud and Internet connection are disrupted. Within the edge layer, when a single edge device lacks sufficient resources for specific offloaded tasks, such tasks can be further offloaded to other edge devices with available resources. This "buffer" layer formed by the edge can make services more resilient.

The third factor concerns energy. Compared with mobile and IoT devices, typical edge nodes are often equipped with power supply such as lamppost, traffic light systems and base stations. Since battery life on many mobile and IoT devices (e.g., smart cars, wearables) is a hard constraint, offloading heavy duty computing tasks to the edge can help extend the battery consumption on those end devices.

Definition: In this book, edge computation offloading represents the process to utilize the edge layer to support computational tasks for cloud and end devices. In addition to conventional offloading practice, i.e., offloading from end devices to the cloud, cloud can also offload computing directly to the edge layer to fulfill specific service requirements [170].

How to Offload: Edge computation offloading can be performed over different strategies, including cloud-to-edge, edge-to-edge, and end-to-edge. The "edge-to-edge" and "end-to-edge" offloading strategies share the similar mechanism with device-driven offloading practices as discussed in Section 6.1.

Cloud-to-edge is a major trend in edge offloading, which aims to strike a balance across performance, resilience, and energy saving. As illustrated in Figure 6.2, to cater for the demands of processing data generated from end devices, the "cloud-to-edge" represents a paradigm where computation is dispatched from the cloud or datacenter servers to computing devices deployed at the network edge as close to the data sources. The tasks offloaded to the edge can take the form of full and partial offloading as discussed in Section 6.1.3.

For edge computation offloading, collaboration and orchestration across edge, cloud and end devices are crucial, given the distributed resources deployed in diverse environments [209]. In the meantime, the overhead of such management (e.g., orchestration) shall be lightweight and scalable to fit the three-tier of device-edge-cloud setting.

6.2.3 Enabling Technologies for Edge Offloading

The technology supports from communication, networking, and computing are essential for edge offloading to be efficient and scalable. For readers looking for comprehensive knowledge, we highlight the enabling technologies from three perspectives.

- **Communication**: Advancement in wireless communication has empowered edge offloading to perform in an efficient and on-demand manner. The flexibility of connectivity is important especially for mobile scenarios such as drone and vehicular applications. Two major technology branches come from IEEE and 3GPP standard families. In the IEEE 802 wireless standard family, Bluetooth [210] and WiFi [211] are dominant in edge offloading, providing energy benefit (e.g., Bluetooth Low Energy) and high throughput (e.g., WiFi version of 6 and 7). In the Mobile Broadband Standard 3GPP family, 4G and 5G [107, 212] can provide low latency and comprehensive coverage for edge offloading.
- **Networking**: To better organize edge offloading in dynamic environments, networking protocols like QUIC [213] and multi-path transport control protocol (TCP) [214] can provide low-latency connection establishment, network path migration, higher throughput and improved resilience. For data offloading scenarios, CDN [204, 205] and Information Centric Networking (ICN) [215, 216] can support better content distribution and mobility.
- **Computing**: For highly diverse and resource constraint edge computing environments, lightweight virtualization such as docker container [217] and unikernel [218], is essential for edge layer to efficiently and mutually cooperate with cloud based services and IoT/end devices [78]. For offloading management, platform solutions such as Cloudlet [7, 219] and Oakestra [209] can further support computing offloading in a scalable manner.

In addition, dedicated hardware and accelerators have been facilitating edge offloading from hardware angle. We will dive into the hardware innovations in Chapter 9.

6.3 Edge Offloading in Practice

This section presents use cases and proposals of edge offloading (Section 6.3.1). To shed light on future development, AI-driven edge offloading is illustrated as a representative trend (Section 6.3.2).

6.3.1 Use Cases of Edge Offloading

Edge offloading has be explored in several application domains, including drones, AR, road safety, and pollution monitoring.

- **Drone**: For the advantages of high mobility, unmanned aerial vehicles (UAVs) have been explored in recent years for edge offloading by acting as "edge" node. By having a mobile UAV with computational resources to process on-demand tasks, we can handle the cases of emergency response in disasters and object detection/recognition (e.g., surveillance purpose) for hard-to-access locations. In such drone context, energy and trajectory planning are essential factors to consider [220].

- **Augment Reality**: For more interactive and immersive experience, AR applications are exploring edge offloading for energy saving, low latency and high accuracy [221]. Since AR applications demand real-time responsiveness on resource constraint mobile devices, several heavy duty tasks such as pose estimation, object detection, and depth estimation can benefit from the resources on nearby edge node to share the costs of computing and battery consumption.
- **Road Safety**: For road safety and better support of upcoming self-driving vehicles, edge offloading is utilized together with roadside infrastructure for real-time assessment of road context [206]. With edge offloading, it becomes feasible to harness the sensing and computing capability of road infrastructure for collaborative road condition assessment and hazard detection (e.g., black ice). In such scenario, the sensing limit of a single car can be extended beyond what can be captured by its on-board sensors.
- **Air Pollution Monitoring**: For real-time insights of air pollution, edge offloading has been explored together with UAVs and field sensors of the infrastructure to perform local data processing [78, 222]. Such localized operation not only enhances data privacy but also reduces data volume and potential processing latency. By deploying new mobile edge nodes (e.g., UAV) on demand in various locations, we can scale up the pollution monitoring coverage without disruption to the existing infrastructure.

6.3.2 AI-Driven Edge Offloading

Given the vast growth of connected IoT devices, AI and data driven automation has become a promising direction to achieve more accurate edge offloading strategies and reducing the costs of decision. Compared with the conventional approaches relying on static logic, which are difficult to cope with to highly dynamic operational conditions, AI-driven edge offloading can facilitate the decision process by utilizing advanced machine learning (ML) methods together with collaborative computing frameworks such as KubeEdge (details Chapter 10).

Being one of the latest research frontier in MEC^2, the AI-driven approach has the potential to make edge offloading scalable, adaptive, and efficient. We illustrate here two representative AI-driven approaches for readers to grasp the new development in edge offloading.

- **Deep Learning**: As a major branch of machine learning, deep learning (DL) [163] can facilitate the learning of data representation based on artificial neural networks. DL can extract essential data features along with the accumulation of training input. For edge offloading, a supervised learning method is proposed to model the computation offloading. The proposed Deep Supervised Learning (DSL) method considers the local overhead on end devices, limited communication and computation resources of the network, for minimizing the overhead [223]. In the Deep Contextualized Compressive Offloading for Images (DCCOI) framework [224], a lightweight neural network is designed to compress the image data in variable-length. By allowing an image region to be encoded into different numbers of feature values, this spatial-adaptive approach offers more optimal performance than the conventional fix-length coding methods. For application domain of video analytics, the Dynamic Adaptive Offloading (DAO) framework jointly considers the dynamics of network bandwidth and video content for maximizing the analytics accuracy [225]. The DAO design utilizes its optimization formulation to adapt the bit-rate and resolution dynamically to strike a balance among bitrate, resolution, analytics accuracy and video content.

- **Reinforcement Learning**: For dynamic decision-making, reinforcement learning (RL) [226, 227] has been explored in edge offloading for optimal or near-optimal offloading strategies that can adapt to different application scenarios. To handle unknown load level dynamics at the edge nodes for long-term cost minimization in a decentralized setting, deep reinforcement learning is proposed allowing each device to determine its offloading decision without knowing the task models and offloading decision of other devices [228]. To enable cloud-edge collaboration, RL is also explored to support dynamic model selection in video analytics [229] and eXtended Reality (XR) application [230]. The ModelIO design jointly considers the decision-making for video pre-processing, Deep Neural Network (DNN)-based model selection and local inference to maximize performance [229]. By taking into account XR requirements such as latency and energy consumption, the proposed Deep Reinforcement Learning-based offloading scheme for multi-access edge computing-supported eXtended Reality (DRLXR) design trains and derives the close-to optimal offloading decision through optimizing a utility function dedicated for XR devices [230].

6.4 Summary and Outlook

Edge offloading is one of the key enablers for MEC^2. Among open challenges in edge computing [3, 162], there are three future directions that deserve our further attentions for edge offloading.

- **Service (Target) Discovery for Edge Offloading**: Due to the complexity and heterogeneity of edge environments in terms of device capacity, OS and protocol stack, and battery life, it is important to explore practical service and target discovery mechanisms. The discovery design shall handle device heterogeneity and also edge network dynamics since devices can join and leave the edge network frequently. The availability of services and computing resources can change rapidly. As the number of IoT and edge devices keep growing, we need to address the scalability challenge since overhead and latency in discovery process will increase. Pro-active and publish-subscribe paradigms can be explored to scale efficiently with the constant growth of devices and services in the future computing environment.
- **Programming Models for Edge Offloading**: For the success of MEC^2 applications, we need to go beyond existing programming models, which still lack necessary abstraction for developers to simplify develop and deploy applications that utilize edge offloading in the future cloud-edge collaborative infrastructure. In addition, diverse requirements from applications, hardware architectures, and operating systems will demand highly flexible and adaptive programming models for edge offloading, especially allowing task partition, migration and orchestration to be both adaptive and of low overhead.
- **Safety and Reliability of Edge Offloading**: As edge devices are often deployed in distributed and uncontrolled environments, it is essential to ensure data privacy

is guarded during the edge offloading process. The design shall balance user trust, usability, and regulation compliance. In addition, by facing edge specific uncertainties in terms of device/network failures and environmental disruptions, edge offloading shall embed robust fault tolerance mechanisms in its design, which is important to meet the user expectations.

To tackle the challenges in those future directions, interdisciplinary efforts spanning computer science, network communication, systems engineering, and cybersecurity are necessary.

7

Edge Storage and Caching

Edge storage and caching brings near-real-time access to storage and caching at the edge of the network. This will facilitate emerging applications that require fast access to data with very low latency.

The evolution of telecommunication technologies, especially the dawn of 5G, have enabled a multitude of advanced applications including, AI, virtual reality and gaming, intelligent healthcare, smart cities, autonomous vehicles, industrial Internet of Things (IIoT) and virtual worlds known as the metaverse [231]. These applications demand high computational power and massive storage, accessible at a very low delay, for massive numbers of connected devices.

The applications can harness the high computational capabilities of massive data centers in cloud computing platforms. Yet, these data centers are usually not located in close proximity to end devices, such as IoT devices, that collect and generate data. Communicating and processing these data at cloud computing platforms will eventually lead to high latency [232]. Therefore, banking on traditional cloud computing platforms will not be sufficient to facilitate emerging applications that require near-real time access to powerful computational resources. These requirements cannot be delivered through traditional cloud computing solutions due to the requirements such as, extremely low latency, fast backup, high accessibility [233]. Hence, such demands are satisfied through edge computing, which brings powerful computing and data storage capabilities of the cloud to the edge of communication networks, as edge devices are in close proximity to users, as illustrated in Figure 7.1 [234]. Furthermore, the number of connected devices at the network edge and the amount of data generated by these devices are expected to increase significantly over time [235]. Hence, storing, managing, processing and securing these data has become a significant challenge. Thus, efficient, robust and secure data storage and caching in the network edge is a key enabler for emerging applications and services [236].

Furthermore, emerging applications and services, including mobile networks, are expected to heavily utilize Artificial Intelligence (AI) and Machine Learning (ML) technologies to analyze data at the network edge. However, this requires storing historical data at the edge of the network to be used for accurate predictions. Yet, this comes as a challenge due to storage limitations at the network edge [237].

This chapter explores edge storage and caching, which is an integral component of emerging Edge computing platforms.

Mobile Edge Computing and Communications: Driving Forces, Technology Foundation, and Application Areas, First Edition. Aaron Yi Ding, Chamitha De Alwis, and Madhusanka Liyanage.
© 2025 John Wiley & Sons Ltd. Published 2025 by John Wiley & Sons Ltd.

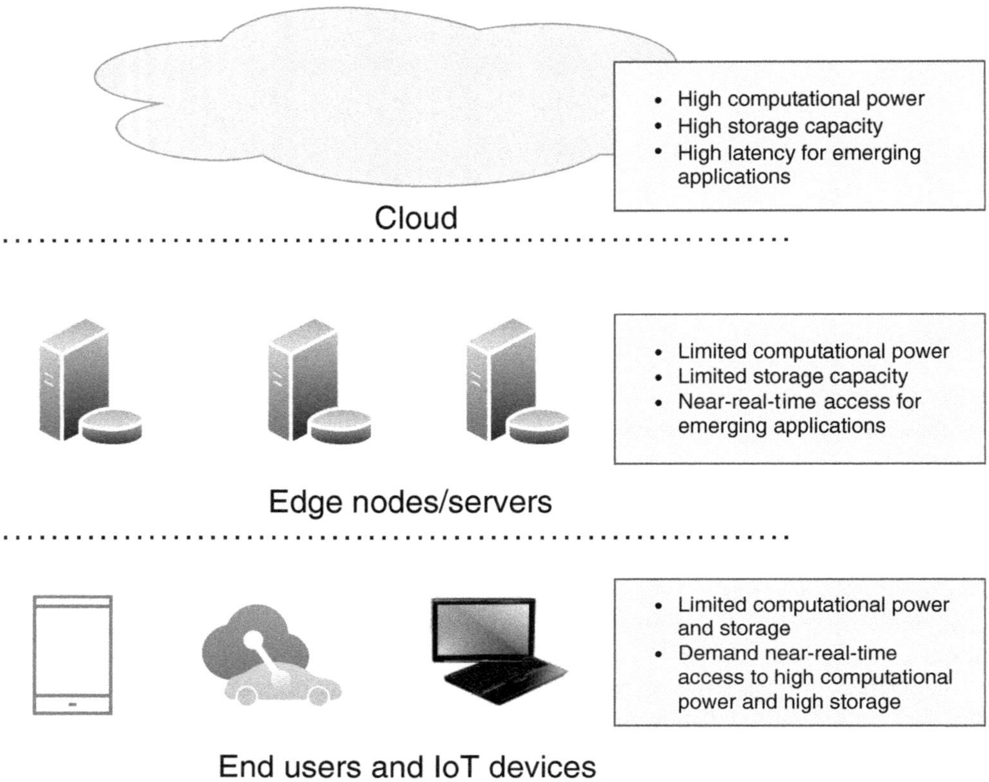

Figure 7.1 Edge Storage. Source: Tumpa/Adobe Stock.

7.1 Edge Storage

The high storage capacity, that is required to store massive amounts of data generated and required by emerging applications, is provided by edge storage systems comprising of large numbers of distributed edge servers [238].

Storing data at the network edge comes with several challenges.

- **Edge Storage Design:** The design of an efficient, robust and secure edge storage solution consisting many distributed edge servers is a challenge.
- **Data Robustness:** Edge storage solutions consist of large numbers of edge servers. A server failure may disrupt the functionality of the edge storage solution and might be difficult to repair due to the distributed nature of servers. This requires fault-tolerant techniques to maintain the availability of data.
- **Data Security and Privacy:** The distributed nature of edge storage solutions consisting many servers may expose data to unintended parties. This requires strong measures to ensure data security and privacy protection. The protection of data is challenging in IoT

environments when users can both generate and consume data. Hence, effective access control mechanisms should be in place to establish a secure edge storage system.

Hence, effective mechanisms should be in-place to ensure data availability through robust data storage solutions, and data security through effective access control mechanisms.

7.1.1 Edge Storage Design

The design of an edge storage system should not only be efficient but also robust and secure. For instance, in edge storage solutions, the storage capacity of edge servers are limited. Thus, edge servers are not capable of independently providing all the storage demanded by applications. This requires edge servers to offload data to other edge servers in case of limited storage resources. Hence, edge servers can collaboratively serve users and applications through shared storage resources. However, these edge servers may belong to multiple edge infrastructure provides. This requires to establish trust and an incentive towards collaboratively sharing data and serving user applications. In order to tackle this issue, a decentralized blockchain based collaborative edge storage solution, namely CSEdge, is proposed in [239]. In this system, edge servers are able to issue data offloading requests to other edge servers, which are evaluated and the winning edge servers where data is offloaded are selected based on their reputation. The offloaded data will be stored by the winning edge nodes and they will receive an incentive for performing the task. The performance of the edge server during the storage of offloaded data will be recorded via a distributed consensus on a blockchain for future decisions on evaluating reputation.

Furthermore, it is evident that managing and processing data at the network edge is a challenge, especially when the amount of data generated and required to be processed by edge network nodes are increasing rapidly. This demands novel solutions to manage and process massive amounts of data at the network edge. Nicolaescu et al. [235] proposes a framework, namely Store Edge Networked Data (SEND), for in-network storage management through data repositories at the network edge. This method considers many criteria including the popularity of data, and proximity of data from different processing functions, towards deciding how to place raw data and processed data at the network edge. The decision is based on labels, which is a system-wide data content identifier system for data context. This method helps achieving efficient data insertion, data lookup and efficient user request completions in edge storage environments.

In addition, many a real-time applications, especially in IoT environments, perform data analytics in the network edge. This will facilitate making near real-time decisions in close proximity to the user. However these data analytics applications require access historical data with minimal delay to function properly. Yet, the amount of data that can be stored in the edge is limited due to storage limitations. Hence, the amount of historical data that should be saved to make high quality near real-time decisions needs careful consideration. This question is take in to account in [232], to propose an architecture to manage data storage in the edge. This adaptive algorithm can facilitate to find the minimum amount of data required make highly accurate real-time decisions in storage limited edge network environments.

7.1.2 Data Robustness in Edge Storage Systems

Conventional applications and services are designed to be bank on cloud services. Moving these applications towards the network edge is challenge due to the distributed nature of edge computing and storage due to the resource limitation.

Edge storage solutions are required to provide robust storage at the edge of the network [238]. This means, edge storage should be robust against events such as, disk damages, server downtimes and hardware faults. Usually edge storage systems. Thus, edge storage systems are required to adopt fault-tolerant mechanisms to keep the stored data safe. This can be performed through storing replicas of data across multiple storage systems. This way, in case of the failure of a storage node, all the services can be switched over to another node containing replicas of the lost data. This can enhance the robustness of data against any possible data loss. For instance, Sonbol et al. [240] proposes a decentralized storage system named, EdgeKV, to be deployed in the network edge. The proposed system provides fast and reliable edge storage utilizing data replication with strong consistency guarantees. Furthermore, EdgeKV is location-transparent and provides a interface-based design. In addition, EdgeKV is also scalable with a heterogeneous edge node system.

However, it should be noted that maintaining replicas of the data has several concerns. For instance, maintaining replicas of data is not storage efficient. This is more prominent in edge storage, where storage is not abundant. In addition, maintaining data replicas in multiple locations increases the possibilities of data leakage attacks. Such attacks can compromise data security and privacy at the network edge. Applying encryption on data replicas can enhance the security of data. However, it is still possible for attackers to crack the encryption key, leave having replicas of data in multiple storage systems to be vulnerable for attacks.

In contrast to storing replicas of data in multiple storage systems, storing data using erasure code based systems can be utilized as a more secure and efficient solution to enhance the robustness of a storage system [238]. In erasure code based systems, a storage node may hold redundant data combinations that can be used to generate lost data in case of a data loss [241]. This is more efficient than holding entire replicas of data. Furthermore, the possibility of data leakage attacks can also be limited due to storage nodes only holding a portion of data. However, in erasure code based systems, the data recovery process may be more complex and time consuming.

It should be noted that network edges are often owned and maintained through multiple entities. These entities might have different capabilities to operate and maintain edge storage nodes. Furthermore, the edge nodes can also be disabled through hardware failures or attacks. Moreover, edge devices can be heterogeneous, making it difficult to have a uniform pricing structure. As a solution, Chen et al. [233] proposes SmartStore, which is a blockchain and smart contract based auction mechanism to allocate edge resources. This method uses Bayesian game theory to analyze how find how data owners and edge network prices can both maximize the respective benefits. This will facilitate data owners to utilize edge network resources in a cost efficient fashion.

7.1.3 Data Security and Privacy in Edge Storage Systems

While offering efficient bandwidth utilisation and low latency, it is a challenge to offer security-aware data robustness together with secure and adaptable data sharing among multiple uncertain users. This is due to the lack of robust fault-tolerant and access control mechanisms. Storing data at the edge of the network demands strong measures to protect data to ensure data security and privacy.

There are many approaches towards ensuring data security and privacy. In addition to data encryption, one prominent approach towards ensuring data security and privacy is only allowing security key owners or specifically assigned users to access data through a virtualized interface [238]. However, this approach can be too aggressive and result in limiting functions such as IoT data sharing. Another approach towards securing data is the implementation of Access Control Lists (ACLs). The ACL specifies the list of users who can access data and should be provided by the data owner. Users who are included in the ACL will be able to access the data using their secret keys. ACLs can also be implemented as Role-Based Access Control (RBAC) or Attribute-Based Encryption (ABE). RBAC allows access to data for users with appropriate roles to decrypt encoded data. On the other hand, ABE controls data based on attributes, which makes it more flexible to allow data access. However, these conventional approaches may not be suitable for edge storage systems in IoT environments, which are highly unpredictable and has a rapid population flow. Hence, a more suitable approach would be conducting access control through trust-oriented evaluation between the owner of the data and third parties based on predefined rules. Yet, this too requires trusting third party agents up to a certain level.

In addition, the work presented in [238] proposes a secure and trust-oriented storage model. This model can strengthen the fault-tolerance and access control issues in IoT environments. The paper proposes a robust and secure edge storage (RoSES) model using a totally local reconstruction code method with high data robustness, security and lightweight computation in edge nodes. The proposed RoSES method was further strengthened using a trust-oriented data access (TODA) method towards facilitating legitimate data access requests from uncertain users in IoT data sharing applications. Hence, this model is able to achieve efficient data storage, data recovery and data sharing along with low storage cost, latency and data leakage probability.

Connected Autonomous Vehicles (CAVs) is one of the key application envisaged to be present in the beyond 5G era [231]. The realization of CAVs demand Intelligent Transportation Systems (ITS) to have a unified and distributed data storage architecture. This is mainly due to the lack of trust between ITS systems and edge servers in the process of storing data. In response, Qiao et al. [242] proposes a distributed trustworthy storage architecture with reinforcement learning for edge services such as ITSs. The proposed method is capable of dynamically storing data. This is using reinforcement learning that is based on popularity and trustworthiness. This way, the resource scheduling and storage space allocation of edge storage can be improved. Furthermore, the work proposes an identity authentication protocol based on trapdoor hashing towards securing the transportation network access.

7.2 Edge Caching

7.2.1 Introduction to Caching

The term caching in computing refers to the usage of hardware and/or software to store data or a portion of data in order to facilitate requests for that data in a faster and efficient manner. Hence, caching brings data or a subset of data to a high speed storage in order to facilitate faster access to data compared to when data is accessed from its primary location [243]. Caching takes place in many places, including, Central Processing Units (CPUs), Graphics Processing Units (GPUs), Digital Signal Processors (DSP), disks, web browsers, cloud storage, and Information Centric Networks (ICNs).

For instance, web caching lowers the utilization of backhaul link capacity of a network and increase the performance of retrieving data to facilitate modern web applications and services. Cache data is usually stored in faster (hence more expensive) storage media or in close proximity to user. Therefore, the storage capacity is usually limited. Hence, caching usually stores a subset of data transiently to optimize the performance of its intended application. One of the metrics used to measure optimize the performance of a caching system is the cache hit probability. This is the ratio between the number of cache files requested and the number of total cache files. It is also evident that the caching system design should consider many parameters, including, cache size, backhaul capacity and cost.

7.2.2 How Edge Caching Works

Caching at the mobile edge is an important use case of multiaccess edge computing (MEC). For instance, a mobile edge server can be placed at the edge of the network or in edge nodes to cache frequently used popular content. This will overcome the need for remote servers to repeatedly push the same content to multiple users and duplicate traffic that would result in network congestion. In addition, caching content at the wireless edge is becoming more viable as the cost of storage becomes lower [244].

Caching at the edge of the network is an efficient way to minimize delays while increasing energy efficiency and enhancing spectral efficiency [236]. With the number of mobile network users increasing, it is vital to seek efficient and effective ways of caching at the wireless network edge, as illustrated in Figure 7.2.

Compared to caching in the cloud or the network core, caching at the Macro Base Stations or Small Base Stations will help minimize the utilization of backhaul links, which will in return minimize the congestion in backhaul connections. Furthermore, by increasing the cache size of a base station, the cache-hit probability can be increase, which will in return reduce the required backhaul capacity and increase the network performance.

Furthermore, caching can also be performed at users. Content can be pre-cached in user devices, such as mobile phones and personal computers towards enhancing the user Quality of Experience (QoE). In addition, wireless traffic can also be offloaded to user devices. This can be performed by a base station broadcasting popular content to users. Furthermore, the base station can unicast content to users understanding their user patterns and preferences. Furthermore, if the cache hit probability of broadcast content is high, the requirement to unicast content to individual users will be decreased, resulting in creating less interference

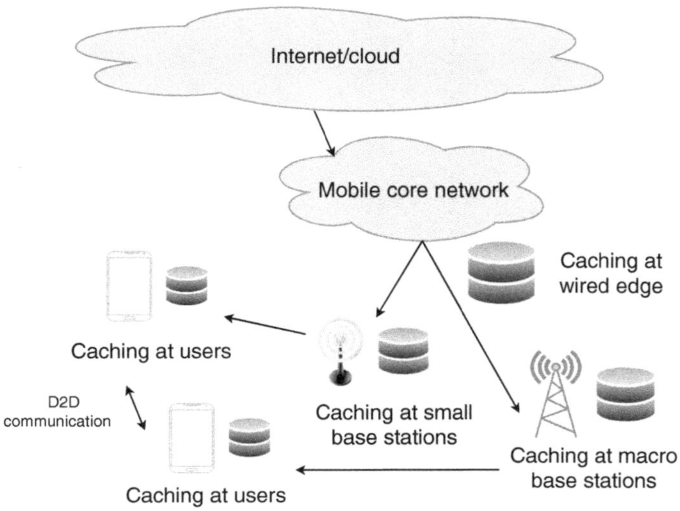

Figure 7.2 Edge Caching.

that would improve the network performance and QoE of other users. However, caching demands storage available at the user end and increasing the cash hit probability and motivating users to cache more content requires more research.

In addition, in case a file is cached at a nearby user device, it is also possible to retrieve the file through Device-to-Device (D2D) communication. Cache supported D2D communication demonstrates significant performance increases and the network throughput increases linearly with the number of users in the cache-enable D2D network. The challenges to overcome are reaching a high offload ratio while reducing the energy consumption of users, especially when user devices have limited battery capacity.

7.3 Edge Caching Towards Information-Centric Networking

The advances MEC and mobile networks can take edge-caching to the next level towards realizing Information-Centric Networking (ICN). The ever-growing demand for mobile applications and services cause significant challenges to mobile networks. These challenges include catering to the exponentially increasing requirement for data, multimedia traffic, and bandwidth, while keeping the networks cost and spectrum efficient. These developments have made existing Content Delivery Networks (CDNs) and Peer-to-Peer (P2P) overlays to evolve towards a novel architecture, knows as, ICN. ICN replaces the existing client-server model and puts information at the center to introduce a new publish-subscribe network model. Hence, ICN together with MEC provides many advantages, including, accelerated delivery of content, improved edge storage and edge caching, high mobility support, better support for real-time and context-aware applications, and content-based security and privacy [245].

ICN provides a novel architecture for the Internet by replacing the host-centric model with the content-centric model, as illustrated in Figure 7.3 [245, 246]. This is through

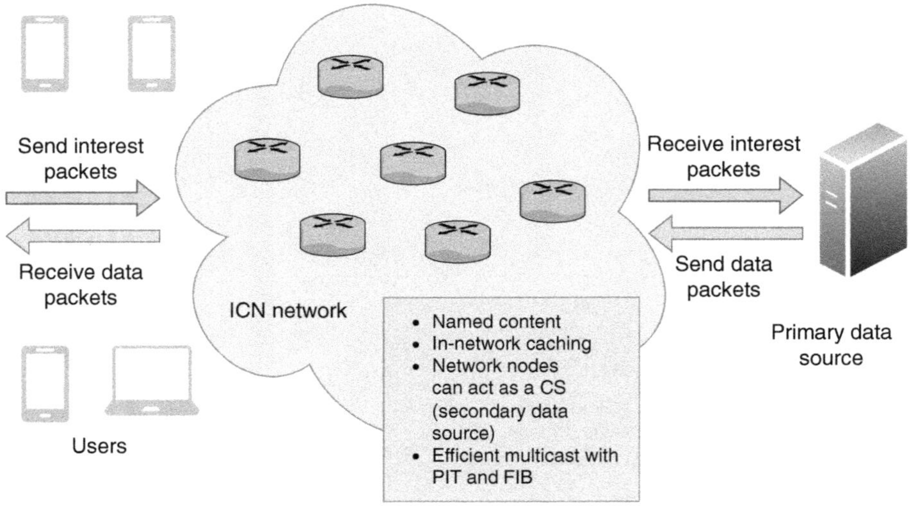

Figure 7.3 Information-Centric Networking.

introducing features, such as, named content, in-network caching, data-level security, and name-based and multi-path routing.

7.3.1 Interest Packets and Data Packets

As a significant move towards ICN, Content-Centric Networking (CCN) proposes the use of two types of packets for communication, namely, Interest packets and Data packets [247]. Interest packets are requests for specific content sent by users. In response to interest packets, data packets are sent and these packets contain the requested content. In order to efficiently use network resources, CCN uses an interest packet window, such that, multiple Interest packets can be sent back-to-back without waiting until the corresponding data packets are received. The flow balance is achieved through adjusting the interest packet window size in real time [245].

Furthermore, it should be noted that CCN uses a pull-based communication mechanism in contrast to the push-based communication in transport control protocol (TCP)/Internet protocol (IP). In this pull-based mechanism, one interest packet only pulls one corresponding data packet. Hence, the pull-based mechanisms reflects the concept of ICN where content is at the heart of the communication network. This method also enables achieving flow-balance, which is instrumental in avoiding congestion and spamming. What is more, the hop-by-hop model involves intermediate nodes towards achieving the flow-balance, unlike with end-to-end models in traditional network architectures.

7.3.2 Named Content

A name in ICN refer to an identifier, which is used instead of an IP address to locate the content [245]. These named-content have their name decoupled from the content location, as content is identified and located using the name. This name also allows network nodes to identify the content that traverse through them. Therefore, named-content also enabled in-network intelligence within networks to allow other interested users to share the content in ongoing data flows, which also results in efficient bandwidth utilization at the network edge. Additionally, as nodes know name of the content, they can replace several interest packets reaching a node with a single packet. Once the response to the interest is received, the data packets are replicated such that they are received by all requested nodes. This enables efficient content multicast in ICN enabled networks.

7.3.3 In-Network Caching

In-network caching is another core ICN feature that takes caching to the network core. This is achieved through performing caching in the network layer. Since ICN enables content to be named, ICN can enable caching more efficiently than traditional web-caching. As network nodes are aware of the content, caching can be performed in a content and content flow aware fashion, resulting in more efficient caching in networks [245].

7.3.4 Data Structures

In CCN, data structures are implemented in several ways. The Pending Interest Table (PIT) holds a list of interfaces that have sent interest packets requesting the same named content. On the other hand, Forwarding Information Base (FIB) keeps the list of interfaces that can be used to send the interest packets to retrieve content. Content Stores (CSs) are network resources that are capable of storing named-content.

In contrast to low capacity data buffers in traditional IP network nodes, CSs in CCNs keep named-content throughout the period that they are found to be useful. Therefore, named-content does not depend on the content locations as any CS node cache the named-content as a secondary data source. Thus, by means of storing content in high demand in CS nodes, ICN can achieve a high cache hit rate, which would make caching in future networks to be more efficient [245].

7.4 Conclusion

It is evident that storing content and caching content at the network edge is becoming a prominent solution towards facilitating emerging applications and services. This chapter covers edge storage, edge storage design, data robustness in edge storage, data security in edge storage and edge caching. Furthermore, the chapter explores the advents of ICN, which will harness the concept of caching at the network layer towards realizing efficient content distribution in future communication networks. However, efficient means of edge caching by increasing the cache hit probability, minimizing energy consumption and enhancing the network performance and user QoE, and the practical implementation of ICNs have many opportunities for research.

8

Edge Communication, Hardware and Enabling Frameworks*

MEC^2 brings near real-time access to powerful computational and storage resources from the cloud to the network edge to facilitate emerging applications and services. However, MEC^2 demands seamless, high speed and near real-time connectivity to communicate between edge computing/storage resources and user devices. This chapter discusses advancements of communication technologies and the evolution of mobile communication networks to realize efficient edge Communication in 5G and beyond networks.

Computing practices have seen a transformation from the use of software-defined networks to the development of Edge artificial intelligence (AI) practices. These deployment and implementation techniques have gained additional popularity in applications and use cases ranging from agriculture, automated vehicles, healthcare, IoT, and smart cities because of several factors that may come in the form of automating tasks, energy efficiency, interoperability, privacy, reliability and security. Behind the success of these techniques, the rapid development and emergence of dedicated hardware, such as application-specific integrated circuits, Edge accelerators and computing systems, play a vital role. In this chapter, a dedicated categorization of such devices and computing systems on the basis of functionality, resources, computing abilities and applications has been provided. This chapter provides fundamental learning to the readers and researchers on the domain-specific relationship between the design considerations of edge hardware, enabling frameworks and AI models, ultimately envisioning a dynamic ecosystem of this technological landscape.

8.1 Advancement of Communication Technologies

The development of science and technologies have paved the path to realize many concepts as technologies that can empower future communication networks. Hence, it is interesting to explore many a advancements of communication technologies that are pivotal towards realizing edge enabled beyond 5G communication networks.

*This chapter is partly contributed by Dewant Katare (TU Delft, Netherlands).

Mobile Edge Computing and Communications: Driving Forces, Technology Foundation, and Application Areas, First Edition. Aaron Yi Ding, Chamitha De Alwis, and Madhusanka Liyanage.
© 2025 John Wiley & Sons Ltd. Published 2025 by John Wiley & Sons Ltd.

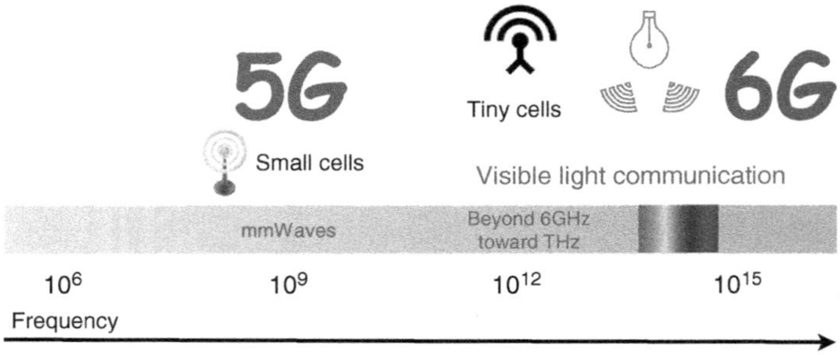

Figure 8.1 Towards THz Communication.

8.1.1 THz Communication

The ever increasing wireless data traffic demand data communication to move towards the Tbps range, which demands exploring higher frequencies beyond mmWave (up to 300 GHz) systems [248]. In response, Terahertz (THz) frequency band (0.1–10 THz) is explored to be utilized in emerging beyond 5G and 6G communication networks, especially to achieve high data rates within short distances, as illustrated in Figure 8.1. The THz band lies between mmWaves and the optical frequency band. Considering the high frequency, THz signals are also ideal to be utilized in tiny cells [249]. The usage of THz links to replace cables in data centres are also explored to achieve less clutter and improved performance. However, the utilization of THz demands solutions to problems, such as, high attenuation, atmospheric absorption, and accurate channel modelling before being utilized in future wireless networks.

8.1.2 Zero Touch Network and Service Management

Zero touch network and Service Management (ZSM) aims to realize a fully automated network. This includes performing functions, such as, self-configuration, self-optimization, self-monitoring, and self-scaling [250]. In a framework proposed by the European Telecommunications Standards Institute (ETSI), seven categories have been identified towards realizing ZSM, namely, end-to-end network and service management, Network-as-a-Service (NaaS), analytics and ML, collaborative service management, security, testing, and tracing [251]. With ZSM, networks are expected to achieve better resource utilization, efficient and coordinated operation, higher security and ultimately provide a better Quality of Service and Quality of Experience to users. It should also be noted that realizing ZSM heavily relies on AI/ML capabilities of a network, especially at the network edge.

8.1.3 Blockchain

Blockchain is a form of Distributed Ledger Technologies (DLTs) that can facilitate the decentralized operation of future edge enabled communication networks. Blockchain can store transactions in to chronologically ordered blocks that are sealed cryptographically [252]. This provides blockchain with a unique set of advantages, including, immutability, proof of provenance, integrity, and pseudonymity [253]. Finance and banking have already adopted blockchain, however, MEC^2 enabled future communication networks are also envisaged to rely on blockchain technologies for efficient, and secure operation.

8.1.4 Federated Learning

The advancements of AI has enabled many a applications, specially in the area of IoT. Processing the massive amount of data generated by IoT devices require AI based techniques. In line with this requirement, Federating Learning (FL) is emerging as a technique to leverage the un-utilized or under-utilized processing power user and network devices.

FL introduces a decentralized approach towards training machine learning models. The raw data can be used at the edge to train a local model, which will be shared with global servers to generate a global model, as illustrated in Figure 8.2. Data privacy can be preserved with FL as raw data is not required to be shared with global servers. FL can be classified in to three categories, namely, horizontal FL, vertical FL, and federated transfer learning [254]. Horizontal FL is when the feature spaces of various datasets are the same. When the feature spaces of data sets are different, it is vertical FL. On the other hand, federated transfer learning has various datasets having different feature spaces and sample distributions [231].

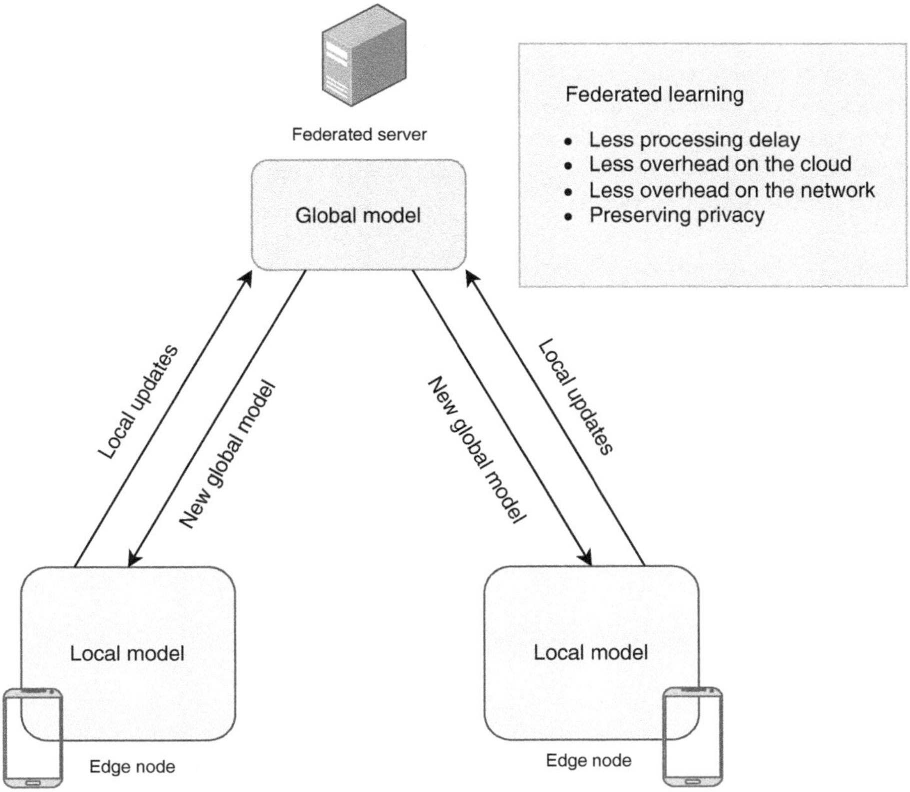

Figure 8.2 Federated Learning.

8.1.5 Large Intelligent Surfaces

Large Intelligent Surfaces (LIS) aims to utilize large surface, such as walls, building, and ceilings as transmitters [231]. This would enable a more controlled and efficient transmission of radio signals. LIS can overcome the challenge of maintaining communication when line of sight communication is not possible. LIS are designed by incorporating large arrays of small, low-cost and passive antennas or meta material in to structures.

Existing methods proposed for LIS deals with changing the phase of the signals. However, by means of harnessing the AI/ML and deep learning capabilities in the network edge, LIS can be developed form a concept to become a fully-fledged technology realized in future networks.

8.1.6 Visible Light Communication

Visible Light Communication (VLC) is a method of using visible light for communication in short-ranges, operating in the frequency range of 430–790 THz [255]. VLC uses Light Emitting Diode (LED) and Light Amplification by Stimulated Emission of Radiation (LASER) diodes for transmitting information while the receives consist of photodetectors (e.g., photodiodes).

VLC enables the utilization of the unlicensed visible spectrum and the available bandwidth is significantly higher compared to the radio spectrum [256]. Furthermore, the frequencies are reusable considering that visible light can be easily blocked. Furthermore, VLS has a low energy consumption, is highly secure due to unidirectional propagation, less costly compared using radio frequencies, and has no health implications for humans. However, efficient utilization of VLC is still a key research area towards realizing next generation communication networks.

8.1.7 Quantum Computing and Communication

Quantum computing aims to harness principles of quantum theory and quantum physics in computing applications. Quantum computing is a rapidly developing technology and is expected to become a key enabler of future computing application to solve complex problems. Quantum communication is a way to transmit quantum states from a sender to a receiver [257]. Quantum computing and communication is still in early stages of development. However, the growth of quantum computing is envisaged to completely change the paradigm of computing, communication and security of future communication networks.

8.2 Evolution of Communication Networks

The advancements of communication technologies as the ever growing demand for advanced communication services fosters the evolution of communication networks. Mobile networks are on their fifth generation, 5G, with a new generation of mobile networks emerging every 10 years [231]. Each new generation brought new capabilities to mobile networks to make humans more capable and connected. The evolution of mobile

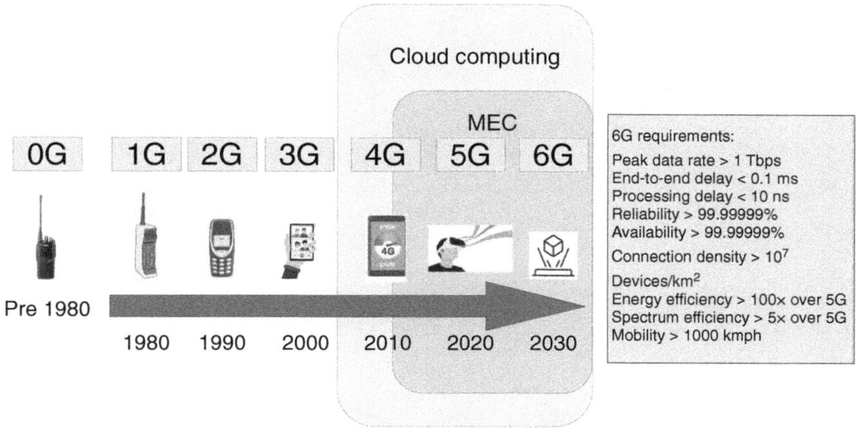

Figure 8.3 Evolution of Mobile Networks Towards MEC^2.

networks towards MEC^2 enabled communication networks is portrayed in Figure 8.3. The zeroth-generation (0G) mobile communication networks only offered basic communication capabilities. The first-generation (1G) of mobile networks introduced cellular mobile networks and analog voice communication. As communication technologies developed, the Second Generation (2G) of mobile networks emerged making a significant transition from analog communication technologies to digital technologies. 2G also brought basic data services to mobile networks. Once such example is Short Message Services, which is widely known as SMS. The third-generation (3G) of mobile networks introduced broadband capabilities to mobile networks. Furthermore, 3G introduced functionalities, such as, video calling, Multimedia Message Services (MMS) and mobile television. The fourth-generation (4G) of mobile networks brought faster mobile broadband to facilitate the streaming of Ultra High Definition (UHD) video. Moreover, 4G also introduced all-IP communication to mobile networks.

The dawn of the fifth-generation (5G) mobile networks enables enhanced Mobile Broadband (eMBB), ultra Reliable Low Latency Communication (uRLLC), and massive Machine Type Communication (mMTC). eMBB offers data rates up to 10 Gbps, uRLLC ensures delays of 1 ms, while mMTC supports over 100× more devices per area compared to 4G. Over 99.999% network reliability and availability is expected from 5G networks [231]. The network architecture utilizes Network Function Virtualization (NFV) and Software Defined Network (SDN) technologies to shift from dedicated hardware to softwarized networks.

The enhanced connectivity offered through 5G provides the communication infrastructure demanded by MEC^2. However, providing seamless MEC^2 along with Edge AI demands extremely high data rates, extremely low delays and latency, and extremely high reliability and availability, which are beyond 5G network capabilities [231]. In response, beyond 5G networks towards the sixth generation (6G) are expected to meet the communication requirements to facilitate MEC^2 and Edge AI. 6G networks are expected to debut by 2030 and are expected to be capable of delivering extreme data rates, extremely low delays, extreme reliability and availability, massive scalability, extreme power efficiency, and

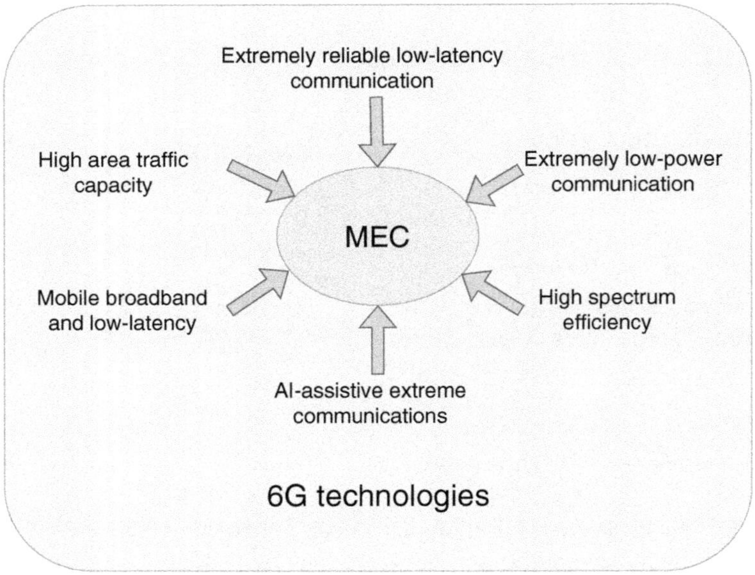

Figure 8.4 6G Technologies for MEC^2.

extreme mobility [231]. These capabilities of 6G will enable seamless MEC^2 and Edge AI to facilitate future applications.

In [231], the authors propose several 6G technologies that would enable future applications and services. Out of the technologies highlighted in [231], the rest of this chapter will focus on the technologies that would be key enablers of MEC^2 and MEC^2 based applications and services. These technologies are, namely, Further enhanced Mobile Broadband, extremely Reliable Low-Latency Communication, Extremely Low-Power Communication, High Spectrum Efficiency, High Area Traffic Capacity, Mobile Broadband and Low-Latency and AI-Assistive Extreme Communications, as illustrated in Figure 8.4.

8.2.1 Further Enhanced Mobile Broadband

enhanced Mobile Broadband (eMBB) evolves on top of 4G Long Term Evolution (LTE) networks to enhance the data rates to reach the gigabits per second range. However, many applications on the horizon, including, metaverse and industry 5.0, demands higher data rates over eMBB [258]. In response, 6G is expected to increase the data rates up to the range of Terabits per second with the dawn of Further enhanced Mobile Broadband (FeMBB).

FeMBB will be able to communicate extreme volumes of data among MEC^2 nodes. Moreover, FeMBB can provide Super-Fast Hot spots with very high data rates to crowded public spaces, high speed transportation mediums such as super fast trains, and smart cities. In addition, FeMBB will enable extremely high bitrate media content including multi-dimensional video and gaming [259].

Many new technological advancements will be required to see the light to realize FeMBB to increase the data rates of edge communication. One such technology is THz communication. Wide radio bands including millimeter-waves can fulfill the data rates required for

5G applications and services. The THz spectrum is expected to offer extreme capacity to provide wide-channels that would in-return facilitate extremely high data rates [260].

In addition to THz communication, VLC is another technology that would act as an enabler of FeMBB. Furthermore, the utilization of AI/ML at the physical and MAC layers (e.g., AI/ML based modulation and coding, AI/ML assisted spectrum sharing, and adaptive beamforming) will also act as enablers of FeMBB towards facilitating fast edge communication in 6G networks [261].

8.2.2 Extremely Reliable Low-Latency Communication

uRLLC is a key 5G feature that enables many a applications that can operate in the range of 1ms latency and 99.999% reliability [262]. Yet, many application including MEC^2 demands extreme reliability and low-latency to provide near-real-time access to powerful computational and storage resources to emerging applications and services. For instance, MEC^2 resources can be utilized to perform an Augmented Reality (AR) enabled surgery. It is evident that this application demands extreme reliability and extremely low-latency, which cannot be satisfied by pre-5G or 5G networks.

6G envisages to provide extremely Reliable Low-Latency Communication (eRLLC) with extreme reliability ranging towards 99.99999% and latency levels running beneath 0.1 ms. This will facilitate a plethora of future applications, including Internet of Healthcare (IoH), that would facilitate holographic tele-presence and AI-Healthcare [249]. However, enabling eRLLC in 6G networks demand many technological developments that would result in fast end-to-end turnaround time, intelligent framing and coding, and efficient resource management [262].

8.2.3 Extremely Low-Power Communication

Enabling MEC^2 in beyond 5G communication networks demand energy efficient hardware and even self-powered devices.

Existing 5G communication networks are not designed to match extremely low-power communication (ELPC) requirements. For instance, large-scale antenna arrays (e.g., Multiple-Input Multiple-Output, MIMO) in 5G consume high power consumption [263]. Beyond 5G and 6G communication networks envisage to deliver seamless connectivity while reducing energy consumption.

It is envisaged to achieve Extremely Low-Power Communication (ELPC) banking on many emerging technologies. For instance, LIS can use less antennas and reduce hardware complexities in transceivers, which would result in lower energy consumption [263]. Furthermore, the usage of passive artificial arrays will also result in reducing the energy consumption. Furthermore, Intelligent Reflecting Surfaces (IRS) do not utilize power amplifiers, keeping the energy consumption to be low [264].

8.2.4 High Spectrum Efficiency

Achieving a high spectrum efficiency is another enabler of MEC^2 in beyond 5G communication networks. It is envisaged that a large number of devices with diverse computational and storage capabilities will be required to connected with MEC^2 resources in envisaged 6G networks [265]. This will cause a disturbance in the wireless spectrum for wireless devices as the more devices connected, the more will be the wireless interference.

In response, 6G and beyond networks are envisaged to deploy sub-THz and THz bands. These frequency bands will can cater to the demand for high spectrum efficiency network based applications [265]. Furthermore, novel techniques, including location based information assisted beamforming techniques can also be utilized to achieve high spectrum efficiency to facilitate MEC^2 in beyond 5G networks. Furthermore, novel channel access schemes, such as Non-Orthogonal Multiple Access (NOMA) that will be available in beyond 5G networks are also identified to be more spectral efficient compared to prevailing pre-5G and 5G schemes.

8.2.5 High Area Traffic Capacity

Realizing MEC in the beyond 5G era demands a High Area Traffic Capacity (total traffic throughput served per geographic area (bit/s/m^2)) for edge nodes to seamlessly interact with MEC^2 nodes [266].

The 10 Mbps per square metre in 5G will not be sufficient for 6G applications and services to efficiently utilize MEC services. Therefore, 6G is envisaged to facilitate over 10 times the area traffic capacity of 5G [267]. Accordingly, the area traffic capacity is expected to reach 1 Gb/s/m with 6G to facilitate the high area traffic capacity demanded by 6G applications and services [267].

8.2.6 Mobile Broadband and Low-Latency

Another technology envisaged with the dawn of 6G towards realizing edge applications and services is Mobile Broadband and Low-Latency (MBBLL). For instance, a VR application that can demand resources for powerful processing and high storage demands. VR applications demand extremely-low-latency to provide an immersive VR experience and enhance QoE [48]. For the human eye to perceive perfect motion, the Motion-To-Photon (MTP) (time within a moment and the pixels of a picture frame that shows a new field of view) should fall beneath <15–20 ms. In addition, this type of VR applications demand high data rates, with downlink peak data rates reaching >1 Tbps. In addition, operating with real-time voice commands demand latency to fall beneath (<0.1 ms) while supporting high mobility applications with speeds over (>1000 km/h).

8.2.7 AI-Assistive Extreme Communications

AI-assistive extreme communications (AEC) is another 6G technology that will be a key enabler of MEC^2 [231]. AEC can monitor, coordinate and operate trillions of network elements while controlling parameters, such as, bandwidth allocation, routing, network planning, network operation, and spectrum management.

Furthermore, AI and Federated Learning (FL) are envisaged to play a key role in the design of 6G networks, as the network nodes are expected to exploit massive mobile data banking on MEC^2 capabilities. Furthermore, ZSM aims to automate all network functionalities, including network configuration, optimization, monitoring, testing and scaling. Intelligent decisions related to ZSM are made owing to the capabilities of AI/ML. Quantum computing and communication is yet another promising area to boost the performance of AEC in future 6G networks.

8.3 Edge Computing Landscape

The computing landscape is undergoing a profound transformation, driven by the convergence of technical advancements and evolving paradigms such as quantum computing, and sustainable/green computing. This transformation is exemplified by the progression of edge computing, a paradigm that decentralizes data analytics and processing, in contrast to the conventional practices of cloud computing. The decentralization is achieved by bringing computing resources closer to the source of data, thus transforming the previous computational processing practices and bringing significant advantages as compared to the traditional practices [219]. The key benefits of using Edge AI such as privacy, low latency, and energy efficiency can be seen in Figure 8.5.

Edge hardware, encompassing a spectrum of devices from mini and tiny IoT nodes to powerful edge servers, has risen to prominence as a response to the escalating demand for real-time processing, reduced communication bandwidth and latency, secure systems and enhanced privacy [172, 268]. A brief categorization of chip manufacturers, edge nodes/units, computing devices and servers is visualized in Figure 8.6, according to the

Figure 8.5 Advantages of Using Edge AI.

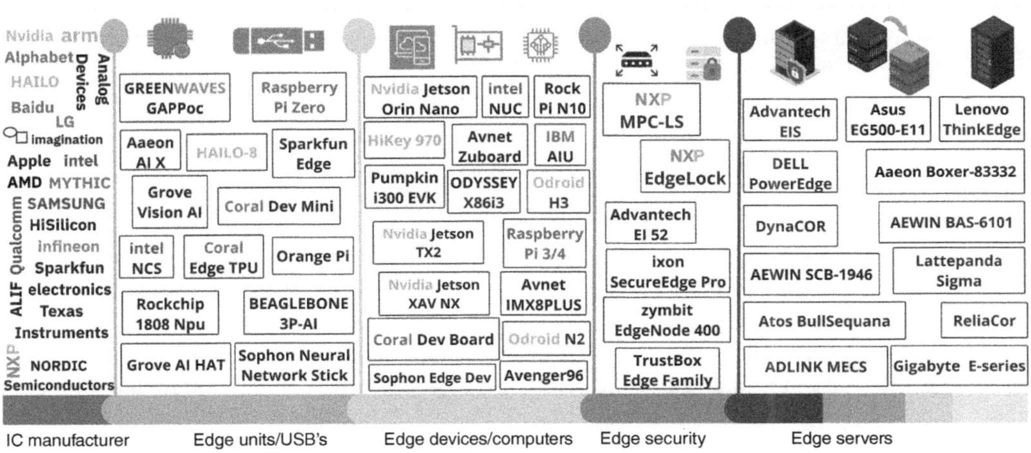

Figure 8.6 Popular Edge AI Chip Manufacturers and Devices Categorized According to the Computational Capability i.e., Node, Computing System, Servers.

popularity within the community [269–271]. In parallel to the growth of edge hardware, software frameworks, models and architectures have also evolved to facilitate ease in deployment for the challenges posed within these cyber-physical environments [271, 272]. Moreover, the convergence of Artificial Intelligence (AI) models and edge computing has created automation and deployment opportunities for several use cases. The computing abilities of Edge hardware to support AI workloads make it a trending approach for deploying machine learning and deep learning models at the edge [273–275]. As AI models increasingly find their applications in diverse domains such as autonomous vehicles, healthcare, and smart cities, the demand for efficient edge hardware and accelerators continues to escalate [268, 274].

Through an exploration of cutting-edge hardware devices and accelerators, this chapter offers fundamental knowledge and insights into the architectural innovations and design considerations that drive the evolution of edge computing. Additionally, it examines the current trends in AI model deployment at the edge, showcasing how edge hardware is instrumental in addressing the challenges of latency, bandwidth, and data privacy. The concluding remarks ultimately underline the relationship between edge hardware, software frameworks, and AI models, through a dynamic ecosystem that can reshape the current technological landscape.

8.3.1 Software Development Timelines and Landscape

The software development timeline and landscape for edge devices is dynamic and continuously evolving to meet the growing demands of edge computing applications. This section delves into the rapid emergence of edge software development over the decades, providing information on their crucial role in shaping the current promising techniques of computation. Below is a general overview of the software development timeline and the evolving landscape of edge device software.

8.3.1.1 Early Stages of Edge Computing (2000s–2010s)

- **Embedded Systems:** During this period, edge devices primarily relied on embedded systems and firmware for their operation. These systems were typically highly specialized and had limited computational capabilities.
- **Custom Software:** The development for edge devices involved developing custom applications and firmware tailored to specific device functionalities. These applications often had to be highly optimized for resource-constrained hardware.

8.3.1.2 Emergence of IoT and Edge Computing (2010s–2020s)

- **IoT Platforms:** With the rise of IoT, cloud-based IoT platforms emerged. These platforms offered cloud-based services for device management, data storage, and analytics. Software development for edge devices began to include IoT platform integration.
- **Edge Gateway Software:** Edge gateways, which serve as intermediaries between edge devices and the cloud, gained prominence. Edge gateway software development focused on data preprocessing, local analytics, and protocol translation.
- **Containerization:** The adoption of containerization technologies like Docker and Kubernetes started influencing edge software development. Containers enabled easier deployment and management of applications on edge devices.

8.3.1.3 Latest Edge Software Trends (2020s–Present)

- **Edge AI and Machine Learning:** Edge AI and machine learning have become central to edge software development. Frameworks like TensorFlow Lite, ONNX, and Edge TPU are used to deploy AI models on edge devices for tasks such as object detection and image recognition.
- **Edge Computing Frameworks:** Developed frameworks like AWS Greengrass, Azure IoT Edge, and Google Cloud IoT Edge provide edge computing capabilities, enabling developers to deploy and manage applications at the edge.
- **Edge-Optimized Operating Systems:** Specialized operating systems like Microsoft Azure Sphere and Google's Android Things are designed for edge devices, emphasizing security and connectivity.
- **Edge-to-Cloud Integration:** Edge software development now involves seamless integration with cloud services, allowing for bi-directional data flow and real-time synchronization between edge devices and the cloud.
- **Edge Device Management:** Software for managing edge devices, including updates, configuration, and monitoring, is crucial. Solutions like IoT device management platforms have emerged.
- **Edge Security:** Security remains a top concern. Edge software development focuses on implementing security measures like secure boot, encrypted communications, and device identity management.

8.3.1.4 Future of Edge Software Development

The software development timeline and landscape for edge devices reflect the increasing complexity and capabilities of these devices. As the technology continues to advance, software developers will need to adapt to new tools, frameworks, and paradigms to harness the full potential of edge computing.

- **5G Integration:** The rollout of 5G networks will further enable edge computing. Edge software will need to harness the low-latency and high-bandwidth capabilities of 5G.
- **Edge-as-a-Service:** Edge computing may evolve towards a service-based model, where developers can deploy and manage applications on distributed edge infrastructure seamlessly.
- **AI at the Edge:** Edge devices will become more capable of running complex AI models locally, expanding their use cases in areas like autonomous vehicles and robotics.
- **Standardization:** Industry standardization efforts will likely continue to streamline edge software development and interoperability among different edge devices.

8.3.2 Hardware Development Timeline and Landscape

The hardware development timeline and landscape in the context of edge applications have witnessed significant advancements over the years based on primary design considerations, also shown in Figure 8.7. The hardware development landscape for edge applications is driven by the need for increased processing power, AI capabilities, connectivity, and ruggedness [276, 277]. As edge computing continues to transform industries, hardware developers will focus on innovations that empower edge devices to perform complex tasks efficiently

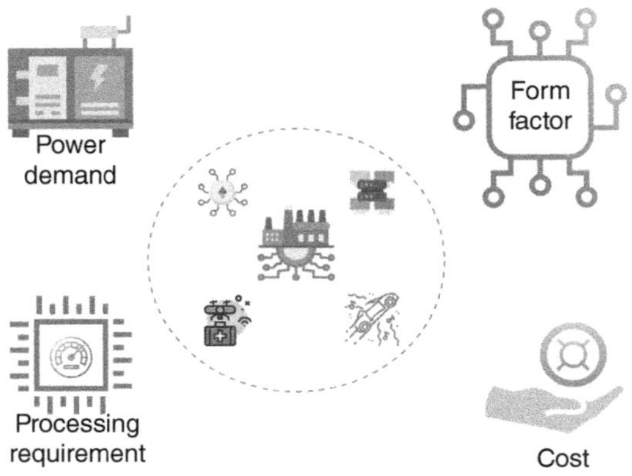

Figure 8.7 Primary Metrics for Edge HW Design Considerations. Source: Adapted from [272].

and reliably. Below is an overview of the hardware development timeline and the evolving landscape of edge devices:

8.3.2.1 Early Development of Edge Hardware (2000s–2010s)

- **Embedded Systems:** Edge devices in the early stages of development were often based on embedded systems. These systems were designed for specific, limited tasks and had minimal processing power and memory.
- **Single-Board Computers (SBCs):** The emergence of SBCs like the Raspberry Pi and Arduino opened up new possibilities for edge computing. These compact yet capable devices provided a foundation for edge applications.
- **Specialized Hardware:** Some early edge applications required specialized hardware, such as industrial controllers, PLCs (Programmable Logic Controllers), and field devices. These were often proprietary and highly specialized.

8.3.2.2 Rise of IoT and Edge Computing (2010s–2020s)

- **Mini PCs and Microservers:** As edge computing gained momentum, mini PCs and microservers emerged. These devices offered more processing power, memory, and connectivity options compared to embedded systems.
- **Edge Gateways:** These devices became essential components in edge computing, and initial tasks include collecting, aggregating/filtering the data received from source. Edge gateways provides connectivity, data preprocessing, and protocol translation capabilities, and are often featured with robust specifications.
- **AI Accelerators:** With an increasing demand of AI at the edge, specialized accelerators such as GPUs (Graphics Processing Units), TPUs (Tensor Processing Units), and FPGAs (Field-Programmable Gate Arrays) were integrated into edge devices to support AI workloads.

8.3.2.3 Latest Edge Hardware Development (2020s–Present)

- **High-Performance CPUs:** Edge devices now often feature high-performance multi-core CPUs, enabling them to handle complex computations and real-time processing.

- **Edge-Specific SoCs:** System-on-Chip (SoC) designs optimized for edge computing have become more prevalent. These SoCs may include dedicated AI accelerators, secure elements, and power-efficient cores.
- **Compact and Rugged Designs:** Edge devices are designed to be compact, rugged, and capable of operating in harsh environments. These designs cater to applications in manufacturing, agriculture, transportation, and more.
- **Edge AI Hardware:** Edge AI hardware, such as Nvidia Jetson and Google Coral devices, provides dedicated AI acceleration for tasks like computer vision and natural language processing.
- **5G-Ready Hardware:** Edge hardware is increasingly designed to harness the capabilities of 5G networks, enabling low-latency, high-bandwidth communication between edge devices and the cloud.

8.3.2.4 Future of Edge Software Development

The software development timeline and landscape for edge devices reflect the increasing complexity and capabilities of these devices. As the technology continues to advance, software developers will need to adapt to new tools, frameworks, and paradigms to harness the full potential of edge computing.

- **Quantum Edge Computing:** It is an emerging area of research, and there are efforts to bring quantum capabilities to the edge. Quantum edge hardware will have the potential to solve complex problems in real-time.
- **Customizable Edge Hardware:** Edge hardware may become more customizable, allowing organizations to tailor devices to their specific needs. This could involve modular designs and flexible configurations.
- **Security Hardware:** With cybersecurity concerns on the rise, hardware security features like secure boot, hardware-based encryption, and trusted execution environments will continue to evolve.
- **Energy Efficiency:** As edge devices are often deployed in battery-powered setting or remote locations, energy-efficient designs will be a priority to extend device lifespans.

8.4 Edge Computing System Solutions

8.4.1 Legacy Systems

Legacy edge computing systems represent earlier-generation technologies and approaches that were used for edge computing before the advent of modern edge computing paradigms. These legacy systems are characterized by their limitations in processing power, scalability, and compatibility with contemporary edge computing requirements. They often possess limited processing capabilities, making them ill-suited for resource-intensive tasks such as real-time analytics or AI inferencing that are central to modern edge applications. Furthermore, many legacy edge systems were built on proprietary hardware architectures, challenging their flexibility and making upgrading and integration with other systems complex. Furthermore, these systems often lacked standardization, resulting in compatibility issues and difficulties in interoperability with other devices and systems, thereby hindering seamless data exchange and collaboration in edge environments.

In terms of connectivity, older edge systems may have offered limited options, impairing their ability to efficiently collect and transmit data, particularly in IoT applications that rely on robust, high-speed connections. Energy efficiency was not a primary concern in many legacy systems, resulting in higher power consumption and operational costs-a significant drawback in an era where sustainability and energy efficiency are paramount considerations. Moreover, many legacy edge systems may lack modern security features, rendering them vulnerable to cybersecurity threats in a world increasingly characterized by connectivity. These systems were often designed in an era when the threat landscape was less complex and the need for robust cybersecurity measures was not as pressing. Consequently, they may not meet the stringent security requirements of contemporary edge applications.

Examples of legacy edge computing systems include traditional industrial automation systems, older embedded systems and microcontrollers, and outdated SCADA (Supervisory Control and Data Acquisition) systems used in utilities and infrastructure monitoring. While these legacy systems may have served their intended purposes in the past, organizations are increasingly recognizing the need to transition to more modern, capable solutions to fully harness the potential of edge computing. Modern edge devices, servers, and software platforms offer improved performance, scalability, security, and energy efficiency, empowering organizations to achieve their edge computing objectives effectively and securely in today's digital landscape.

8.4.2 Application Oriented Systems

8.4.2.1 Scalys

Security and privacy are paramount concerns in the rapidly evolving landscape of edge computing and IoT. Two innovative approaches, represented by the Scalys: secure Edge and IoT are the TrustBox Edge Family and the Enclave Device Blueprint, which are shaping the future of secure edge and IoT devices. The TrustBox Edge Family, crafted by Scalys, stands as a beacon of cyber-secure edge devices tailored to the demands of IoT and Secure OT environments. Built on the robust NXP Layerscape platform, these devices incorporate a host of security features. Key among them is the Trusted Platform Module (TPM) 2.0, providing an unassailable root of trust, guarding against unauthorized access. The Secure Boot process adds an additional layer of protection by ensuring that only trusted software can be booted. Firmware updates undergo stringent verification before installation, rendering tampering virtually impossible. Hardware-rooted trust further bolsters security.

The TrustBox Edge Family is characterized by its exceptional adaptability, with various form factors catering to diverse applications. The TrustBox Edge Nano, a compact, low-power device, finds its niche in battery-powered applications. In contrast, the TrustBox Edge Micro delivers robust performance, ideal for video surveillance and industrial automation. For high-performance edge computing and network security, the TrustBox Edge Gateway steps in. This family's adaptability, scalability, and wide operating temperature range make it suitable for a wide range of deployments, from smart cities to healthcare.

8.4.2.2 Enclave Device Blueprint

On the other front, the Enclave Device Blueprint, a collaboration between Arm® Technologies and Scalys BV, addresses the burgeoning need for confidential computing in the IoT and edge computing realms. With the proliferation of intelligent edge devices and autonomous IoT ecosystems, privacy and safety ascend to critical importance. Confidential computing, particularly leveraging Trusted Execution Environments (TEE) or enclaves, emerges as the solution. Enclave devices, although pivotal, pose engineering complexities, limiting the full potential of edge computing. The Enclave Device Blueprint endeavours to streamline enclave device development and confidential application deployment in IoT. Confidential computing, an extension of traditional computing, augments security for computing workloads and data. In the IoT landscape, enclave devices demand meticulous engineering and a focus on hardware-based trust. The Enclave Device Blueprint navigates these complexities, emphasising security, leveraging a collaborative, open-source approach. It encompasses projects, resources, and guidance to simplify development and enable scalable deployment of confidential IoT applications. This blueprint bridges the gaps, ushering confidential computing into the mainstream IoT domain [278].

TrustBox Edge 201, an Azure IoT Edge-certified, tamper-resistant enclave device developed by Scalys, is an example of a secure edge device. Harnessing Azure cloud services and the blueprint's software components, a robust pipeline for deploying confidential IoT applications was orchestrated. The blueprint's modular design ensures adaptability across technologies, operating systems, and cloud platforms. The TrustBox Edge Family and the Enclave Device Blueprint are pivotal in shaping the future of secure edge and IoT devices. They stand as beacons of security, privacy, and scalability in an increasingly interconnected world. These innovations lay the foundation for a seamless, unified experience across the cloud and the edge, reinforcing privacy and safety in this dynamic landscape [278].

8.4.2.3 Nokia

Edge solutions, encompassing both hardware and software innovations, offer a holistic approach to meet the evolving needs of the telecommunications industry. These solutions include the Nokia AirFrame Open Edge Fronthaul Gateway, designed for seamless connectivity in 5G and 4G networks, the compact yet high-performance Nokia AirFrame Open Edge Server, and the Nokia AirFrame Data Center Manager for efficient infrastructure management. Coupled with the Nokia Edge Automation Tool, this suite empowers Communication Service Providers (CSPs) with comprehensive edge automation capabilities. It streamlines edge planning, deployment, and operation while aligning with customer expectations of reduced operational costs and faster time-to-market for new applications and capabilities at the edge. Nokia's workload-agnostic edge automation tool ensures CSPs can effectively manage thousands of sites while realizing the anticipated total cost of ownership savings. Above mentioned solutions and products are covered in detail below:

- **Open Edge Fronthaul Gateway:** The Nokia AirFrame Open Edge Fronthaul Gateway (FHGW) is a network component that is designed to provide a solution for deploying and managing 5G and 4G mobile networks by enabling the connection between radio units (RUs) and mobile networks. It plays a pivotal role by offering fronthaul connectivity, acting as an aggregator for multiple RUs, and enabling protocol conversion between various fronthaul protocols. This device ensures efficient communication between RUs and mobile networks while enhancing performance and efficiency. Key components of the FHGW include the Common Public Radio Interface (CPRI) interface for high-speed RU connections, the 46 BBU Airscale for data processing, the RF section for signal conversion, and the 5G/4G RRH for radio signal transmission and reception. The interface also connects the FHGW to the mobile network, and the FPGA chip accelerates data processing. The FHGW supports evolved Common Public Radio Interface (eCPRI), a high-bandwidth, low-latency version of CPRI, and offers interfaces for connecting to a Virtualized Distributed Unit (VDU) for data processing from RUs. For 4G connectivity, it interfaces with LTE BBU, another baseband unit. Moreover, it features the optional capability to pass through traffic from eCPRI native RUs without protocol conversion.

- **AirFrame Open Edge Server:** This is a high-performance, compact server platform designed for deployment in edge data centers. This server is tailored to support a wide array of applications, including Virtual RAN (vRAN) for cloud-based radio access network functions, Multi-access Edge Computing (MEC) for applications like real-time analytics and IoT data processing, and IoT data processing itself. At present, it is available in two variants. Key components of the server include the CPU for processing, memory for data and instruction storage, storage for inactive data, networking interfaces, power supply, and cooling fans. As part of the Nokia AirFrame Open Edge platform, this server provides the necessary performance and scalability to meet the demands of intensive edge applications. It is equipped with a 4th Gen Intel Xeon processor, DDR5 RDIMM memory, ample storage options, and expansion slots for NIC and accelerator cards. Security features include an onboard TPM2.0, while management is facilitated through Redfish and an onboard BMC. Its compact form factor and support for edge data centers ensure low-latency, high-throughput data processing, contributing to improved edge application performance.
- **Data Center Manager:** Nokia's AirFrame Data Center Manager streamlines data center operations with an intuitive interface and real-time monitoring capabilities. It offers a single-pane dashboard for tracking critical systems, system firmware version control, and hardware configuration management, facilitating centralized oversight of computing and networking resources. With features like power, thermal, and utilization monitoring, it ensures optimal performance while providing configurable alerts for proactive management. Access controls enhance security, and troubleshooting is simplified with quick data retrieval from faulty nodes. This solution supports integration with external systems through a RESTful API, allowing for seamless connectivity with operations support systems and cloud controllers. Remote control features enable efficient hardware management, and it's adaptable to multisite deployments, effortlessly scaling to manage thousands of nodes.

  In terms of hardware requirements, it mandates a minimum of 500 GB of disk space, an Intel CPU with at least 6 cores, and 32 GB of RAM for managing up to 300 nodes. For larger deployments of up to 3000 managed nodes, it recommends 2 TB of disk space (SSD recommended), an Intel CPU with 24 cores, and 256 GB of RAM. Importantly, in multisite deployments, the central instance demonstrates remarkable scalability, effectively handling tens of thousands of nodes, making it an ideal choice for modern data center management, where efficiency and security are paramount.
- **Edge Automation Tool:** This tool is designed to streamline the operations of edge infrastructure, automating various tasks from planning to deployment and ongoing management. Integrated with the Nokia AirFrame Data Center Manager, this cloud-native application optimizes edge infrastructure operations, eliminating complexity and significantly reducing operational costs. It provides real-time insights into system health and

configuration through a unified dashboard, supports simultaneous system upgrades, and enables central management of user interfaces and APIs. This tool simplifies infrastructure management, consolidating system information and enhancing overall efficiency. It empowers communication service providers (CSPs) with a robust cloud computing infrastructure, allowing virtualized software and cloud-native applications to be deployed at the edge. Additionally, it offers flexibility for on-premises deployments, catering to diverse customer needs.

The Nokia Edge Automation Tool recommends CentOS 7/8 and RHEL 7/8 with the latest updates as the preferred operating system and environment for optimal performance. However, it is also compatible to be installed on other recent Linux distributions, providing flexibility in deployment. Regarding hardware requirements for common deployment alongside the Nokia AirFrame Data Center Manager, it's recommended to have a system with 2 TB of disk space (SSD recommended), an Intel CPU boasting 24 cores, and a substantial 256 GB of RAM. These specifications ensure the tool operates smoothly and efficiently, meeting the demands of modern edge infrastructure management.

- **MX Industrial Edge:** This is a cloud-native, mission-critical industrial edge solution tailored for supporting industries such as manufacturing, energy, and transportation. This solution is powered by the Nokia Digital Automation Cloud; it offers seamless management of applications and private wireless networking in a single user interface. This solution enhances operational efficiency and reduces costs, energy optimization, and new service development. In practical terms, it combines hardware and software technologies. For hardware, it consists of ruggedized server, driven by Intel Xeon Scalable processors, which ensures high performance and reliability. Meanwhile, the cloud-native platform supports various industrial applications, featuring container orchestration for easy deployment, real-time edge computing, and robust security measures.

8.4.2.4 Amazon

Amazon's edge AI solutions are tailored and varied according to the applications. The variety can be envisioned, from smart speakers, edge IoT gateways to rugged servers deploying AI applications at the edge. They are adaptable to a wide array of processors and hardware providers in terms of hardware compatibility. The hardware selection primarily depends on the particular use case and the computational requirements. This encompasses a spectrum of options, ranging from CPU, and GPU instances to FPGA and specialized AI inference chips like AWS Inferentia. Similarly, the software solutions are highly adaptable and are designed to be compatible with diverse environments and applications supporting real-time image processing at the cameras to speech recognition using natural language processing. For example, AWS IoT Greengrass enables local device processing in various IoT applications, while AWS Deep Learning AMIs and AWS Inferentia provide robust AI capabilities for edge devices, allowing organizations to harness the power of machine learning in edge computing environments. Brief summary of popular solutions is described below:

- **IoT Greengrass:** AWS IoT Greengrass is a software service designed to run on a variety of edge devices, including those equipped with ARM-based processors (e.g., ARM Cortex-A and ARM Cortex-M) and x86-based processors, thus making it compatible with a wide range of devices. Its processing power largely depends on the capabilities of the underlying hardware, which can vary from low-power microcontrollers to more powerful multicore processors. This flexibility allows IoT Greengrass to adapt to the specific computational needs of different edge devices and use cases [279].

- **Sagemaker:** Amazon SageMaker is a fully managed machine learning service that Amazon Web Services (AWS) offers. It is designed to build, train, and deploy machine learning models at scale. SageMaker provides a comprehensive set of tools and capabilities for every step of the machine-learning workflow, including data preprocessing, model training, hyperparameter tuning, and model deployment using machine/deep learning frameworks such as TensorFlow, PyTorch, and MXNet [279].
- **Deep Learning Amazon Machine Images:** AWS offers a range of Deep Learning AMIs supported by deep learning frameworks such as TensorFlow. These AMIs are optimized for both CPU and GPU instances. These AMIs are compatible with Intel and NVIDIA hardware, among others. They support Intel Xeon processors for CPUs, while for GPUs, they can be used with NVIDIA GPUs like Tesla V100 and P3. These instances provide significant processing power for training and inference tasks in deep learning applications. The choice between CPU and GPU depends on the specific use case and the need for parallel processing capabilities [279].
- **Inferentia:** AWS Inferentia is an AI inference chip designed by AWS. It's optimized for high throughput and cost-effective inference. AWS Inferentia chips are integrated into Amazon EC2 instances and Amazon SageMaker instances, offering a choice between different instance types. These instances use AWS Inferentia chips to accelerate inference tasks. While the specific processors in these instances may vary, they often feature Intel Xeon CPUs alongside AWS Inferentia chips to provide a balanced mix of general-purpose computing and AI acceleration [280].

8.4.2.5 IBM

IBM offers a range of edge devices, software solutions, and services designed to support edge computing and IoT (Internet of Things) applications. These solutions are aimed at helping organizations process data closer to the source, reduce latency, and make real-time decisions in edge environments. Here are some key aspects of IBM's edge computing offerings:

- **Edge Devices:** IBM has partnered with various hardware manufacturers to develop edge devices optimized for edge computing workloads. These devices often feature hardware acceleration for AI and analytics and robust security features. Some of these devices are designed for specific industries such as manufacturing, retail, and telecommunications.
- **Edge Software:** IBM provides software solutions designed to manage and orchestrate edge computing deployments. This includes edge application management, data management, and security solutions. IBM's software offerings help organizations deploy and scale edge applications efficiently.
- **Edge Computing Services:** IBM offers a range of consulting and deployment services to help organizations implement edge computing solutions effectively. This includes services for designing, deploying, and managing edge environments and integrating edge solutions with existing IT infrastructure.
- **Cloud Services for Edge:** IBM's cloud offerings include edge computing capabilities, allowing organizations to extend their cloud resources to the edge. This enables seamless

data integration and processing between edge devices and cloud services, supporting hybrid cloud-edge architectures.
- **IoT and AI Integration:** IBM's edge solutions often include integration with IoT platforms and AI technologies. This allows organizations to collect data from IoT sensors and devices at the edge, process it locally, and apply AI and analytics for real-time insights.
- **Security and Data Privacy:** IBM strongly emphasises security and data privacy in edge computing. Their solutions include encryption, identity management, and threat detection capabilities to protect data and devices at the edge.

8.5 Edge Hardware and Devices

Hardware platforms for edge AI applications can vary depending on the form factor or specific design requirements [281]. Edge devices in this case can range from tiny miniature devices operating in micro-watts to industrial Edge servers supporting terabytes of data. Some applications also require collaboration amongst clusters of devices present in the cyber-physical ecosystem. Some applications may require collaboration between Edge-Fog-Cloud to achieve high performance. A possible correlation and collaboration between Edge devices arranged as tiers/level is shown in Figure 8.8. Popular hardware platforms categorization for Edge devices currently includes:

- **Edge Accelerators:** These are specialized hardware devices that are designed for AI applications. They can provide significant performance improvements over traditional CPUs and GPUs. For example, the NVIDIA Jetson Xavier NX is designed for AI applications in the smart city, smart devices, healthcare, and mobility as a service.
- **End Devices/Nodes:** These are the devices closest to the data source. They may not have the same processing power as edge accelerators, but they can still be used for approximate computing. For example, a smart traffic light could use an end device with an edge accelerator to detect and track vehicles in real-time.
- **Edge Computing Devices:** These are devices that are located between the end devices and the cloud. They can provide additional processing power and storage for edge AI

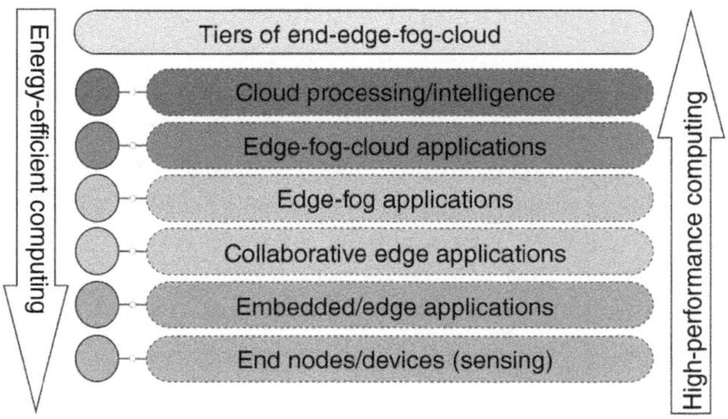

Figure 8.8 Hardware Correlation in Edge Applications. Source: Adapted from [274].

applications. For example, a fog node could be used to process the data from a fleet of autonomous vehicles.

- **Base Stations:** These devices connect end nodes/devices to the cellular network. They can also be used for edge AI applications, especially for low latency applications. For example, a base station could process the data from a video surveillance camera rather than transmitting it to the cloud.
- **Fog:** This is a computing layer located between the edge and the cloud. It can be used to distribute the workload of edge AI applications and improve their performance and efficiency. For example, a fog node could process the data from a smart city traffic management system.
- **Cloud:** The cloud often serves as the ultimate repository for processed data. The current role encompasses functions such as data storage, AI model training, and the deployment of AI applications. For instance, in medical imaging system; the cloud can be leveraged as the central hub for archiving the extensive volumes of imaging data generated, and can also be used for processing and deploying large models at the cloud while supporting local models deployed within Edge devices.

8.5.1 The Jetson Family—GPU Oriented

Nvidia Jetson offers development and deployment to various computing requirements by categorizing performance and capabilities. From beginners exploring AI to professionals developing advanced edge solutions, these devices provide a platform for real-time AI processing, fostering innovation across industries and reshaping the landscape of edge and cloud computing.

- **Nano:** The Nvidia Jetson Nano is an entry-level edge computing device with a compute capability of around 0.5 TFLOPs (FP16). It features a quad-core ARM Cortex-A57 CPU, a 128-core Maxwell GPU, and 4 GB of LPDDR4 RAM. The Jetson Nano is popular for robotics enthusiasts, smart camera projects, home automation, and educational purposes. Its capabilities include object detection, image classification, and basic AI processing.
- **Xavier NX:** With a compute capability of approximately 6 TFLOPs (FP16), the Nvidia Jetson Xavier NX offers substantial processing power. It boasts a six-core ARM Cortex-A57 CPU, a 384-core Volta GPU, and 8 GB of LPDDR4x RAM. This compact and energy-efficient device facilitates AI applications in resource-constrained edge environments. It finds applications in robotics, drones, industrial automation, and smart cities, performing tasks ranging from object recognition to intelligent decision-making.
- **TX2:** The Nvidia Jetson TX2 provides a compute capability of about 1.3 TFLOPs (FP16) and is equipped with a dual-core Denver 2 CPU, a quad-core ARM Cortex-A57 CPU, a 256-core Pascal GPU, and 8 GB of LPDDR4 RAM. This edge device caters to applications requiring higher AI performance, including object tracking, video analytics, and autonomous navigation. It is commonly integrated into drones, robots, and industrial equipment for real-time data processing.
- **AGX Xavier:** At the forefront of Nvidia's Jetson series, the AGX Xavier boasts a compute capability of approximately 10 TFLOPs (FP16). This powerhouse features an

octa-core ARM Cortex-A57 CPU, a dual-core Denver 2 CPU, a 512-core Volta GPU, and a substantial 16 GB of LPDDR4x RAM. Primarily designed for demanding AI applications, such as autonomous vehicles, industrial AI, and advanced robotics, the AGX Xavier enables real-time AI inferencing in complex and resource-intensive environments.

8.5.2 Coral Platforms

Coral edge devices empower edge computing with AI capabilities via several low and medium compute-capable devices. From the Coral USB Accelerator's simplicity to the comprehensive Coral Dev Board, each device offers a unique entry point into the world of AI-powered edge computing. As a result, these devices are revolutionizing industries by enabling real-time AI inferencing, privacy-conscious processing, and unlocking new horizons for intelligent edge applications.

- **USB Accelerator:** It is a compact and versatile device that connects via USB and incorporates the Edge TPU coprocessor. This hardware gem is designed to quickly and affordably add AI capabilities to existing systems. By harnessing the power of the Edge TPU, it enables devices to perform AI inferencing locally, unlocking real-time insights without relying on cloud processing. This device serves as a gateway for a wide range of applications, from enhancing computer vision in cameras to enabling voice recognition in various IoT setups.
- **Dev Board:** It is a comprehensive single-board computer tailored for AI development at the edge. Powered by the Edge TPU and built around the NXP i.MX 8M SoC, it provides both the hardware infrastructure and the software tools necessary to experiment and innovate. Its primary purpose lies in prototyping and building AI applications that require real-time processing. As a versatile platform, it facilitates the development of applications spanning from robotics to intelligent industrial automation.
- **Dev Board Mini:** Designed for compact edge AI projects, the Coral Dev Board Mini packs the power of the Edge TPU in a smaller form factor. Based on the MediaTek i500 SoC, this diminutive device retains the capabilities of its larger counterpart while catering to space-constrained applications. Its compactness makes it suitable for projects where real-time AI processing is essential, but size is a critical consideration, such as wearable devices and small-scale robotics.
- **System-on-Module (SoM):** This device encapsulates the prowess of the Edge TPU in a compact module format, designed for seamless integration into custom hardware designs. This module empowers developers to embed AI capabilities directly into their edge devices, fostering the creation of intelligent systems across various industries. By utilizing the Edge TPU's AI acceleration, the SoM enables edge devices to execute complex AI tasks locally (inference), minimizing latency and ensuring privacy.

8.5.3 Raspberry-Pi Series

These devices are a series of single-board computers known for their affordability and compatibility and play a significant role in the edge computing ecosystem. As edge devices, they enable real-time processing and analysis of data closer to the data source, making them ideal for IoT applications, home automation, and basic AI inferencing. Additionally, their potential as lightweight edge servers allows them to perform data aggregation and local processing and act as gateways, enhancing the efficiency of edge computing networks. While lacking specialized AI accelerators, Raspberry Pi devices excel in their accessibility, community support, and adaptability, positioning them as valuable tools for various edge computing scenarios.

- **Zero:** The smallest and most basic model, Raspberry Pi Zero, is designed for compact projects with minimal requirements. It features a single-core ARM1176JZF-S CPU and 512 MB of RAM. While it's not as powerful as other models, it's well-suited for lightweight edge tasks.
- **Pi 4 Series:** The latest model in the series, Raspberry Pi 4 offers improved processing power, better memory options, and enhanced connectivity compared to previous models. It features a quad-core ARM Cortex-A72 CPU and up to 8 GB of RAM. It's suitable for a wide range of edge computing tasks, including IoT applications, home automation, and lightweight AI processing.
- **Pi 3 Series:** The predecessor to the Raspberry Pi 4, Raspberry Pi 3 features a quad-core ARM Cortex-A53 CPU and up to 1 GB of RAM. It's still widely used for edge computing projects that require moderate processing power and connectivity options.

Compared to Nvidia Jetson or Google Coral, the Raspberry Pis are not designed for a dedicated AI acceleration chipset but are mostly used as general-purpose computing units. Their accessibility, versatility, and vibrant community support make them valuable assets for various edge computing applications.

8.5.4 Intel Devices

Intel's edge devices and technologies provide a comprehensive ecosystem for edge computing solutions. Their wide range of hardware and software empowers developers and businesses to create intelligent edge devices capable of real-time data processing, AI inferencing, and enhanced decision-making. As edge computing continues to reshape industries, Intel's contribution is pivotal in making edge intelligence more accessible and powerful. Their lineup of edge devices encompasses a range of products designed to deliver performance and scalability at the edge of the network.

- **Intel NUC (Next Unit of Computing):** These are compact and powerful mini PCs that can serve as excellent edge devices. Equipped with Intel Core processors, these devices provide the computing power needed for data processing and analytics at the edge. They are suitable for a wide array of applications, including industrial automation, digital signage, and IoT gateways.
- **Movidius Neural Compute Stick:** For AI at the edge, the Intel Movidius Neural Compute Stick is a compact yet potent device. It features a dedicated VPU (Vision Processing Unit) optimized for AI workloads, making it ideal for tasks like image recognition, object detection, and natural language processing. This device is particularly valuable in applications such as smart cameras and robotics.
- **FPGA (Field-Programmable Gate Array):** Intel's FPGAs offer reconfigurable hardware acceleration, making them versatile edge devices for specific workloads. They can be

customized to accelerate AI inferencing, data compression, encryption, and more. FPGAs are instrumental in scenarios where specialized hardware acceleration is required, such as in network appliances and edge servers.

- **Edge Servers:** Intel provides a range of powerful edge servers, including the Intel Xeon D processors. These servers offer robust computing capabilities, multiple cores, and extensive memory support. They are well-suited for demanding edge computing applications like autonomous vehicles, industrial automation, and large-scale IoT deployments.
- **OpenVINO Toolkit:** Intel's Open Visual Inference and Neural Network Optimization Toolkit complements their edge devices. It's a software framework designed to optimize and deploy AI models on Intel hardware, including CPUs, GPUs, VPUs, and FPGAs. This toolkit simplifies the development of AI-powered applications at the edge.

8.5.5 Beelink

Beelink devices, powered by Intel processors, are compact yet powerful computers tailored for edge computing applications. Their versatility and robust performance make them suitable for various applications across smart cities, industrial automation, retail, healthcare, and transportation sectors. Here's a closer look at the current lineup:

- **Beelink SER:** Equipped with an Intel Celeron N4120 processor, 4 GB of RAM, and 64 GB of storage, this device is a reliable choice for handling demanding edge computing tasks. With Intel UHD Graphics 600 and comprehensive connectivity options, including Gigabit Ethernet, Wi-Fi 5, and Bluetooth 5.0, the Beelink SER is ideal for digital signage, kiosk displays, and retail point-of-sale systems. Its USB and HDMI ports support various peripherals and displays, enhancing its deployment.
- **Beelink GTR:** This device is powered by an Intel Atom x5-Z8350 processor and offers a balance between performance and energy efficiency. With 4 GB of RAM and 64 GB of storage, it's well-suited for IoT gateways, lightweight edge analytics, and smart home applications. The integration of Intel HD Graphics, Gigabit Ethernet, and wireless connectivity options ensures seamless data transmission and processing, making it a practical choice for edge nodes in healthcare monitoring systems or industrial IoT setups.
- **Beelink MINI S:** Supported by an Intel Celeron N3450 processor and designed for low-power, compact edge computing solutions, the MINI S is particularly effective in sensor data aggregation, basic AI inferencing, and as a controller for automated environments. Its small form factor and connectivity capabilities, including Wi-Fi 4 and Bluetooth 4.0, make it an excellent choice for space-constrained environments like smart home applications and small-scale industrial control systems.

8.5.6 NXP

They are known for its wide range of products tailored for edge computing, IoT (Internet of Things), and embedded systems. NXP offers a variety of microcontrollers and processors that are commonly used in edge devices and IoT applications.

- **i.MX Processors:** NXP's i.MX series of processors are widely used in edge devices such as industrial automation systems, smart appliances, and automotive infotainment systems.

These processors offer a range of features, including ARM Cortex-A and Cortex-M cores, multimedia capabilities, and robust security features.

- **EdgeLock Secure Enclave:** This is designed to enhance the security of edge devices. It provides a hardware-based security foundation for protecting sensitive data and ensuring the integrity of edge applications.
- **EdgeVerse Edge Compute:** It is a comprehensive suite of edge computing solutions. It includes processors, connectivity solutions, and software tools to enable developers to create intelligent edge devices for various applications.
- **IoT Connectivity Solutions:** These devices provide a range of connectivity solutions such as Wi-Fi, Bluetooth, and Zigbee for IoT applications. These solutions are essential for connecting edge devices to the Internet and other devices.
- **Automotive Edge Devices:** NXP is a major player in the automotive industry, offering edge devices and processors for advanced driver assistance systems (ADAS), infotainment systems, and vehicle-to-everything (V2X) communication.
- **Industrial Edge Devices:** NXP's processors are commonly used in industrial edge devices for tasks like data acquisition, control systems, and edge analytics. These devices are essential for Industry 4.0 and smart manufacturing applications.

8.6 Edge Accelerators

Edge computing has become a widely adopted computing practice in various industries by enabling real-time data processing and decision-making closer to data sources. To achieve the desired levels of performance and efficiency, edge accelerators have emerged as essential computing components. Among these accelerators, Tensor Processing Unit (TPU), Field-Programmable Gate Arrays (FPGAs) and Application-Specific Integrated Circuits (ASICs) stand out as compact yet powerful units [270, 282]. An example of these processing units corresponding to flexibility, efficiency, generality and performance is shown in Figure 8.9. This section explores the current developments in TPU, FPGA and ASIC units.

8.6.1 FPGA

Field-Programmable Gate Arrays (FPGAs) have gained popularity in edge computing use cases because of their flexibility and efficiency. Their re-programmable nature allows for the execution of specialized tasks with optimized performance. They also benefit from their ability to process data in real time, thus making them necessary for applications requiring rapid decision-making. The significant features which make FPGAs suitable for edge acceleration are:

Devices: Xilinx Versal AI Edge Series [283], leveraging 7nm MOSFET technology, combines high performance with power efficiency and are equipped with integrated AI Engine for hardware-accelerated AI inferencing, further making them suitable for smart cameras, industrial automation, and robotics applications. Built on Intel's 10nm Enhanced SuperFin technology, FPGAs from Intel Agilex M/MP Series [284], offer high logic density and integrated AI accelerators, suitable for network security, video analytics, and autonomous vehicles. An example of cost-effective FPGAs are designed for low-power, low-cost edge applications is the Lattice ECP5/ECP6 Series [285]. These series feature embedded Digital Signal Processors (DSP) blocks and efficient processing capabilities, ideal for sensor data pre-processing and wearables.

Applications: FPGAs are increasingly being utilized for edge AI and machine learning applications. Their ability to accelerate machine learning inference tasks while maintaining low power consumption makes them ideal for scenarios where energy efficiency is critical. In telecommunications, FPGAs are employed to improve network performance and flexibility through Network Function Virtualization (NFV). They enable the virtualization of hardware functions such as routing and encryption, significantly enhancing network adaptability and efficiency [283, 285].

Deployment: FPGAs significantly reduce latency by processing data locally, eliminating the need to send it to distant data centres. This benefits autonomous vehicles, industrial automation, and augmented reality applications. Developers can tailor FPGA designs to match the exact requirements of an application. Whether it's a machine learning model, encryption algorithm, or data filtering, FPGAs can be fine-tuned for optimal performance, thus complementing battery-powered edge devices and remote deployments.

8.6.2 Tensor Processing Units

TPUs are specialized hardware accelerators designed specifically for high-speed machine-learning computations. Developed by Google [286], TPUs are engineered to optimize the performance of TensorFlow, an open-source machine learning framework. Since their introduction, TPUs have evolved through several generations, significantly improving processing speed and efficiency for AI applications. For training large ML models, cloud TPUs are also available to accelerate AI applications and deployment using Google compute engine [287]. Current applications include enabling intelligent decision-making, real-time insights and enhanced operational efficiency in edge computing scenarios [288].

Devices: TPUs (TPU v1) were initially developed to process large volumes of low-precision computations, such as 8-bit operations while enabling higher input/output operations per joule compared to the graphics processing units (GPUs), which are commonly used for matrix operations. With a performance of up to 92 TOPS (tera-operations per second) on 8-bit integer calculations. This first version was primarily used internally by Google to accelerate its own products, such as "Search and Translate." TPUs were developed to complement the diverse ecosystem of processors by correlating the suitability of different processors, which varies across machine-learning models. For example, as per operations and computing suitability, TPUs excel in convolutional neural networks (CNNs), whereas GPUs show higher performance for fully connected neural networks, and CPUs are often preferred for recurrent neural networks (RNNs) [289].

Second Generation (TPU v2): Introduced in 2017, TPU v2 added support for training deep learning models, not just inference. It significantly increased computational power, offering up to 180 teraflops of performance and 64 GB of high-bandwidth memory per TPU device. TPU v2 devices were networked together into "pods" that could provide up to 11.5 petaflops of computational power, demonstrating a massive leap in the ability to train complex models more quickly. Third Generation (TPU v3): Google announced TPU v3 in 2018, further improving its predecessors' capabilities by offering even more computational power and memory. Each TPU v3 pod could deliver over 100 petaflops of performance, highlighting the rapid pace of development in processing power. Enhanced cooling systems were also introduced to manage the increased power consumption and heat generation. Fourth Generation (TPU v4): Revealed in 2021, TPU v4 continued this trend, boasting more than double the matrix multiplication capabilities of TPU v3, and significantly increasing the amount of on-chip memory. TPU v4 pods can deliver exaflop-level computational performance, marking them as some of the most powerful computing systems for AI workloads.

Applications: TPUs find applications in a wide range of edge AI scenarios. In autonomous vehicles, for instance, TPUs facilitate critical functions such as real-time object detection, image recognition, and sensor data integration [290]. This enables vehicles to interpret complex environmental inputs, making instant decisions for navigation and safety. In the industrial sector, particularly in smart manufacturing, TPUs drive quality inspection and predictive maintenance advancements. By analyzing data directly at the manufacturing line, TPUs help identify defects, predict equipment failures, and optimize production processes, thereby significantly improving operational efficiency and minimizing unplanned downtimes. In healthcare, using TPUs for medical image analysis transforms patient care by accelerating the interpretation of diagnostic images, such as MRIs and CT scans, enabling quicker, more accurate diagnoses and treatment plans [291].

Deployment: TPUs deployment at the edge requires strategic consideration of power consumption and form factor. These processors are integrated into various edge computing devices for diverse applications, including edge servers, industrial control systems, and gateways, to facilitate local, real-time AI processing capabilities. For applications in challenging conditions, such as remote monitoring stations, outdoor infrastructure, or manufacturing, TPUs are incorporated into devices designed to withstand rugged conditions, ensuring reliable performance. The inherent efficiency and computational power of TPUs are crucial for applications demanding immediate data processing and decision-making, underscoring their indispensability in achieving the goals of edge AI, such as reducing latency and increasing the throughput of data-driven insights [292].

8.6.3 ASIC

Application-Specific Integrated Circuits (ASICs), with their application-specific design, offer unmatched performance and efficiency in edge computing tasks. From enhancing customer experiences in smart retail to advancing autonomous technologies in vehicles and robotics, ASICs provide the necessary computational power and efficiency. Their deployment strategy focuses on maximizing the advantages of their custom design, emphasizing energy efficiency, adaptability to harsh environments, and cost-efficiency at scale. Together with FPGAs and TPUs, ASICs complete the ecosystem of edge accelerators, each contributing unique strengths to address the diverse requirements of edge computing applications [270, 293].

Devices: Intel Movidius Myriad X VPU [294], a low-power, high-performance vision processing unit developed for deep learning inference at the edge, supports a broad range of AI frameworks and models. The current applications involve smart cameras, drones, and industrial automation. In a similar context, NXP RT6000 Series, which features dedicated neural processing units (NPUs), offers low latency and high throughput for demanding

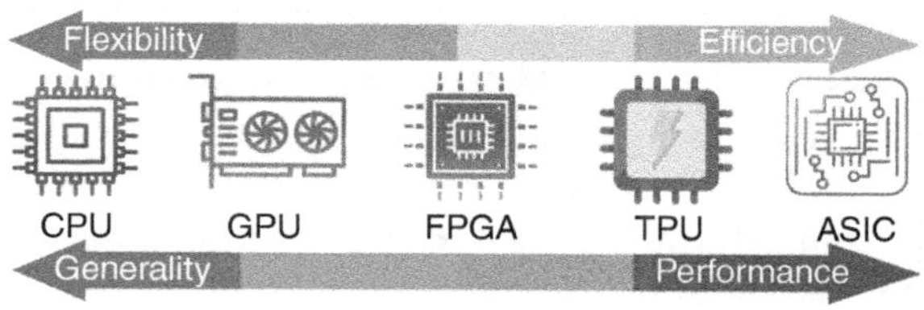

Figure 8.9 Processing Units and Alternatives for AI Model Deployment at Edge.

AI workloads, targeting autonomous vehicles, robotics, and AR/VR applications. For joint processing, Cambricon M1300 Series [295], offers acceleration for edge and cloud applications, supporting diverse AI workloads with flexible programming options suitable for smart retail, healthcare, and industrial automation.

Applications: ASICs are purpose-built for a specific workload, delivering unmatched performance because of their highly optimized nature. ASICs have been recently used for edge applications that require intensive computations, such as deep learning inference or cryptographic operations.

Deployment: ASICs' small form factor and low energy requirements make them ideal for deployment in energy-sensitive environments. This includes wearable devices, IoT sensors, and mobile devices, where long battery life is crucial. Also, these devices are adaptable to harsh environments in the form of custom-designed ASICs, which are found in devices deployed in industrial sensors and outdoor monitoring equipment. Their robustness ensures reliable operation despite environmental challenges. While ASICs require a significant initial investment in development, their long-term deployment in high-volume applications becomes cost-effective. The efficiency and performance gains justify the upfront costs, especially in large-scale deployments where their optimized performance can substantially save power and operational costs [294, 295].

8.6.4 Key Takeaway: The Future of Edge Acceleration

Edge computing's rapid evolution demands performance, efficiency, and adaptability. FPGAs and ASICs are at the forefront of this transformation, enabling edge devices to process data faster, make decisions smarter, and conserve energy. Comparison based on flexibility, efficient processing, generality and performance between CPU, GPU, FPGA, TPU and ASIC is shown in Figure 8.9. As industries continue to embrace edge computing, the role of these accelerators will become more prominent [282, 296]. These technologies represent the bridge between data generation and actionable insights at the edge, ushering in a new era of intelligent, responsive, and efficient edge computing applications [162, 297].

8.7 Industry Frameworks

8.7.1 KubeEdge

KubeEdge is an open source framework, that is built on Kubernetes, to support containerized application orchestration and device management in the network edge [79]. KubeEdge supports networking, development of applications and provides synchronization between the cloud and the edge network. KubeEdge also supports Message Queuing Telemetry Transport (MQTT), which is a machine to machine network protocol for remote locations with resource and bandwidth constraints.

KubeEdge framework extends cloud computing capabilities to the network edge [298]. Accordingly, their cloud architecture incorporates computing and storage resources not only at centralized data centres but also at the network edge, as distributed resources. Both the cloud and edge components are open-source. KubeEdge framework connects and coordinate centralized and distributed resources towards realizing a better performance and user QoE.

KubeEdge allows cloud services to be adopted at edge. This is through providing the same network protocol infrastructure and runtime environment of the cloud in the edge network. Hence, cloud applications and services can run seamlessly in the network edge. Thus, KubeEdge is a edge enabling framework that can support many emerging applications and services in the beyond 5G era. KubeEdge can process high data volumes locally at the edge in a efficient, low-latency, secure and privacy preserving manner while causing network backhaul links to be less overloaded.

8.7.2 Edgent

Edgent is a framework that is designed to bring the capability to perform Deep Neural Network (DNN) based tasks in the network edge [299]. This is through synergetically utilizing edge nodes to perform DNN collaborative inference. Edgenet design follows two approaches. In the first approach, Edgent partitions the computational overload and coordinates cloud resources and edge resources in the close proximity to perform real-time DNN inference. In the second approach, DNN right-sizing is performed, where the computing latency is reduced trough performing early existing inference in an intermediate DNN layer. Hence, DNN does not complete the full set of layers of DNN inference.

In addition, the design of Edgent support both static and dynamic network environments [299]. In case of a static environment with slow bandwidth changes, Edgent finds the most appropriate configuration through regression based prediction. In case of a dynamic environment with varying bandwidth, Edgent can generate the best execution plan banking on online change point detection algorithm towards identifying the best configuration for that state of bandwidth at given time.

The Edgent framework operates in three stages, namely, offline configuration stage, online tuning stage and co-inference stage, as illustrated in Figure 8.10.

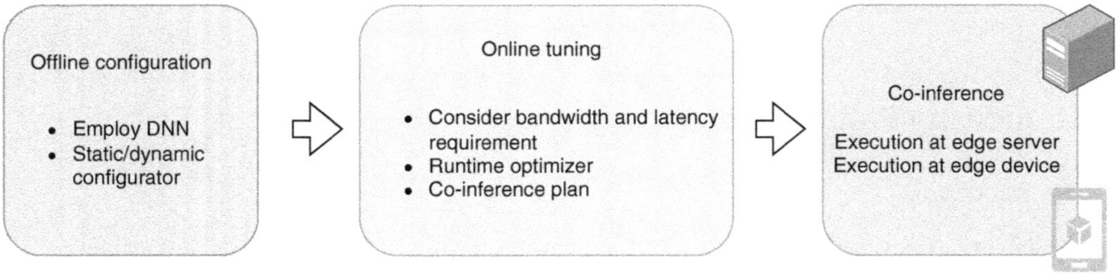

Figure 8.10 Edgent Framework.

The offline configuration stage incorporates the Static/Dynamic Configurator that selects the suitable configuration for online training, where Edgent performs DNN partitioning and DNN right-sizing. Finally, at the co-inference stage, the layers before the partition point will be executed on the edge server while the remaining will be on the device.

8.7.3 GreenGrass

Amazon Web Services (AWS) Greengrass brings Amazon's computing, communicating and ML inference capabilities to edge computing [300]. AWS Greengrass mainly comprises of two AWS components. They are, AWS Lambda, which is an AWS function module that is relation to function computing, and AWS IoT Core, which is a cloud computing processing centre.

AWS Greengrass also incorporates device shadow functionality, which can act as virtual backup of real devices. Device shadow can cache local devices states, keep track of and record edge device state, and synchronizes these states and data to the cloud.

Another feature of AWS Greengrass is that it enabled edge devices to operate even when they are offline or not seamlessly connected to the cloud. Once connected, AWS Greengrass seamlessly synchronizes the latest device status with the cloud.

It should be noted that most of the data generated at through IoT devices do not require analysis at the cloud. Hence, through operating at end devices, AWS Greengrass can carefully select the data that should be sent to the cloud to minimize transmission costs and minimize over utilizing backhaul communication links. AWS Greengrass can also deploy their cloud-run ML modules at the local network edge to minimize data transfer costs, latency and boost performance, when performing activities, such as, event processing, ML, and image recognition.

8.7.4 Azure IoT

Azure IoT is a platform designed using a range of services that run on the cloud and the edge towards controlling and monitoring large numbers of IoT devices [301].

Azure IoT combines Microsoft-managed cloud services, edge nodes, and SDKs to provide various services, including, operating systems, security, and data analytics. These services intend to support building and managing IoT businesses.

8.7.5 Private Edge Service Providers

An organization can host a private cloud within their organization, especially if they are operating at a large scale. These cloud service providers are extending their services towards the network edge, in order to facilitate emerging applications and services demanding near-real-time access to powerful computational and storage resources.

For instance, estimating that over 70% of enterprise data are created and processed at the edge by 2025, Dell Technologies Cloud provides services spanning from public clouds to private clouds, and the edge [302]. Hence, this type of cloud services can be used towards automation, operation, optimization, analytics and securing of the complete data life-cycle in industries and other cloud and edge computing applications. Harnessing the capabilities of such platforms, applications and services, such as industries can harness advanced computing and storage resources from centralized cloud to the edge considering the processing load, storage load, backhaul link availability, backhaul link cost, and sensitivity to delays and latency. These capabilities are accessible for users as Data Center as a Service (DCaaS).

8.8 Concluding Remarks

This chapter explores the diverse landscape of edge communication, hardware and enabling frameworks, showcasing the technological breadth from compact, versatile USB sticks to sophisticated systems engineered by leading corporations like Intel, Nvidia, IBM, and NXP. We investigated emerging beyond 5G and 6G communication technologies. The chapter focuses on the advancement of communication technologies and then explores several envisaged 6G technologies, such as, Further enhanced Mobile Broadband, extremely Reliable Low-Latency Communication, Extremely Low-Power Communication, High Spectrum Efficiency, High Area Traffic Capacity, Mobile Broadband and Low-Latency and AI-Assistive Extreme Communications.

Besides communication advancement, we also explored the specialized domain of hardware accelerators, including Tensor Processing Units (TPUs), Field-Programmable Gate Arrays (FPGAs), Application-Specific Integrated Circuits (ASICs), and vision processors, which are specifically designed to provide unique computational benefits to edge environments, illustrating the synergy between hardware innovation and practical utility. The practical applications of these technologies span a broad spectrum of industries, marking a transformative shift in digital capabilities. In smart cities, IoT innovations are enhancing operational efficiencies and further fundamentally elevating the urban experience, making cities more livable and responsive. The healthcare sector benefits from remote monitoring to AI-enhanced diagnostics, showcasing how edge devices are important in advancing patient care outside traditional settings. Moreover, the Mobility as a Service (MaaS) paradigm is evolving and is significantly propelled by edge computing to offer more agile, sustainable, and user-centric transportation solutions. In this evolving scenario, scalability, security, and interoperability remain challenging. As edge computing becomes more embedded in critical infrastructure and services, addressing these challenges head-on will be crucial for realizing its full potential. The future of edge computing is not just about technological advancements but also about fostering an ecosystem where innovation, security, and sustainability complement each other to drive the digital transformation of our global society. This chapter, therefore, serves as an indication of the current state of edge computing and further highlights the path towards a future where edge computing continues to redefine the boundaries of what's possible and further paving the way for a smarter, more connected world.

Part III

Integration and Driving Forces

As the system perspective of this book has a strong inclination on practical deployment, this part of the book will illustrate the integration and deployment aspects of MEC^2, including the driving forces in Chapter 9, security in Chapter 10, privacy in Chapter 11, telecom and cloud integration in Chapter 12.

Mobile Edge Computing and Communications: Driving Forces, Technology Foundation, and Application Areas, First Edition. Aaron Yi Ding, Chamitha De Alwis, and Madhusanka Liyanage.
© 2025 John Wiley & Sons Ltd. Published 2025 by John Wiley & Sons Ltd.

9

Driving Forces for Edge Integration

Emerging applications and services, such as connected autonomous vehicles, AR, VR, small and big data analytic, rely on seamless connectivity and processing extreme volumes of data in near-real-time harnessing the capabilities of AI and ML. On the other hand, these applications generate most of the data in the network edge and demands powerful processing and storage in the network edge. For instance, it is estimated that nearly 850 ZB of data will be generated at the network edge in 2021, whereas the global data center traffic will only reach 20.6 ZB. Hence, mobile edge computing (MEC) is considered to be a significant enabler of many emerging applications [303]. The rest of the chapter continues to explore the requirement for edge integration in beyond 5G mobile communication networks, as illustrated in Figure 9.1.

It is a challenge to facilitate near-real-time access to powerful computational and storage resources to process massive amounts of small and big data [231]. This demands the convergence of communications, computing, control, localization, and sensing (3CLS) in future communication networks [249]. Therefore, 3CLS is envisaged to be a key driving trend toward the realization of beyond 5G mobile communication networks. This will enable beyond 5G mobile communication networks to have intelligence at the network edge [249]. Accordingly, it is evident that having powerful computing and storage solutions in the network edge is a significant enabler to realize future mobile communication networks and their applications and services [231]. The rest of this section explores the growth of Internet of Things (IoT), the growth of small data, the development of AI/ML based communication technologies, and the advent of novel technologies and applications, as key driving trends toward the integration of edge computing resources with future communication technologies.

9.1 Internet of Things Toward Internet of Everything

The number of IoT devices are increasing exponentially and is expected to reach 24 billion by 2030 [231]. What is more, the total IoT market is also expected to hit 1.5 trillion USD in 2030 [304]. The growth of IoT will pave the path toward the development of Internet of Everything (IoE). IoE is envisaged to connect things with people and data [305]. IoE is expected to connect various devices and platforms toward realizing many emerging applications and services [306].

Mobile Edge Computing and Communications: Driving Forces, Technology Foundation, and Application Areas, First Edition. Aaron Yi Ding, Chamitha De Alwis, and Madhusanka Liyanage.
© 2025 John Wiley & Sons Ltd. Published 2025 by John Wiley & Sons Ltd.

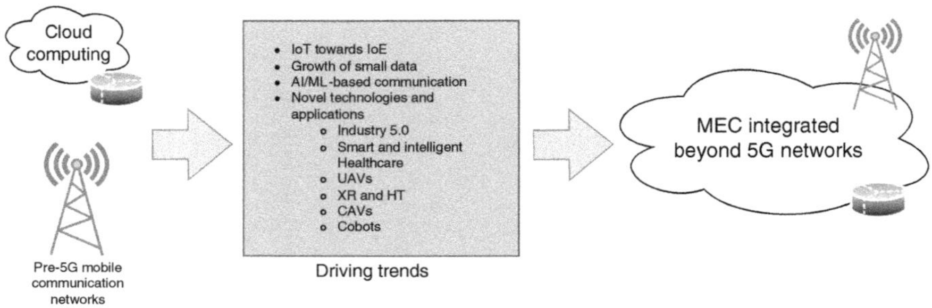

Figure 9.1 Driving Trends Toward MEC Integrated Beyond 5G.

It should be noted that the growth of IoT toward IoE generates large amounts of data in the network edge. These data should be efficiently communicated and processed in the network edge to enable IoE applications. However, processing such high volumes of data demands powerful processing and storage capabilities. This is due to emerging applications banking on AI and ML technologies. This demands future communication networks to integrate MEC as an integral component of the network.

9.2 The Growth of Small Data

The growth of IoT toward IoE leads to the generation of high volumes of Small Data. Small Data is simply a limited set of data in a niche area that can provide meaningful insights [307].

The term Big Data refers to large sets of historical data. Such Big Data can be efficiently processed utilizing existing cloud computing resources. On the other hand, Small Data are small sets of data that is real-time or of a limited life-span. Small Data is a key enabler of Industrial IoT (IIoT). This is due to processing Small Data being instrumental toward operating and maintaining machinery and devices in the IIoT environment [231]. It should be noted that the collection of Small Data sets and processing them demands utilizing AI/ML in the network edge. This requires seamless communication of large amounts of data from massive number of connected devices and processing this data at the network edge.

In addition, the process of offloading massive amounts of Small Data to edge computing facilities should also be performed in an energy efficient manner. The integration of MEC with future communication networks will enable many IoE nodes to seamlessly communicate small data with edge computation resources in near-real-time. This would facilitate the utilization of Small Data to enable applications, such as, IIoT and the retail industry (to collect data from sensors and wearables to provide personalized services).

9.3 AI/ML Based Communication Technologies

The advancement of communication technologies from beyond 5G toward 6G demands harnessing the capabilities of AI/ML based technologies. In addition to converging 3CLS and utilizing Small Data for network node operation and maintenance, there are many other areas where AI/ML will be utilized in future communication networks. For instance, growing on the concept of multiple-input-multiple-output (MIMO), large intelligent surfaces (LIS) are electromagnetically active surfaces that can perform the functionality of a large antenna array. These are made using meta-materials that are placed surfaces such as walls, roads and buildings [308].

In addition, novel channel access schemes, such as non-orthogonal multiple access (NOMA) that can allow same orthogonal resources to be simultaneously allocated to multiple users, are more spectral efficient compared to orthogonal multiple access schemes. Moreover, utilizing frequencies beyond millimeter Wave (mmWave) toward THz frequency bands to facilitate high bandwidth to facilitate faster data rates also demands advanced transceiver techniques and beamforming. In addition, communication networks will move from small cells toward tiny cells in order to support the high-frequency bands in THz spectrum. Furthermore, multi-mode base stations will facilitate network operations in a wide range of spectra while visible light communication (VLC) will mark the dawn of a new era in communication harnessing the same frequency range that is perceived by the human eye. VLC utilizes optical wireless communication (OWC) technologies toward providing short-range communication [255].

On the other hand, Zero touch network and Service Management (ZSM) envisages to establish fully automated network management. ZSM will enable future networks to be able to perform many functions autonomously, including, self-configure, self-optimize, and self-monitor.

However, it should be noted that the technologies mentioned above are designed to harness the capabilities of AI/ML. For instance, LIS transceivers need low-complexity channel demodulation. Furthermore, ZSM requires AI/ML based analytic, which provides to be yet another driving force to integrate MEC with future communication networks. This needs utilizing various AI/ML techniques, including deep learning (DL) [309]. Processing AI/ML in near-real-time demands access to powerful computational resources.

9.4 Novel Technologies and Applications

The ever increasing requirement of humans to make life and work easier has given rise to the dawn of many advanced technologies and applications. Following technologies and applications are explored as key driving trends toward edge integration in future communication networks.

9.4.1 Industry 5.0

Industry 5.0 is a new era of industries where people will work along robots, smart machines and devices to efficiently utilize the creativity of humans together with the efficiency of machines [310]. With Industry 5.0, the use of machines and humans together AI can elevate human capabilities toward personalize autonomous manufacturing [310, 311]. Therefore, it is evident that a large number of IIoT machines and devices will be connected with Industry 5.0. These will be connected using 5G and beyond communication technologies. Moreover, the data generated by these devices are required to be processed in near-real-time using AI/ML, harnessing the computational capabilities of edge computing. This demands MEC to become an integral part of 5G and beyond communication networks.

9.4.2 Smart and Intelligent Healthcare

The emergence of AI-driven smart and intelligent healthcare applications are expected to enhance human quality of life (QoL) [231]. This is through weaving a network of Intelligent Wearable Devices (IWD), Internet of Medical Things (IIoMT), and Hospital to Home (H2H) services [231, 312, 313]. IWDs and IIoMT will connect health monitoring devices and sensors in real-time to communicate health data with powerful computational resources at the network edge. This data can be used for diagnosing health conditions and even perform medical triage to facilitate remote diagnosis by medical professionals. Furthermore, the extreme communication capabilities envisaged by 6G networks will pave the path toward tele-surgery and remote patient caring in natural disasters, emergency situations, rural areas or during pandemics. It should be noted that smart and intelligent healthcare demands extreme communication capabilities together with powerful computational capabilities at the network edge. Hence, the paradigm shift toward smart and intelligent healthcare is a key driving trend toward MEC integrated beyond 5G networks.

9.4.3 Unmanned Aerial Vehicles

Unmanned aerial vehicles (UAVs) are becoming a key technology toward realizing many futuristic applications and services. UAVs are expected to grow up to a 45.8 billion USD industry by 2025 [314]. For instance, autonomous drones are increasingly being used in defense applications [315]. Moreover, UAVs are used in many other civilian applications, ranging from, large scale agriculture projects, disaster recovery activities, and traffic monitoring [316].

UAVs will also play an integral part in beyond 5G communication networks to provide enhanced mobility, rapid base station development, non-terrestrial networks, and providing connectivity to remote and isolated locations [231]. Furthermore, UAV-to-Everything (U2X) network is a concept that is being explored to utilize the capabilities of UAVs to enhance future communication networks. However, operating UAVs with limited battery, payload, and flight duration has many challenges. This demands heterogenous UAVs to operate as swarms to operate in a coordinated manner [317].

Hence, the operation of UAV swams demand sophisticated algorithms that can harness the capabilities of AI/ML to perform activities such as path planning [231]. This in-return demand seamless connectivity and access to powerful computational resources, requiring future communication networks to have integrated MEC capabilities.

9.4.4 Extended Reality Toward Holographic Teleportation

Extended reality (XR) provides a combination of real and virtual worlds. XR can be an enabler of many emerging applications, including smart education, smart healthcare and brain computer interfaces (BCI) [231]. On the other hand, holographic teleportation

(HT) is expected to project real-time three dimensional images of people and objects with full-motion. HT requires capturing surrounding objects and movements to transmit these information across mobile networks. In order to provide users with a satisfactory QoE, HT demands the network connectivity to have a latency of 0.1 ms and data rates of multi Gbps [318]. However, it should also be noted that it is required to not only communicate the high volume of data in XR and HT applications but also processed efficiently at the network edge. Hence, the advancement of XR and HT technologies is also another driving trend toward MEC integrated communication networks in the beyond 5G era.

9.4.5 Connected and Autonomous Vehicles

Autonomous driving vehicles are evolving toward connected and autonomous vehicle (CAVs) that will harness the capabilities of emerging communication networks, such as, Internet of Vehicles (IoV) and Vehicular Ad hoc NETworks (VANETs) [319, 320]. Together with the advancement of AI/ML and next generation communication technologies in beyond 5G communication networks, such as, intelligent V2X communications, CAVs will provide efficient, reliable and safe autonomous vehicles for the future society. This demands extremely low latency, extremely high reliability and availability, and near-real-time access to powerful computational resources, which can be provided through MEC integrated beyond 5G communication networks.

9.4.6 Collaborative Robots

Collaborative robots (Cobots) are robots that will work collaborate with humans. Cobots can undertake difficult, risky and repetitive tasks to promote health and safety in work environments while boosting the productivity [321]. Hence, cobots will also play a significant role in Industry 5.0. Cobots are required to process high data volumes toward making decision in real-time. This demands extreme storage capacity, and computing and communication connections. Therefore, the dawn of cobots will also demand MEC integrated beyond 5G networks.

9.5 Conclusion

This chapter explores many driving trends toward MEC integrated beyond 5G and 6G communication networks. The driving trends discussed in this chapter demands extreme communication capabilities with extremely low latency, extremely high reliability/availability, extreme data rates, and seamless connectivity together with near-real-time access to powerful computational resources to have extremely low processing delay. These driving trends will force future communication networks to have MEC as an integrated and integral component. This also raises the requirement for beyond 5G and 6G standardization efforts, such as, European Telecommunications Standards Institute (ETSI) [322], Next Generation Mobile Networks (NGMN) Alliance [323], Alliance for Telecommunications Industry Solutions (ATIS) [324], and Next G Alliance [325] to deeply incorporate MEC as an integral component of future mobile network technologies, applications and services.

10

Edge Security*

10.1 Introduction

Security is a broader concept that extends to information, cyber-, forensic, and network security. Information security was defined as the preservation of confidentiality, integrity, and availability (also referred as the CIA triad) of information under the standard ISO/IEC 27002 in 2005 [326]. The information contextualized under this definition applies to physical or electronic/digital forms of data that are subjected to be documented, stored, in transition or conversed. Forensic security covers acts committed against the laws and disciplines in the governing domain. However, digital forensic methods are used in the IT domain to ensure security. A nascent definition for cyber-security is presented in [327] as the approaches and actions associated with security management processes followed by organizations and states for protecting the CIA of data and assets in cyber-space. Though, the latest requirements of cyber-security exceed CIA aspects. Traceability, authentication, authorization, anonymization, granularity, localization, and trust are novel requirements for systems where cyber-security is applicable [328, 329].

Initially, network security was specified to secure communication networks from possible intrusions and vulnerabilities. Those attacks and threats were limited to intervening and masquerading attacks such as man-in-the-middle (MitM), relay, and spoofing. With adequate encryption and cryptography primitives, probable attacks were plausible for mitigation. Though, novel communication services prioritize the network's data rate to serve more subscribers. Thus, cumbersome cryptographic primitives are imprudent. Moreover, softwarized approaches of software defined networks (SDN), network function virtualization (NFV), and network slicing are contriving more requirements for security assurance as presented in [330–332]. Most emerging systems are cyber-physical systems (CPS) that integrate computation, networking, and physical processes to contrive an environment extending to cyber and physical spaces [333]. Thus, security for a CPS poses an extensive domain applied for cyber, information, forensic, and network security contexts.

10.1.1 Importance of MEC Security

The MEC system, placed at the edge of the mobile network, is the ingress access point to all the services emanating from the RAN. This critical juncture is one of the weakest

*This chapter is partly contributed by Pasika Ranaweera (University College Dublin, Ireland).

Mobile Edge Computing and Communications: Driving Forces, Technology Foundation, and Application Areas, First Edition. Aaron Yi Ding, Chamitha De Alwis, and Madhusanka Liyanage.
© 2025 John Wiley & Sons Ltd. Published 2025 by John Wiley & Sons Ltd.

points of the entire network regarding security. Most Internet of Things (IoT) devices in the market are produced with economically manufactured circuitry that employs weak encryption/encoding schemes and other security measures for maintaining an affordable price range to meet the competition [9]. Most such devices are vulnerable to cloning and physical tampering, imperiling the entire mobile network for countless attacks. Verifying the credibility of these devices at the edge is a major concern. In addition, distributed nature of the MEC paradigm is broadening the avenues for adversaries due to the migration of storage and processing service infrastructure to a proximate radio access range. Even a service-impeding attempt is intimidating the purpose of MEC for attaining ultra-low latency and higher bandwidths.

Emerging applications and services demand handling personal credentials and information at the network edge to meet service demands. Privacy, integrity, and trust management assurances are prime requirements with MEC deployments, despite the attributed locational and contextual awareness facilitated for the users. Virtualization technologies are vital for realizing the MEC paradigm and creating a serviceable platform with dynamic resource allocation capability. Security of the virtualized platforms is still a gray area due to lesser deployments. The vulnerabilities and attacks plausible on virtual machines (VMs) are unique and cause significant consequences to the MEC system.

Like cloud computing, outsourcing MEC subscriber data to a remote storage and processing environment creates a predicament regarding privacy rights. Establishing boundaries regarding the extent of authorized conduct on service providers' capabilities is imperative for guaranteeing the trust of MEC consumers. Considering all these facts, security and privacy are important for realizing a pragmatic MEC paradigm deployment.

10.2 Security Threats in MEC

A threat vector (TV) is a specific path, method, or scenario that can be exploited to break into an information system [18]. In this section, we identify the TVs that exist in the considered MEC deployment scenario derived in compliance with the ETSI standardized architecture presented in Figure 1.4 of Chapter 1. Figure 10.1 illustrates the TVs applied at different levels in the MEC structure. The TVs are categorized into four areas based on their scope of intrusion. They are access network, mobile edge network, core network, and architectural.

10.3 Threat Vectors Related to the Access Network

The access network (AN) represents the RAN defined by the 3rd Generation Partnership Project (3GPP) and the access infrastructure related to non-3GPP networks such as; wireless local area network (WLAN), worldwide interoperability for microwave access (Wi-Max:IEEE 802.16), or code division multiple access (CDMA) 2000 networks [334]. In any communication system, threats are mostly originated in the access network. The prime rationales for those threats are the diverse technologies deployed in the AN scope. The scalability of the proposed TVs based on AN is a great concern for the applicability due to the heterogeneous nature of the MEC-enabled services. We classified the TVs related

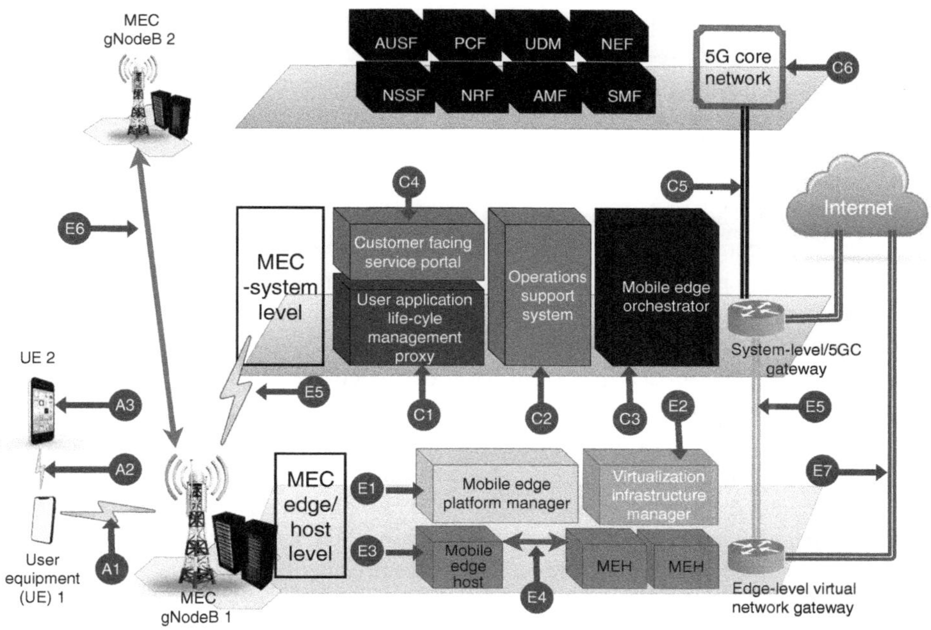

Figure 10.1 Identified Threat Vectors in an MEC Deployment. Source: Ranaweera et al. [10]/IEEE.

to AN into three generic categories, described in the following subsections A1, A2, and respectively A3.

10.3.1 A1: Link Between the User Equipment and a Base Station

The connection from UE to the BS is the most typical communication link and the most vulnerable to threats in a mobile communication system. Since A1 applies to the most exposed part of the mobile communication network, an adversary could intervene in the communication link or introduce a malicious device to compromise the BS. The concept of mobile delegation is concerned with offloading computationally intensive tasks to the edge servers due to resource constraints of the UEs. The bulk storage and information traversing intended for processing in an offloading scenario are elevating the network traffic through the air interface [21]. Additionally, novel technologies such as massive multiple-input-multiple-output (MIMO), interference-aware receivers, advanced coding/modulations, millimeter Wave (mmWave), carrier aggregation, Wi-Fi offloading, LTE and license shared access (LSA) have been introduced with the purpose to improve the spectral efficiency in the access network [37]. The connectivity of the UE through these heterogeneous technologies is raising concerns over compatibility and interoperability factors that adversaries could exploit.

The broadcasting nature of the wireless medium in the air interface linking the UE to the BS is prone to attacks such as eavesdropping or jamming [335]. Moreover, intrusion-based attacks such as MitM, relay, DoS, distributed DoS (DDoS), advanced persistent threat (APT), sybil, spoofing and privilege escalation are probable. The MitM attacks are plausible between the 3G network and the non-3GPP WLAN networks that target to compromise the internal virtualization infrastructure entities of the edge level [9]. All these attacks would intend to gain access of the ME Apps operating in MEHs and manipulate the virtualization infrastructure of the MEH to exhaust its resources while infecting other MEHs connected to the infiltrated one. The UEs connecting to the BS do not possess the intrinsic requirements for generating effective cryptographic primitives to ensure security and privacy [336]. The diversity of the UEs connecting to the BSs would present an issue of managing the compatibility among mobile devices produced by different vendors and the communication protocols of varied UEs or UE Apps. In such circumstances, malicious nodes could be injected to the system exploiting the vulnerabilities in the UE devices, such as susceptibility to device cloning, less secure wireless protocols (WPA/WPA2), proneness for hardware Trojans, predictable access control credentials (PIN/pattern) or alleviated resiliency against malicious software agents acting disguised as partial entities of UE Apps such as spyware, adware, Trojans and malware. Such vulnerabilities could feed malicious content or counterfeited information to the ME Apps through the BSs; which would manipulate the services offered by the MEH as mentioned in TV E3. However, these attacks could be mitigated by employing an effective authentication mechanism embedded with a trust verification scheme for the UE and BS connectivity.

10.3.2 A2: Ad Hoc Connectivity Between User Equipment

The threats on A2 are associated with the links established between UEs in an ad hoc manner. These links employ short-range communication channels used for data-transferring purposes under the influence of specific UE Apps. The connectivity is device-to-device (D2D), which focuses on establishing a direct communication link between two devices without requiring any BS for connectivity [337]. The short-range communication technologies such as Bluetooth, Bluetooth Low Energy (BLE), near field communication (NFC), ZigBee, Wi-Fi direct, narrowband IoT (NB-IoT), SIGFOX or any technology which could form a mobile ad hoc network (MANET) are capable of deploying connections between UEs [9, 21, 338]. Moreover, FlashLinQ and Proximity Services (ProSe) are other techniques capable of forming D2D communication platforms. FlashLinQ, developed by Qualcomm, facilitates content sharing, gaming, and social networking features to proximity devices [339]. ProSe is a standardization published by the 3GPP for enabling proximity discovery and direct communication for future AN-based deployments [337].

The vulnerabilities of this TV are limited to the communication channel established between the UEs. Though, attacks such as eavesdropping, impersonation, forging, free-riding, DoS and privacy violation are probable [337]. Most of such attacks are feasible due to the nature of the communication protocols embedded with short range communication technologies. These technologies are prioritizing leveraging the bandwidth for D2D execution over employing security measures. The threats originating in a UE do not directly influence the intrusions into MEC systems in this TV. A UE infiltrated by a D2D based attack could use its connectivity with the BS to infect the MEC servers for various manipulations of the MEH explicated in E3.

The method of offloading cellular traffic to the UEs by the MNO is an exemplification of a D2D instance [337]. In this approach, MNO is only transmitting the content to specific

UEs considered as cluster heads and those UEs are multi-casting the content to respective UEs in the scope of the cluster. Moreover, use cases such as extending the coverage through D2D connectivity for rural areas and establishment of critical communication channels for disaster or terrorist situations (where the cellular network is disabled) are envisaging future potential for D2D based services. In these use cases, connectivity between a UE is considered as a cluster head and the MEC servers is maintained for content sharing under the supervision of the MNO. Thus, this scenario opens up new possibilities for adversaries to exploit the cluster head UEs to manipulate the service they offer.

10.3.3 A3: User Equipment (UE)

UE can be considered any device, such as a mobile, personal computer, CCTV camera, wearable sensor, or sensory system, which can be in direct contact or connected through a gateway device to the BS. A variety of technologies attributed by a UE on the aspects of operating systems (Android, iOS, Windows, Symbian, BlackBerry, and WebOS), memory management (SD, micro-SD, and HDD), communication (RF, RFID, NFC, Bluetooth, Wi-Fi, and Ethernet), physical design and structure contribute to the fact of improbable deployment of a generic security solution for UEs in a holistic extent [340]. The UE contains information related to various aspects of the daily life of a person such as private information (photos, medical reports, medical statistics, and CCTV footage), location (GPS), daily routines (shopping and transportation), enterprise information, critical infrastructure information (energy consumption, financial, banking, and emergency service status) and online account statistics; where exposing such divulge credentials and parameters could be fatal for one's well-being [21]. Thus, mobile users' privacy is a threat of great concern [23, 341]. The resources embedded in a UE regarding processing power, storage capacity, and battery life are the most significant factors for this threat vector [336]. Certain softwarized and virtualized attacks require minimal resources to launch in an executable platform [342]. Thus, enhanced processing and storage resources in current UEs are improving the possibility of launching such attacks capable of hindering detection by conventional means.

A UE could instantiate the threats with or without the user's knowledge. Even a genuine user is capable of activating a malicious software agent unintentionally. This risk of UEs being vulnerable to physical and remote attacks makes this threat vector extremely critical. The UEs are vulnerable to physical damage, Side channel attacks (SCA), malicious code injection, and hardware Trojans, while all other attacks explicated in TV A1 and TV A2 apply to the communication interfaces. Physical damages are the most common type of attack for this TV, where they lead the attacker to re-configure the affected device to convey misdirecting information to ME Apps [23]. These misinforming attacks lead the MEC system to interrupt its services by feeding fake but calculated information to edge devices. A typical scenario would be the reconfiguration of ME Apps to execute continuously (without termination and commandeering maximum resources) to exhaust its resources. The attacker's intention of an SCA is to extract the cryptographic parameters by cryptanalytic means. Security protocols engaged in communication channels are exposed with such revealed credentials.

The offloading of services due to mobile delegation could result in UE experiencing various offloading mechanisms, such as full and partial offloading. These offloading mechanisms are prone to attacks [51]. An infected UE App or a UE, when offloading partial or complete executable or passive content to the MEC server can inject a malicious agent to be activated in the corresponding ME App in the MEH. Such an attack is directly contributing to the TVs E3 and E4. In case of a UE acting as a machine type communication gateway (MTCG) for applications like e-health or any other MTC deployments, the malicious content could be generated at one of the sensors or actuators where the UE is acting as an intermediary ingress point to the intrusion. Resource allocation and scheduling of the ME App is controlled by the mobile edge platform (MEP) in each MEH, where MEP communicates with the UE App in case of a mobile delegation circumstance [14, 51]. An infected UE App could influence the MEP to allocate inessential resources at the MEH to cause a service interruption. The capability of a UE to be deployed in a tamper-resistant manner, employing lightweight but effective cryptographic primitives with high resiliency, relies on the device's design and manufacturer.

10.4 Threat Vectors Related to the Mobile Edge Network (MEN)

The entities located in the edge network or the host level of the MEC paradigm, such as the mobile edge platform manager (MEPM), virtualization infrastructure manager (VIM) and the mobile edge hosts (MEHs) are investigated in this section for establishing threat boundaries. The placement of MEC servers is localized in comparison with conventional data centers. Thus, the MEC edge (i.e., host) level is prone to tangible or physical security attacks [34].

10.4.1 E1: Mobile Edge Platform Manager (MEPM)

The MEPM is the entity that monitors the MEH activities, using the connectivity that it maintains with the mobile edge platform (MEP) contained by the MEH. This entity performs resource allocation and monitoring functions with the connections it maintains with VIM, MEO and OSS. As MEPM is the highest level entity in the MEC host level, it performs the following functions: managing the ME App life-cycle, traffic steering, recording fault and performance metrics from VIM. Moreover, MEPM reports the holistic monitoring statistics of the host level entities to the system level.

The placement of MEN entities at the edge limits physical tampering based attack vectors. However, the risk is higher compared with conventional CC. MEPM could be furnished with fake information regarding resource allocation based configuration data. Such a threat could originate within the AN or from an infiltrated ME App. An infected ME App could force the corresponding MEP to lead the MEPM for allocating undesired resources for intending service disruption. An already exposed UE App or a communication channel in the AN can mislead the MEPM for allocating more MEHs for processing which leads to resource depletion. As the MEPM is connected to the OSS, MEO and VIM, such false information is disseminated to the system-level entities destabilizing the entire MEC system. VM based attacks such as VM manipulation, Domain Name System (DNS) amplification, VM escape, virtual network function (VNF) location shift, and security log troubleshooting attacks are probable for this threat vector.

10.4.2 E2: Virtualization Infrastructure Manager (VIM)

VIM executes the integral task of facilitating ME Apps of virtualized infrastructure resources in every MEH operated within a single edge vicinity. The connections of VIM toward MEO and MEPM are feeding the statistics to manage and monitor Local Area Data Network (LADN) deployed in MEHs. As this is the main entity assigned for facilitating the virtual resources at the edge, the role of the VIM is similar to a hypervisor for MEHs. Thus, VIM performs the functions of allocating, managing, releasing, and performance monitoring of virtualized resources [14].

As VIM is responsible for allocating resources, it could be subjected to attacks such as VM manipulation, VM escape or any malicious attack targeted for virtual deployments. These attacks would exhaust the system resources in various methods such as allocating inessential processing and storage facilities for a single ME App in the virtualization infrastructure (VI), allocating excessive amount of ME Apps to process a single application or blocking resources or service interruption to a particular ME App. The system level entities could be confiscated for privilege escalation or service interruption threats due to their connections with VIM. If an attack compromises the VIM, malicious misconfiguration exploits could be launched by an attacker [343].

10.4.3 E3: Mobile Edge Host (MEH)

MEH is the main host-level functional entity that performs the computational, storage, or networking operations guaranteed by the MEC paradigm. A MEH consists of an MEP, Virtualization Infrastructure (VI), and a data plane or Local Area Data Network (LADN) which maintains the local connectivity among ME Apps. Additionally, the User Plane Function (UPF) is a 5G access network inside the MEH for integrating the 5G core network into the LADN. As MEH is the only entity that stores the content conveyed from UE Apps, the risk of being exploited is high.

MEHs are the main target of any attack from the AN toward the MEC system. As they are the main functional elements which support the service processing, storage and computations; any attempt to penetrate the MEC system through a malicious act should be directed toward the MEH in the attackers' perspective. The threat of a MEH being subjected to an infection of a malicious adversary is highly reliant on computational offloading processes as discussed in the AN threat vectors. Once infiltrated, it can mislead the MEP and VI for resource allocation and service continuation. The attacks applicable to any VM based deployments are prone to ME Apps as they are deployed in a VI. Moreover, the false statistics conveyed by the affected ME Apps could be causing mis-configurations in the VI and the MEP; that could be exploitable by privilege escalation type attack. Apart from resource depletion and service interruption, a malicious entity could exploit the connectivity to the UPF to misconfigure the 5G core network entities.

10.4.4 E4: Connectivity Between Mobile Edge Hosts

ME Apps might require to connect with one or several MEHs for processing high end applications; in which single MEH does not possess the resources to perform the entitled function. This requires the connectivity between ME Apps operated under MEHs, which are established through MEPs. This is probable for high end applications such as Industrial IoT (IIoT), surveillance, or critical infrastructure services. As the VI and LADN of two MEH entities are accessed, respective VIM and UPF should be notified in addition to the subscriptions in the MEPM.

A malicious ME App that emerged from the methods explicated in E3 can infect other connected ME Apps and MEP elements in addition to other entities inside a MEH. Though the connections among MEHs are internal and obscured to adversaries. Thus, intrusions are limited to malicious injections traversed from the access network. However, the allocation of MEH for services is provisioned by MEPM. An infected ME App could manipulate the MEPM for resource depletion while depleting the resources of each MEH that a malicious agent manages to propagate. A mechanism for detecting and terminating such malicious agents would be a viable countermeasure. An intrusion detection system (IDS) operating under the MEPM at the host level that monitors the anomalies of the VM operations would be a suited directive for detecting such exploits.

10.4.5 E5: MEC Platform Connectivity Between the Edge and the Core

The connection between the mobile edge system level entities and the host level entities is a critical link in the MEC paradigm. The registration process for a particular MES requested by a UE App is established through this link [14]. UEs connecting to a MEC system should first register at OSS through the connection extended from the BS to the UALCMP entity in the core [344]. This is the initial interfacing of UE Apps with the MEC platform. The nature of the UE App (whether it is operated by a trusted or a malicious entity), authenticity, and content transmitted from the UE App to ME App are factors to be investigated in this threat vector.

This bidirectional connectivity conveys the control information between the MEC system and host-level entities. As the location separates these two levels, the connectivity could be established from long range communication links using the technologies such as microwave (MW), fiber optic (FO), satellite or RF. Thus, a resourceful adversary could manage to intervene the communication link. RF links are vulnerable to attack vectors discussed under A1. Electromagnetic pulse (EMP) based, Sybil, DoS, and DDoS attacks are probable with MW links [345–347]. FOs are vulnerable against fiber tapping and hidden pulse attacks [348, 349]. Satellite communication (SATCOM) links face the threats of kinetic, jamming, and cyber-attacks [350]. Even though the possibility of being intervened is less with long range communication links, the exposure of the control information could allow the attacker to exploit the MEPM, VIM and MEH host-level entities as desired. Similarly, if the attacker managed to alter the status updating parameters conveyed from the host level to the system level, MEO and OSS could be subjected to service interruption attacks.

10.4.6 E6: Connectivity Between Mobile Edge Apps Operated Under Mobile Edge Hosts at Different Base Stations

In this situation, the UE App related to the user is accessing ME Apps operated under MEHs at different mobile edge host levels residing in two BS locations controlled under a single MEO (or ME system level). A crowdsourcing or smart grid application that deploys two instances of the same ME App operating at different locations exemplifies such a scenario. Even such a ME App instances operate at geographically dispersed edge levels, they are governed by a singular trust domain contrived by an OSS and a MEO.

The connectivity between two host levels governed by the same system level could be prone to intervening attacks, as explicated in E5. The service migration from one host level to the other probes the possibility of infecting two host levels through malicious content. As these applications are only probable with upscale use cases, the effects of the attacks could be incisive. However, a malicious ME App is not in a position to request a service migration from the OSS. OSS approves the allocation according to the specifications requested through the subscriptions of UE Apps.

10.4.7 E7: Connectivity with the Mobile Edge Host and the Cloud Servers

In case of services hosted to third party consumers, MEH maintains a connection to a centralized cloud or a server platform as a typical massive IoT based cloud implementation. In that scenario the MEH (or MEHs) is pre-processing the content at their possession before conveying them to the cloud service. The cloud service typically hosts the edge services as a function as a service (FaaS) implemented through a cloud wrapper MEC application [15]. The connection to the cloud server is instigated through the LADN of the MEH.

The channel between the cloud service and the MEH is vulnerable to packet sniffing and location based attacks from geo-location leakage [23]. Moreover, this channel type's security policies and rules are rarely investigated. CSPs are susceptible to such attack vectors. Thus, interoperability issues might be plausible. An attack launching from this channel due to its bi-directional connectivity and exposed nature could instill a malicious agent in the MEH to enable E3 based threats. Moreover, the cloud platform or centralized servers are prone to malicious attacks from an infected MEH. Intervening attacks such as MitM, relay or impersonation attacks are plausible for the communication channel extending from MEH to the cloud platform.

10.5 Threat Vectors Related to the Core Network

The core network expands from the MEC system level devices such as UE App Life-cycle Management Proxy (UALCMP), Customer Facing Service Portal (CFSP), Operations Support System (OSS), and Mobile Edge Orchestrator (MEO) to the backhaul network that extends till the internet connectivity.

10.5.1 C1: User Application Life-Cycle Management Proxy (UALCMP)

This entity is the initial contact point for any UE App that intends to subscribe MEC services. The main function of the UALCMP is handling multiple UE App requests while determining their life cycle. As this entity includes proxy functionality, internal addressing function is facilitated for the MEC system to link UE Apps to their corresponding ME App or ME Apps operated at MEN.

Request handling nature of this entity pertains the possibility of DoS, DDoS, or masquerading attacks which would entail for service disruption or access granting for malicious intruders. As this entity determines the UE App life cycle, an adversary is capable of furnishing falsified information to obtain an increased life cycle excessive of its requirements. As the OSS depends on the requests and information of the UALCMP, service disruption of UALCMP could directly affect the OSS operations.

10.5.2 C2: Operation Support System (OSS)

The OSS grants the service requests forwarded through UALCMP or CFS portal while instantiating or terminating ME App functions. Additionally, OSS maintains links to MEPM and MEO for extracting control information. OSS approves for subscribers to use ME Apps that are configured for a particular MEC service. Thus, this entity is critical for attackers to gain access to the MEC system.

As the UALCMP is the entity facing the service requests from UE Apps, OSS has to be protected from DoS or DDoS attacks. Since all the UE Apps subscribing for ME Apps should be registered in the OSS, attackers could attempt to inject fake information to impersonate itself as a valid entity for pertaining MEC services. If the attacker was successful, mobile delegation-based operations would permit the possibility for infiltrating the MEC host level.

10.5.3 C3: Mobile Edge Orchestrator (MEO)

MEO represents the core functionality of the MEC concept, which assigns the role of hypervisor for the holistic MEC system. It observes the deployed MEC hosts and resource utilization status at the edge. As the hypervisor, MEO supervises the VMs and underlying hardware configured for virtualization [343]. The main functions of MEO are service migration, mobility management, and traffic steering monitoring. The hypervisor role of MEO is still applicable to scenarios where the MEC system is integrated with other driving technologies such as NFV Infrastructure (NFVI) with NFV Orchestration (NFVO) capability [351].

Though MEO is deployed at the system level where malicious intrusions are improbable, resource allocation and service manipulation attacks such as DNS amplification, VM escape and security log troubleshooting would be highly probable. Due to the effects of such attacks, the configuration of the MEO system could be destabilizing to perform at its optimum level. As in the case of a security log troubleshooting attack, logs of the operations of MEO or any other entity would be altered. Thus, the control statistic could not be conveyed to the corresponding ME entities for optimum operation. Even if the log information is conveyed, the entities will malfunction due to their altered content. The configuration parameters forwarded from the entities such as MEPM, OSS, UALCMP and VI are also vital for the utilization of the MEO. In the scenario of the MEO acting as the hypervisor being compromised, automated network configuration exploits, orchestration exploits, malicious misconfiguration, and SDN controller exploits are probable [343]. Therefore, mechanisms should be employed for detecting such attacks and remedying them.

10.5.4 C4: Customer Facing Service Portal (CFSP)

CFSP facilitates the access of ME Apps to third-party services, where it is capable of recalling service level information from such applications [14]. Car park monitoring, connected vehicles, and IoT big data are applications suited for MEC deployment. These deployments employ sensors that gather enormous amounts of data, which are pre-processed at a MEC edge server and conveyed to a centralized corporate server for further analysis [51]. For most of these services, MEC acts as a low latency aggregation point. Thus, third party consumers are instigating the service requests from cloud service providers (CSPs) end. The role of the CFSP is to approve the deployment of MEC resources for processing of such third-party requests.

As the role of the CFSP is handling requests, it could be prone to service based attacks such as DoS and DDoS. Moreover, proper approval mechanism should be employed by the CFSP for enabling traffic steering of third party applications to the MEC host level entities. Use cases applicable to E7 are exemplifications of extended services approved by the CFSP for third party applications. A compromised CFSP could manipulate the service subscriptions of OSS.

10.5.5 C5: Connectivity of the Mobile Edge Orchestrator (MEO) and the 5G Core Network

A secure interface has to be defined for MEO and 5G core network [352]. The control signals will exchange between the 5G core network and MEO via this interface. Since this is the main interface that interacts with the 5G core network, this is one of the critical interfaces in the whole MEC architecture. It is possible to host both 5G core network elements and MEO in the same physical host. However, it is more likely to implement them in two different physical hosts [353, 354]. In that case, the communication link should be established between MEO and 5G core network via the physical network. This network can be implemented using any network technology such as wireless, wired or optical. Depending on the connecting medium and the deployed location, these entities will face different security challenges.

The interface between MEO and 5G core is yet to be defined. However, several security threats can already be identified related to this interface. It is more likely to use separate physical hosts for 5G core network and MEO [353, 354]. In that case, the control traffic will be transferred between two entities via an open 5G backhaul network. Therefore, the interface between MEO and 5G core will be vulnerable to typical TCP/IP attacks such as eavesdropping, spoofing, DoS, replay and reset attacks. It is necessary to enable proper security mechanisms such as mutual authentication, E2E encryption, and challenge–Response Procedures (CRPs) to mitigate these issues. However, the impact of these attacks will be minimum if both MEO and 5G core deployed in the same physical host. Another challenge is to properly define an interface between 5G Core and MEO [355]. It is challenging to define proper and unified security mechanisms without a standard interface. However, this challenge is somewhat relaxed since ETSI [322] is leading both 5G and MEC standardization tasks.

10.5.6 C6: 5G Core Network

Ultimately, 5G core network controls the entire 5G network [356]. 5G core network will enable the MEC capabilities for the selected services. Moreover, all the control signals will be forwarded to the MEC system via the 5G core network. Therefore, 5G core network can be considered as the vital element to ensure the proper operation of the whole MEC system.

Since 5G core network is the main control entity of the whole 5G network, whatever the attack on 5G core network will have a significant impact on 5G MEC system. In contrast to the pre-5G networks, the 5G networks have a softwarized or virtualized core [357]. Here, all the core network functions are implemented as VNFs. However, several security concerns are observable in the functionality of VNFs. The hardware-based pre-5G core network had natural protection against many attacks due to closed, complex and vendor-specific nature [358]. However, NFV base 5G core network is open and software controllable. It is comparably easy to manipulate a software-based system than a hardware-based system [358]. Moreover, VNFs are now vulnerable to attacks such as Inter-operability issues [359], DoS/DDoS [360–362], VM escape [363], VNF Manipulation [343] and VNF location shift attacks [343].

Different VNFs are developed by different VNF providers and they attribute different levels of security polices. The mismatch between these differences can lead to vulnerabilities when they are deployed in a same system [359]. Variety of DoS/DDoS attacks on targeted services are possible when VNFs are hosted in the cloud, e.g., attack on Bitbucket [360]. The impact of DDoS is even greater for virtualized networks since the attack could propagate to untargeted VNFs that are hosted on the same physical host [361, 362]. Since VNFs are software, they are vulnerable to software flaws which can lead to unintended behavior. For instance, these software flaws can be used to bypass firewall restrictions or perpetrate buffer overflow to execute arbitrary code [361]. A malicious VM can escape out of the virtualization environment and execute arbitrary code within the hypervisor to compromise it [363]. An attacker misuse the privileges of a compromised hypervisor to install kernel root kits in VNF's OS and manipulate the VNF [343]. An attacker can migrate from a compromised VNF to a different location where less security or privacy policies are enforced to gain additional access to the system [343].

10.6 Architectural Threat Vectors

These threat vectors elaborate the vulnerability vectors that could exist in architectural improvements associated with the MEC deployments.

10.6.1 AR 1: Network Slicing (NS)

The concept of network slicing is slicing a physical network into several logical networks to enable the simultaneous usage of a singular physical network at different virtual/logical levels for heterogeneous IoT applications in order to alleviate capital and operating expenditure [364]. Thus, it creates an agile and dynamic on-demand networking platform on top of a physical networking infrastructure [351]. According to the 5G information-centric networking (ICN) model, NS framework has five functional planes. They are: the service business plane, service orchestration/management plane, IP/ICN global orchestrator plane, domain service orchestration/management plane, and infrastructure plane [365]. Compared with other resource-sharing initiatives such as RAN sharing, dedicated core networks (DCN), and enhanced DCN (eDCN); NS offers a higher range, virtualization support, function modularization support, end-to-end connectivity, and better isolation techniques [366]. This threat vector discusses the vulnerabilities associated with NS processes and techniques.

Most of the threats originating for the network slicing technique exist through its vulnerabilities such as: the non-existent mutual authentication schemes between the entities in different slices; insecure communication among Network Slice Instances (NSIs) and Network Slice Managers (NSMs); incompatibility of diverse security protocols and policies at different slices; different security levels at different slices which could permit an attacker to exhaust the resources a secure slice gaining the access from a low-security slice; attachment of UE to multiple slices which arises the tendency for blending of information flaws in case of an infected UE App [367]. These vulnerabilities attract attacks such as impersonation, MitM, tracing, DoS, DDoS and SCA [364, 367]. Invalidation of NFV-based network slices are probable due to server shutdown for maintenance, misconfiguration, and firmware errors [368]. An impersonation attack is highly probable threat as it could target NSMs and NS entities in different scenarios in the interaction [367].

The vulnerabilities and the complexity associated with NS affect the MEC deployment. In order to coordinate services for various applications, MEC system-level and host-level entities are dispersed through different network slices. Thus, the attacks mentioned previously could assist adversaries to gain control over the networking interfaces of the MEC hosts for service exhaustion and injecting malicious entities to disrupt the functioning MEHs. The incompatibility and dissimilarity between security protocols and policies as mentioned earlier would maintain different security levels among different MEHs interacting at various network slices. These risks are susceptible to TVs E3 and E4.

10.6.2 AR 2: Traffic Steering

Traffic steering of the MEC platform is conveyed by the Mp2 reference point (the connection between the MEP and the data plane of the VI in a MEH). The configuration of the data plane is managed by the MEC platform where it conveys traffic steering requests to Policy Control Function (PCF). PCF sends the corresponding traffic steering rules to the Session Management Function (SMF), where SMF handles the corresponding Protocol Data

Unit (PDU) sessions [15, 334]. Moreover, Service Function Chaining (SFC) is a technique that facilitates traffic steering policing by adopting SDN and NFV constructs [369]. SFC forms the middle boxes (MBoxes) or Service Functions (SF) such as firewalls, Deep packet inspection (DPI) entities, Access control lists (ACLs), Intrusion detection/prevention systems (IDS/IPS) or network address translation (NAT) [370]. As SDN and NFV are forming the MEC infrastructure for interconnecting system level, host level, and access network; SFC could be considered as a viable traffic steering technique for MEC deployments. This TV is investigating the flaws in traffic steering rules at 5G core network entities of the MEC architecture and intermediary networking entities.

Application Functions (AFs) could influence the traffic routing through the selection procedure of the UPF or the service request function for configuring the traffic steering rules [15]. The ability of AFs to influence traffic rules could prompt a flaw in this process. If an attacker exploits an AF, traffic-based attacks such as sinkholes, wormholes or Reduction of Quality (RoQ) attacks are applicable to the LADN located at the MEH. Thus, more effective attacks such as service manipulation could be launched with resource allocation failures in the core network. In terms of SFC deployments in the MEC architecture, confidentiality and integrity risks exist due to the possibility of interception, modification, or manipulation of steered traffic with incompatible security and traffic forwarding policies [369]. The SFs could disrupt services due to overloading, misconfigurations, resource scarcity, or security attacks. Moreover, opaqueness attributed in MBoxes through modification of packet headers would lead to inconsistencies in steering with altered addressing parameters. Therefore, attaining compatibility among security and traffic steering policies is a vital requirement for this TV.

10.6.3 AR 3: Service Migration

Service migration is the process of transferring executable content configured to offer a specific MES either between edge levels or between the cloud and the edge [371]. This process could induce unprecedented vulnerabilities and flaws in a MEC environment. In a CC-based service migration, services originally hosted in cloud environments are migrated to the edge servers located proximate to the mobile devices. This results in reducing the latency and improves the capacity of the access network. Thus, the services being executable programs, tools, or software running on a virtualized platform; the code of that particular software should be migrated to the edge in such circumstances. There are four approaches being considered for code migration by Rodrigo et al. [21]: (i) migrating only part of the code; (ii) migrating an exact replica or a clone of the entire execution environment with the memory and CPU images; (iii) migrating mobile agents created by mobile devices to the edge; (iv) amalgamating process cloning and mobile agents at the edge. The MEC services are typically launched as VMs in the VI of MEHs. VM migration is conducted as either live or non-live approaches [372]. In a non-live migration, the entire VM with its running states is encapsulated and transferred to the migrating vicinity while the local operation suspends completely. In live migration, VMs are orchestrated simultaneously at different edge platforms without suspension while multiple VM migrations are plausible via local area network (LAN) or wide area network (WAN) coverage. Other than services, migrating computational processes are viable applications for MEC deployments; where the network controller acts as the resource selector for utilizing the computation power. The computational migration models could be formulated using the Markov decision process (MDP) problem based on a random-walk mobility model or a threshold-based model such as the Lyapunov optimization technique [34, 341].

All the service migration methods explicated in Section 10.2, divulge migrating an entire serviceable platform or a part of it to the mobile edge hosts operating at the edge. Still, the security of the migration process is a grey area due to the diversity of utilized resources and the scope of the services. The migration process instigates a service operated within a single MEH or multiple MEHs. The unauthentic nature of the Internet-based connectivity among MEC edge entities is posing security issues extending to VIM, UEs, and the migration data traversing channels [372]. Moreover, malicious code injection attacks targeting the migration channels are imminent at the edge network. Detecting the malicious code would be improbable once the migration process is completed. The repercussion of this leads to the exploitation of communication links between the edge service infrastructure and the cloud server or core network entities. In the case of employing mobile agents for service migration at the edge, the probability of an intrusion is higher as the code or the service platform migrated from the access network is subjected to threats under A1. Thus, it is imperative to secure the migration processes with proper security mechanisms for mitigating massive service manipulations at the edge of the MEC deployment regardless of the diversity of the services.

10.6.4 AR4: Mobility Management

The term "mobility" can be described as maintaining the connectivity of a UE when roaming from a certain coverage area of a serving BS to another via a handover mechanism for the intention of maintaining service continuity [51]. The coverage of a BS could be varied from multiple macro BS, small cell BS (SCBS), Wi-Fi access points (APs) to standard RAN BSs. Thus, concerns are raised over security with different capacity BSs offering various coverage and handover models. Under this TV, those issues would be investigated to identify the best strategies for countering prevailing vulnerabilities under mobility scenarios.

The factors of the heterogeneous nature of configurations, user-server association policies, and trust domains among different coverage areas or cells are issues associated with mobility that results in severe interference and pilot contamination. Additionally, these issues are degrading the transmission performance and improving the service provisioning latency [34]. These issues are imposing availability threats on UE Apps that are probable for exploitation by an attacker with significant awareness through hijacking the frequency of the roaming channel of UE prior to or while the handshake process is undergoing. Such an attack is plausible with a higher latency due to inefficient mobility. UE mobility in the prospect of 5G core network integration is complicated. Core network entity Network Exposure Function (NEF) and AN entity User Plane Function (UPF) are subscribing mobility services for MEO and MEHs in MEC deployments [15].

In a mobile delegation circumstance where a UE has offloaded its processing to the edge data centers in the previously serving BS, there are two scenarios in which the continuity is established [51]: (i) VM migration where the current VM in the serving BS MEH would be migrated to the roamed BS MEH; (ii) selection of a new communication path to the UE

App and the serving MEH VM. In case of VM migration, VM or the UE App could be prone to VNF location shift attack where the UE App would be configured to instill parameters beyond the legal bounds of the service. The key management protocols are also burdened to renew the credentials in this scenario as the VM is migrated into a different trust domain. In the second scenario, the request for establishing the connectivity to the serving VM or ME App should be conveyed through the UALCMP to be registered in the OSS operated under the same or a different MEC system level. An attacker pretending to be a valid UE App under mobility could launch an impersonation attack which causes OSS to allocate MEH resources at the edge through MEO intervention. An adverse situation is launching simultaneous mobility requests as DoS or DDoS attacks for disrupting the edge services.

Moreover, mobility and roaming events can happen very frequently in the 5G network because of the popularity of local 5G networks or micro 5G operators [373–375]. Such localized 5G networks have a limited coverage. In such a scenario, the possibility of encountering a malicious local 5G network is quite high [376]. Most of these local 5G operators do not have a high level of security similar to the main MNOs. Therefore, it is comparably easy to attack local 5G networks.

10.7 Security Requirements of MEC

A set of requirements is required to design security measures for identified TVs. These requirements specify the focused aspect of security, the scope of its applicability, and the scenario it should be used in. Security cannot be achieved by a single action and require multiple functions conducted to maximize the protection of the secured content. In other words, multiple security mechanisms are required to satisfy multiple security requirements. This section states the identified security requirements for MEC deployments in the context of conventional and MEC-specific scenarios.

10.7.1 Classical Security Requirements of MEC

10.7.1.1 Confidentiality

At the design stages, obscuring information, ingressing, egressing, and traversing within the MEC system is critical. This will prevent disclosing information about intervening and eavesdropping attacks perpetrated at the AN. Mobile networks evolving from GSM to LTE employ encryption algorithms ranging from A5/2 to Evolved Packet System (EPS) Encryption Algorithm (EEA) [377, 378]. Moreover, security specifications published under TS 23.122 and TS 33.210 are guaranteeing the security measures available at access stratum (AS) and non-access stratum (NAS) levels of 3GPP architecture [379]. The insights gained from the prevailing network models have aided in enhancing the security of mobile protocols regarding confidentiality.

For 5G and beyond 5G based RAN, however, communication protocols should be customized according to the deploying use cases and applications. Though, vehicular communication and e-health applications could be employed with standardized encryption schemes available under mobile networks [380, 381]. Slice isolation ensures the confidentiality of UEs sharing the same infrastructure, accessed through the RAN [382].

Non-3GPP-based IoT devices are facing the same threat levels as mobile UEs. However, their threat domain exceeds a typical UE device due to their resource scarcity and inability to launch adequate security measures. Mechanisms such as lightweight security protocols, ECC, PHY schemes are emerging methods for securing Zigbee, Bluetooth, Wi-Fi, and LPWAN-based devices [383, 384].

The information traversing within the edge infrastructure and toward the core network should be encrypted with considerable security. Ramifications of information disclosure at this level impose more damage to the MEC system than the AN. Both signaling and controlling information related to the mobile network and virtualization platform are conveyed through the link between edge and core levels as described under TV E5. To these links and internal connections of the edge infrastructure, tunneling, IPSec, TLS/SSL-based encryption schemes are viable adaptations for securing through End-to-End (E2E) encryption strategy [23]. In addition, Quantum Resistance (QR) or anti-quantum cryptographic methods are researched to limit the vantage of adversaries brimming of resources [385]. Lattice-based, multivariate, hash-based, and elliptic curve schemes are employed for formulating QR algorithms; to overcome exhaustive key searching or brute-force attacks commandeered by quantum computers. These methods guarantee integrity in addition to confidentiality.

10.7.1.2 Integrity

Manipulation and destruction of data perpetrated to mislead the parties engaged in communication are considered integrity violations. Similar to confidentiality, integrity is a widely addressed concept for mobile networks. The EPS Integrity Algorithm (EIA) offers integrity protection to AS and NAS stratum levels in LTE [379]. In 5G, 5G-AKA handles the integrity assurance mechanisms for signaling channels [386]. Moreover, tenant isolation is a prospect that improves integrity protection in a multi-sliced environment [382]. Compared to 4G-LTE, 5G supports integrity at the user plane [379]. This feature enables resource-constrained IoT nodes to utilize integrity verification mechanisms available under 5G RAN. Hash-based and session-key-based encryption schemes could be employed for IoT devices to ensure integrity at the PHY level [387].

Typically, edge level and core level are linked with communication channels that employ existing protocols. The IPSec's encapsulation Security Payload (ESP) attribute is utilized for integrity protection of IPSec-based tunnels, which are adaptable for egressing channels of MEC edge [388]. In addition to traversing information, the integrity of applications or MESs should be validated perpetually. OSS can verify initial integrity while MEO can monitor the integrity in softwarized entities deployed at the edge infrastructure [389].

10.7.1.3 Availability

Availability corresponds to the omnipresence of MEC resources for consumers willing to subscribe to services. This factor primarily relies on the network performance and effectiveness of the network interfaces. Therefore, the performance of the mobile network is paramount for MEC. DoS-resembled attacks that impede services are the main causes of availability disruptions in communication channels. In the existing LTE network model, Radio Resource Control (RRC) connection status creates issues in validating the eNodeB for the UE [390]. Thus, it emanates opportunities for DoS attacks to compromise the availability of eNodeBs.

Since the UE requests are directed to the UALCMP initially, its capacity for serving UE instances should improve significantly to cater the 5G-based applications. Once the service request is granted, UE establishes a direct link with the MEC edge to instigate its service. This channel is formed via the RAN and connects to the MEH via the UPF instance in the LADN. Thus, the N3 interface of the 5G service architecture should feature adequate capacity to handle minimum UE instance requirements [15].

In the edge infrastructure, the placement policy of the virtualized entities plays a key role in maintaining availability parameters. However, MEPM and VIM are static placements. Depending on the service provisioning from VMs as isolated hosts or VMs deployed within a singular host, they represent different availability and cost factors [391]. As the virtual environment formed by the ME Apps within a MEH can be customized according to the service type, various ME Apps should function distinctly to each other, not the same function. Therefore, ME Apps' placement within the MEH directly affects availability.

10.7.1.4 Authentication and Authorization

Authentication is the process of verifying the identity of the parties engaged in communication or resource access [28]. These mechanisms are either performed by a single party or mutually through an extended scenario. Authentication schemes employ various measures of authenticity for validating the entities. Keys are the generic commodity employed for authenticating non-human entities. Depending on the domain where the authentication is instigated, the mechanisms are classified as primary or secondary. The authentication for MEC-based UEs are mostly handled via the air interface of mobile RAN as a primary approach. Thus, heterogeneous IoT devices and services are incurring diverse authentication requirements [392]. Ensuring the confidentiality of keys and authentication credentials is intrinsic for UEs, core-level, and edge-level entities. UE protection can be acquired by enhancing the existing EPS Authentication and Key Agreement (AKA) mechanism employed by LTE [390]. Further, 5G-based AKA and Extensible Authentication Protocol (EAP) AKA are two mandatory authentication schemes proposed under 5GPP phase 1 [379].

For most IoT devices, non-3GPP-based technologies are employed for communication. Wi-Fi is a common technique utilized by most edge computing circumstances due to improving the Wi-Fi range. Methods such as PUF [393], accelerometer data [394], and visible light (referred as Li-Fi) [395] are explored for improving the authentication of Wi-Fi networks. Moreover, novel methods are introduced for securing LPWAN [384, 396, 397], NB-IoT [398–400], RFID [401, 402], and BLE [403, 404] authentication phases. These technologies are utilized for D2D or ad hoc type authentication, which are common for IoT deployments. QR authentication is a directive that would benefit resource-constrained devices due to its cryptographic primitives bearing a lesser overhead [405]. Thus, such schemes are viable for IoT-based technologies that interface the MEC system.

From the MEC systems' perspective, UALCMP and CFSP handle the authentication of MES UEs' to the edge system. These entities at the MEC system level govern the access controlling function, including the authorization. Authorization is granting access to authenticated entities classified under diverse capability levels. Depending on the service type the

UE requests, OSS approves the capability level for the specified ME App at the edge level. In addition, MEPM is responsible for restrictions imposed by the OSS for ME Apps. Thus, MEC already possesses an authorization discipline devised within its architecture. However, a proper authorization framework should be identified from the existing deployments to be compatible with prevailing mobile services.

10.7.2 MEC Specific Security Requirements

In MEC, security is a paramount conundrum due to its intricate interoperability and compatibility concerns emanating from underlying driving technologies of NFV, Software defined networking (SDN), IoT, ICN, and Network Slicing (NS). As these novel technologies are attributing security concerns of their own, amalgamating such techniques creates inconceivable circumstances for forming security protocols applicable to all cases. A holistic MEC system comprises a RAN, an edge network, and a core network. These sub-levels are formed with diverse technologies and protocols that operate discretely.

10.7.2.1 Secure Charging and Billing

The responsibility to launch MEC service infrastructure is granted to the telecommunication operators or MNOs merely due to their authority over the prevailing mobile network with dispersed BSs. As MNOs expect profits for their existence, many MESs would be subscription-based deployments with pay-per-use billing models that resemble cloud computing services. In these circumstances, current billing models applied in conventional cellular networks by tracking the routed traffic traversing through the core network are not applicable. There is a high probability that UE Apps will launch at the edge network as ME Apps not convey their subscription-based consuming status to the core network entities. Displaced operations in the edge infrastructure isolated from the core entities constrict the traceability of the UE Apps subscribed through the OSS. The proposed approach to charging is perpetrated by the Policy Control Function (PCF) Network Function (NF) in 5G core network employing both online and offline charging schemes [15]. Functioning with myriads of potential UE Apps and accounting for their corresponding ME Apps and diverse service types (priority/best effort) for consumption appears to be arduous. Thus, the potential for exploitation with possible deliberated billing and roaming frauds reveals a critical security failure in the MEC charging system. An adversary capable of misleading an autonomous charging process can confer financial losses to the MNO. Therefore, mitigating the inconsistencies in the charging process is a critical security requirement from the perspective of the MNO.

10.7.2.2 Mitigating Service Impeding Threats

The MES subscription functionality performed by the OSS receives requests through UALCMP and CFS Portal. Once the UE Apps from the regular subscribers and third-party subscribers are approved for launching ME Apps, all the MESs are executed in a common virtualized infrastructure without distinction. Such an environment is subjected to resource exhaustion by insignificant processes and would lead to service disruption of priority services. An adversary can infiltrate the MEH infrastructure by leveraging the weaknesses of the access controlling method.

Approved malicious ME Apps can consume excessive resources in the MEH environment resulting in service disruption. This type of infiltration could result from a delinquent access control mechanism that does not feature a scheme for classifying the MES service type or its nature. Hence, a method to separate and distinctly recognize ME Apps based on their emanating service type is intrinsic to MES deployment. Moreover, the directive of "application-aware performance optimization" in MEC allows the applications to influence the configuration of RAN resources according to the customer experience [22]. Denial of service (DoS) based threats are probable for RAN access interfaces from a malicious agent influencing the MEC cell configuration for coverage adjustments.

10.7.2.3 Secure Mobile Offloading Schemes

The mobile offloading processes intended by the MEC edge deployments require caching functionality in the virtualization infrastructure. The multitude of UE Apps delegating to the edge is demanding extended caching spaces that would overrun the VM processor caching resulting in service interruption and major latencies. An adversary could target this vulnerability of caching mechanism for instigating cache poisoning attacks. Managing and monitoring the caches are intrinsic to utilizing the dynamic memory of the VI and ceasing any feasible attacks and ensuring the security of sensitive credentials used in authentication schemes established between the edge and mobile devices are imperative for eliminating exposure of the entire system. Compromised credentials allow the perpetrator to launch spoofing, eavesdropping, and data manipulation attacks on the edge system. In addition, credentials traversing from the edge to the core network are susceptible to intervening attacks due to the dispersed nature of the two networks. Thus, employing adequate levels of cryptic engagements in the communication channels and trust management schemes among entities operating on the edge and the core is critical for protecting the confidentiality and integrity of the secure credentials.

10.7.2.4 Robustness of MEO

The security of individual entities has to be ensured. Prominently, MEO should be equipped with novel and strong cryptographic primitives to avoid any exploitation from malicious entities. An internally or externally influenced compromised MEO might lead to sabotaging the entire platform, endangering the services catered to thousands of individuals and industries. MEO should be equipped with an internal threat detection entity, such as a VMI agent, besides the communication channels. One of the objectives of introducing the MEC paradigm was alleviating the latency for critical IoT applications. To achieve higher speeds with enhanced responsiveness, employing strong security protocols would be cumbersome. Thus, identifying a balance between speed and security is a timely requirement.

10.7.2.5 Secure Virtualized Environment and Orchestration at the Edge

Edge computing paradigms rely on virtualization technologies for deployment. Virtualization offers a dynamic service provisioning functionality that could launch flexible and configurable softwarized executable instances at the edge platform. This flexibility, migration, and offloading scenarios invite virtual attacks [219]. Prevailing virtualization approaches amalgamate hypervisor-based and lightweight virtualization technologies to form edge computing platforms [406]. Containers are deployed within a VM to attribute more dynamic governing capabilities. Even though hypervisor-based and container-based methods provide an adequate level of isolation among their instances, a hybrid system might generate unintended causes that might affect in security and performance perspectives. Moreover, various orchestrating deployment options are available such as OpenNFV, CloudNFV, OpenBaton, Open MANO, Cloudify, and T-NOVA [13]. Interoperability and compatibility among these technologies are vital for creating a virtual ecosystem. These virtual entities are typically networked with virtual gateways and SDN controllers [407]. Therefore, any softwarized attack is capable of exploiting this system. To develop a secure virtualized environment at the edge, the security of these individual technologies and their integration should be validated before launching MEC.

10.8 Possible Security Solutions for MEC

This section discusses general solutions that can be adopted in the MEC edge environment and the system level. These solutions exceed the capabilities of the conventional security solution and are novel to the research community. These directives are state-of-the-art security solutions focused on pragmatically deploying edge computing.

10.8.1 Security as a Service

The security as a service (SECaaS) concept first emerged with cloud computing deployments [408]. The goals of provisioning security mechanisms of intrusion detection and prevention, network security and monitoring, authentication (i.e., trusted third party services), certification authority, identity and access management, security and vulnerability scanning/assessing, and securing the transmission channel or VPN facilities are the expected functions of SECaaS. The flexibility and scalability offered by cloud services are ideal for launching these services. Though, the limitations of cloud computing in the context of IoT as stated in Chapter 1, fades the feasibility on a practical SECaaS implementation. And considering the diverse security requirements of the novel applications, a singular security solution is not adequate to secure their conduct [409]. The edge computing platforms, however, improve the realization of SECaaS potential with the attributed ultra-low latency, higher serving bandwidth, locational awareness, and real-time responsiveness [23]. These features enable the launching of diverse security mechanisms as virtual security instances or ME Apps under a SECaaS MEH, which is configured for serving one type of security service (i.e., either network intrusion detection, VPN, or authentication). These virtual entities are launched in a virtualized infrastructure administered by a MEPM that performs the security orchestration function [17]. Such a SECaaS initiative can be extended to a global context, where the cloud infrastructure can be integrated to perform the analysis, assessment, predictive modeling, and security/privacy policy formation and management tasks. This is becoming a popular directive with emerging 5G-based technologies that offers a comprehensive security solution that can be configured for any situation.

10.8.2 Artificial Intelligence with Machine Learning and Deep Learning Methods

Machine learning (ML) and deep learning (DL) are the methods for paving the path toward artificial intelligence (AI). ML approaches presumably imitate the human approach to learning and adapting through experience. The outcome of the ML approaches depends upon the amount of available prior knowledge and experience derived from them. The prior knowledge extracted from the training data can be utilized for formulating variant ML models or algorithms that can be formalized for different applications according to their requirements. Such models are categorized into supervised, unsupervised, and reinforcement learning. Supervised learning relies on respective outputs or labels of the training data while unsupervised learning methods formulates their data representation structure without the labels; and reinforcement learning methods are ideal for dynamic environments where the goals are pre-specified. DL methods, though they falls under ML methods, are more accurate for classification scenarios in computer vision, bio-informatics, language processing, and advanced gaming based on AI; where neural networks and their variants are the most common approaches at play [410]. The AI approaches are imperative for edge computing-based applications to perform optimization, classification, decision-making, prediction tasks, and security mechanisms. It is the future of beyond 5G networks that pave the path toward 6G [411].

Security service providers can rely on AI to launch autonomous security solutions. Such approaches are evident with malware and anomaly detection adaptations [412]. An autonomous security system is required for MEC deployments that rely on virtualization technologies with minimal human or administrator involvement and is expected to cater plethora of devices that are provisioned with heterogeneous services. Already, there are initiatives taken to launch AI-based security solutions for edge computing environments [413, 414]. Intrusion detection and prevention systems (IDPS) are the most obvious adaptations of the AI approaches for the MEC edge, which can monitor the network behavior and traversing content in the LADN [409]. Adversaries possessing legitimate credentials pose a significant threat to the MEC system, especially when the operations are autonomous and virtualized. Thus, ingress initiatives such as CFSP and UALCMP in the MEC system level are always in the radar of the attackers. With the level of scalability envisaged, assumptions should be drawn that the penetration of the MEC system would be inevitable at some point. Hence, the MEC orchestrators or hypervisors should be monitored for any anomalous behavior at the MEH level. From the UE perspective, AI-based malware detection schemes for smart mobile devices are presented in [415, 416]. Honeypots can be launched at the edge to evade attackers as a cyber defense strategy [417]. Even VMI (i.e., explained in sub-section 10.8.8) services can be benefited from AI involvement to improve their detection accuracy [418].

10.8.3 Network Slicing (NS)

Existing networking infrastructure In NS, a physical network is logically compartmentalized into different slices to enable diversified services to operate simultaneously [364]. With that approach, expenditure for resource utilization could be alleviated significantly.

NS offers virtualization support, function modularization, end-to-end connectivity, and better isolation compared with other resource-sharing concepts [366]. According to this concept, a singular slice is composed of network slice instances (NSIs) and a network slice manager (NSM) [367]. Variant ME Apps provisioned to service diverse UE Apps share the same LADN in a MEH. The NS concept could be applied in such a situation to simplify service provisioning. A simplified networking deployment is improving the chances of security solution adaptations. A MEH or an ME App could be assigned as an NSM for embedding security mechanisms. The Next Generation Mobile Networks (NGMN) alliance is actively working toward NS security for adapting the concept to the 5G use cases [367].

10.8.4 Context Aware Security

Context awareness is drawn from the ambient intelligence plausible with prevailing smart devices that possess advanced sensory capabilities [419]. These features of current UEs enable services customized for personal and locational contexts. Similarly, ambient intelligence could be leveraged to enhance UE security from the security-related contextual data gathered from the sensory installations. Such an approach would be ideal for securing the communication channels discussed under TV 2. Adaptive security protocols could be employed for securing these channels [420]. Moreover, MEHs can deploy context aware security mechanisms to detect ME Apps with anomalous behavior.

10.8.5 Physical Unclonable Functions (PUFs)

PUF is a method of extracting a bio-metric resembled imprint from non-human entities (i.e., devices) considering their unique, unclonable, and instant-specific features attributed to the fabrication process. Various parameters in circuitry, such as RF carrier frequency offset, in-phase quadrature-phase imbalance, and transient effect ring oscillator, are used as UE PUFs. Like bio-metrics, PUFs could be utilized to develop authentication protocols for UEs as presented in [421, 422]. Due to their uniqueness, PUF-based authentication messages do not incur heavy overhead comprised of authentication credentials. Thus, efficient and more secure security protocols could be developed with this adaptation. Such PUF-based Challenge Response Pair (CRP) constructs within the authentication mechanisms are ideal for D2D, M2M, or V2V instances where the environment is completely autonomous [423].

10.8.6 Blockchain

Blockchain is a secure and decentralized data management framework that emerged with the Bitcoin digital currency concept [424]. Blockchain intends to create a tamper-proof ledger that justifies the mere concepts of trust and accountability through secure entries or blocks enabling the verifiable consensus within its access domain [425]. Blockchain supports pseudonymity, eliminating the requirement for a certificate authority [28]. This concept overcomes the limitations (i.e., single point of failure) of centralized data management systems with improved security attributed to the dispersed data fragments/blocks. Even with the locational and context awareness facilitated to the MEC subscribers, outsourcing private data to another party still raises concerns. To overcome these privacy predicaments, Blockchain is an evident solution. From the MEC point of view, Blockchain could be adaptable for securing the anonymity of UEs, service migration channels, mobile offloading channels, and maintaining service status parameters hindered by adversaries [426–429].

10.8.7 Modern Authentication and Key Agreement Schemes for MEC Infrastructure

Authentication is a crucial security measure that can provide security for MEC edge infrastructure. Since edge infrastructures are lying intermediary to the cloud service paths, proper authentication offers intrinsic identity verification to the access network devices, eliminating the requirement for any trusted services employed by the cloud services for authentication. Various authentication schemes are proposed to secure the channel between the UE and the MEC edge base station and the link between edge gNBs. A key agreement protocol for MEC-enabled 5G networks focusing on remote patient monitoring is proposed in [430]. The use of symmetric key encryption makes this protocol efficient and lightweight for the context of this application. A TTP is employed for validation purposes, and a key agreement occurs between a medical IoT device and a MEC-enabled base station. Braeken et al. [140] present an edge-based secure architecture for healthcare applications to perform secure key management. The proposed protocol targets Mobile Augmented Reality (MAR) based applications, where man-in-the-middle (MitM) type, distributed denial of service (DDoS), impersonation, malware injection, and side channel attacks (SCA) attacks are in its scope of defense.

10.8.8 Virtual Machine Introspection (VMI)

Also referred to as hypervisor introspection, VMI approaches detect anomalous patterns in the behavior of VMs in a virtualized environment governed by a hypervisor [18]. The parameters related to the processor, memory, and networking utilization of a VM environment are monitored by the VMI. The monitoring mechanisms can be either passive (i.e., analyses the main memory based on external occurrences) or active (i.e., interferes with the control flow for monitoring). There are various introspection methods, such as process-bound introspection, core-selective introspection, and hypervisor-based switching [431]. As edge environments as in MEC are envisaged to employ virtualized tools for deploying the infrastructure, these VMI approaches are salient; either as a security strategy or a performance-enhancing strategy. LibVMI is such a tool that acts as a host-based Intrusion Detection System (IDS) for VMs [343]. The VMI proposed in [432] is capable of interrupting the VM execution or isolating the running programs if the VM states show any deviation from the normal operational standards. The RapidVMI proposed in [431], is a framework for active virtual machine introspection that enables fine-grained, multi-core aware VMI-based memory access on virtual address spaces. It was designed to overcome the program tracing and process injection shortcomings of VMIs in a production environment. The vProVal [433] was introduced to act as a security tool to detect any hidden processes or rootkits within a VM running in a Kernel-VM (KVM) based cloud environment; that can be adapted for edge platforms. The monitoring is conducted externally at the KVM layer level, where low-level detail extraction is possible. From the MEC perspective, VMIs could be launched at both edge and core levels to inspect the behavior of MEHs and system-level entities via the MEPM and MEO, respectively. The VMI process can be implemented internally for orchestration entities aligned to the VIM hypervisor. Any attack that intends to compromise the virtualized entities can be detected through these approaches, especially following anomaly-based detection methods. Moreover, proper introspection methods can improve the performance of the hypervisors.

10.8.9 Virtual Private Local Area Network Service (VPLS)

VPLS is an interconnection technology for industrial sites that leverages the Multi-protocol Label Switching (MPLS) standards for its provider network [434]. The tunneling nature of VPLS, resembling the effect of a virtual private network (VPN), guarantees the security and integrity of conveyed information through the channel. The rapid label exchanging featured with the MPLS backbone of VPLS is optimizing the network performance by alleviating the latency and jitter. To overcome the tunnel management limitations of VPLS architectures, SDN has been proposed as a solution [435]. MEC envisages deploying an SDN-based networking infrastructure for establishing communication within and among the edge levels. Thus, the software defined VPLS (Soft-VPLS) approach is adaptable for securing the communication channels between edge and core levels of the MEC deployments. These channels transmit mobile network signaling statistics (i.e., 5G), MEC control statistics, MEC service requests, MEC service approvals, user data, and executable content in migration or an offloading circumstance. With Soft-VPLS, each traffic category could be classified and conveyed via different tunnels. Moreover, the Internet connectivity (i.e., TV 7) could be secured with Soft-VPLS adaptations.

11

Edge Privacy*

11.1 Introduction

Privacy, in general, is an elusive concept that can be objectified to the phrases: freedom of thought, solitude over one's belongings, freedom of surveillance/searches/interrogations, and control over one's information and reputation [436]. In other words, privacy is an individual's right to act or behave independently of any records or surveillance activity conducted without their consent. In this digital era, mishandling personal/sensitive data in the absence of proper consent is considered a privacy violation. Moreover, one's actions conducted in the digital domain should be obscured and untraceable to preserve their identity. The extent of the digital domain on prevailing technology invites adversaries to conduct in a sophisticated manner to extract personal credentials on weakly protected systems. Advanced sensory apparatus appended to both human and non-human entities are increasing the privacy leakages in all the layers of the TCP/IP model [219]. From the core network point of view, in the envisaged software defined mobile networks (SDMN), the openness, software-enabled access, and control contrive an advantage for the intruders to exploit the presented exposure and thereof invade the privacy of the conveyed information [407]. Such unlawfully gathered personal data can be outsourced to institutions (i.e., insurance companies, mobile operators, and the entertainment industry), which are dependent on personal intents, behaviors, and interests to expand their commercial market. These acts are recognized as unlawful practices, and novel legislation focuses on mitigating these occurrences.

Impending applications require the functionality to outsource/offload information, including the location, to a storage and processing infrastructure for the sake of enabling improved scalability, performance, compatibility, and interoperability with real-time responsiveness [437]. Cloud computing (CC) was such an intended solution deployed for achieving the above-mentioned goals. Though, CC as a paradigm was not living up to its expectations for the intended applications of 5G and IoT; due to the associated latency, jitter, locational unawareness, limited access, and inefficient data outsourcing inherited from the geographically dispersed cloud servers and their centralized architecture [9, 13]. As it is a concept designed to outsource the user data and services to trusted third party (TTP)

*This chapter is partly contributed by Pasika Ranaweera (University College Dublin, Ireland).

Mobile Edge Computing and Communications: Driving Forces, Technology Foundation, and Application Areas,
First Edition. Aaron Yi Ding, Chamitha De Alwis, and Madhusanka Liyanage.
© 2025 John Wiley & Sons Ltd. Published 2025 by John Wiley & Sons Ltd.

servers, the privacy of the user data was an evident conundrum. With the advancement of cloud hosting applications, a particular user data set is shared among various cloud services without assimilated consent drawn from the users. Even though the users are granting their consent via online agreements prior to subscribing to the service, eagerness to acquire the services clouds the judgment and leads to ignoring the importance of the context in the agreement. As there are myriads of cloud service providers (CSPs) operating nowadays, it is an arduous task for a user to distinguish the ingenuity due to non-existing attestation authorities to verify their service. Moreover, cloud services are not advertising the actual location of the servers. Current application layer or security protocols do not attribute the binding of geo-location of the servers with the identity certificate [23]. This is a clear violation of user privacy rights on being aware of the location in which the data are stored.

Typically cloud services are hosted from a remote data center, where the distance to the user might extend to several continents. In this scenario, employing strong encryption schemes might endure cumbersome circumstances in terms of efficiency in transmission. The absence of a provably secure cryptic protocol, where most cloud services are failing to notify the users, raises the risks of intervening attacks occurring on data traversing channels. The distance between the cloud entity and the user accessing device elevates the probability of this occurrence. In addition, authentication credentials are traversed from the UE to the remote entity. Exposure to the credentials poses an inconceivable situation that is more intimidating than exposure to data. Adversaries are capable of revealing the encrypted data perpetually. As most cloud services are extending their services to mobile devices as an advancement in mobility and flexibility, the above concern is higher due to the alleviated computing and storage limitations in mobile devices that lead to employing lightweight security protocols. In addition, information stored in a mobile UE is more private than the ones stored in user PCs. And to support the mobility and flexibility requirements engaged with UEs, the probability of launching lightweight protocols raises concerns. These factors are exposing the critical privacy issues associated with the existing information systems. 5G-based edge computing paradigms such as fog and MEC are introduced as the solutions to the limitations of cloud computing for preserving the privacy of subscribers [21, 438].

11.2 Privacy in 5G and Beyond

Privacy in 5G networks is a crucial requirement as it transforms the current digital infrastructure to introduce novel daily life applications and access modes of digital services. Also, 5G will bring new enhancements in terms of architectural and service-oriented requirements as compared to traditional mobile networks (3G, 4G), so it will require strong privacy policies and regulations [379].

Privacy in 5G and beyond networks will be vital for whole ecosystems, including users and various other stakeholders [439]. Therefore, in order to have complete public acceptance and adoption of future networks, it is mandatory that privacy issues are well addressed. User privacy of future mobile networks can be divided into three main categories; *data*, *location*, and *identity privacy* [440, 441]. These aspects are applicable for

6G, as the fundamentals of trust are derived from safeguarding the principles of privacy [411]. In this section, we explain each privacy category suited for future mobile-based technologies.

11.2.1 Data Privacy

Data privacy represents the confidentiality and privacy of stored data [439]. As the advancement toward future 5G mobile networks is rapid and concrete, consumers will be giving more preference to mobile networks as compared with traditional Internet-based services. Consequently, 5G mobile networks are expected to have high data speed and low latency, which eventually results in a huge volume of data. For example, critical applications such as healthcare and banking will generate sensitive user data and thus require data protection while storing and utilizing that data.

In 5G, various heterogeneous service providers and operators might be tempted to store and use personal information of consumers with or without their permission. This is because it allows companies to follow the trend of their own product, which helps them in improving the quality of services. Companies also analyze the consumers' data to check whether any additional necessary feature is required in that particular product. In most cases, it is considered good practice for companies to take some of the user's personal data, and based on that, they can make new/useful products and services for consumers. However, the privacy of users' data can be at huge risk if companies are using critical data other than the agreed/predefined reasons. Thus, to ensure data protection in 5G networks, strong regulation is required.

11.2.2 Location Privacy

With the advent of 5G technology, location based services (LBS) are getting more important as users can access useful services based on knowing the location information [442]. Also, nowadays, several gadgets possess the capabilities of positioning and tracking, providing numerous location-based services [443]. For example, online applications on mobile devices can suggest locations of various hospitals, restaurants, shopping malls, etc., that are the nearest to the particular user. Recently various social networking websites such as Facebook introduced location-aware features such as "check-in." These features are very helpful in the socialization of people by knowing the location of nearest friends/relatives. However, along with such kinds of services, users would be continuously tracked by various actors through their personal gadgets or devices embedded in the environment, and that could cause serious concerns for user's privacy.

For example, online applications on mobile devices can suggest locations of various hospitals, restaurants, shopping malls, etc, that are the nearest to the particular user. Service providers/companies are building new applications and services based on the data gathered through analyzing location information.

Consumers are well-known for the benefits of the LBS but unaware of most of the privacy leakages associated with it. Traditionally, location privacy is mostly explored from the network or application layer perspective, but the work in terms of the physical layer is very less or mostly lacking. Thus, all necessary measures need to be taken in order to protect location privacy in 5G systems.

11.2.3 Identity Privacy

Identity privacy refers to protecting the identity information of the subscriber and the device/user equipment (UE). With 5G and IoT, it is expected that billions of devices will be connected to the internet. In such digitalization, every entity (user or device) would be categorized by some identity in order to access or deliver required services. Using identities, services, and personal information can be accessed. For example, online applications such as healthcare require identity to access patient information, and online shopping or banking requires payment modes through some cards, which include identity information. Similarly, the device identity can also lead to leakage of the user's personal information [444, 445]. Thus, it is important to have a secure and efficient identity management mechanism in 5G networks.

Another key challenge in 5G networks is to preserve identity privacy. Along with other information, service providers/companies store and utilize the identity information of the user. However, in most cases, users do not want to reveal their real identity, for example, during online feedback or asking for any information. The consumer prefers to stay anonymous unless they are dealing with their known friends/service providers or are well sure of that particular environment. Nowadays, adversaries steal identity through various means; for example, online shopping or banking requires payment modes through some cards, and thus the real identity can be traced. Identity privacy can further be classified into two categories, i.e., subscriber's identity privacy and device identity privacy.

In addition to the consumer's point of view, privacy will be an important aspect in the perspective of various 5G stakeholders such as network operators, cloud service providers, manufacturers, and third-parties applications developers. 5G will allow these stakeholders to build novel applications, and based on that, they may draft new business models, which may be useful to increase their revenue and provide services to consumers in a much easier and more efficient way [446]. For example, in a future mobile communication environment, various stakeholders will collaboratively provide services to the user. Thus, these new business models, architectural and technological enhancements in 5G will bring new privacy challenges for involved stakeholders. For example, in a shared service environment, these stakeholders would require mutual trust among each other to provide services through a common platform.

11.3 MEC Privacy

MEC and other edge paradigms were introduced primarily to overcome the latency and bandwidth limitations of the existing cloud computing (CC) based information systems. As an unforeseen refinement, provisions for employing privacy protection schemes are inherited from them. Such factors of MEC are listed below:

Improved Location Privacy: Primarily, launching the data storage and processing infrastructure at in-proximity BS guarantees the user the location of their data. However, existing security protocols should be amended to bind the geo-location credentials of the hosting MEH (located at the closest or a remote MEC edge level) for realizing location awareness.

Trusting Service Provisioning Authority: The guarantees given to the users are reinforced by the service level agreements (SLAs) provided by MEC service providers to draw consent from the users [447]. As telecom operators are serving as MEC service providers, user trust is ensured in contrast to CSPs.

Improved Cryptographic Primitive Usage at UE: Alleviating the energy consumption of UEs is one of the objectives of MEC deployment. With this energy utilization, communication channels are provisioned to employ cryptographic primitives for securing the transmission medium. Thus, the penetrating success of intervention-based attacks could be mitigated. Moreover, enhanced power management during an MES operation allows an additional level of encryption within the internal processing stages, ensuring user data privacy within the UE while being processed. A similar approach as Blockchain could be employed constrained to the resource availability [448]. As the user data is traversing only within the edge boundaries on the majority of the time, the probability of being intervened is alleviated.

Privacy Preserving Augmented Data Processing Platform at the Edge: User data is handled by the ME Apps in the edge corresponding with the related UE Apps in the UEs. As ME Apps are enriched with resources facilitated by the VI, maintaining a cryptographic base for Blockchain or homomorphic encryption resembled operations is not an issue. Secure multi-party computation (SMC) is another approach to handling information without exposing them [449]. This method is the highest expectation for enforcing privacy on outsourced user data. The in-proximity MEC deployments that feature ultra-low latency with enhanced bandwidth utilization improve the probability of launching a cryptic base for the holistic system undergoing storage, processing, and communication aspects. In that scenario, user data are obscured and integrity protected in every state of the MEC system.

Improved Privacy via Authentication: Authentication is a critical phase of a security protocol for ensuring privacy. As the initial contact of the UE App service request is forwarded to the UALCMP, threats are only limited to the RAN interfaces of the MEC. The extent of the threat vector is significantly reduced in contrast to cloud computing as the authentication credentials are not leaving the MEC system level. Attributed higher resources in UEs enable the secure cryptographic mechanisms that lead to protecting the authentication credentials during the service instigation.

Monitoring Authority at the Edge: The MEC system-level entities OSS and MEO are ensuring the privacy preservation of the user data with their continuous monitoring of the service platform. The proximate nature and convenient monitoring scope (minimal number of MEC edge levels) granted to the MEC system-level entities guarantee the detection of abnormal occurrences in terms of privacy violations. This feature of MEC is significant in comparison with CC.

Impending applications and services are demanding the handling of personal credentials/information at the edge of the network for realizing the service requisites. Privacy, integrity, and trust management assurances are prime requirements with MEC deployments despite the attributed locational and context awareness facilitated for the users. It is evident that virtualization technologies are vital for realizing the MEC paradigm for creating a serviceable platform with dynamic resource allocation capability. Security of the virtualized platforms is still a gray area due to lesser deployments. Vulnerabilities and attacks plausible on virtual machines (VMs) are unique and cause significant consequences to the MEC system.

Similar to CC, outsourcing MEC subscriber data to a remote storage and processing environment creates a predicament in terms of privacy rights. Establishing boundaries regarding the extent of authorized conduct on service providers' capabilities is imperative for guaranteeing the trust of MEC consumers. Considering all these facts, security and privacy are important for realizing a pragmatic MEC paradigm.

11.3.1 Impact on Data Privacy

The confidentiality of user data that is either stored, processed, or traversed is considered data privacy. Any mishandling of such private data is denoted as a privacy leakage. Enabling Big Data applications facilitated with high bandwidth and ultra-low latency enhancements of MEC are generating massive amounts of personal data in the future. Applications such as healthcare, banking, and crowd-sourcing are brimful of such sensitive data [439]. Moreover, a certain data set of a user is displaced among various operators due to the open service platform of MEC. This poses a confidentiality issue. In the current era, artificial intelligence (AI) based pattern recognition techniques are applied to user data by certain companies for identifying trends and interests. As these activities are outsourcing data without the user's consent, privacy is violated conspicuously. Thus, subscriber privacy preservation is a vital requisite for launching the MEC paradigm.

11.3.2 Impact on Location Privacy

Though location-based services (LBSs) are enabling various applications to the MEC subscribers, exposure of the geo-location is endangering the financial, entertainment, professional, and secrecy aspects of human life that leads to hijacking, blackmailing, or ransoming situations [450]. User location could be revealed to the subscribed service legitimately, either intentionally or unintentionally, as a pop-up service request for location sharing, that the subscriber is consenting without proper assimilation. Moreover, secondary mobile channels broadcasting the wireless transmission apart from the direct channel are bound for monitoring by eavesdroppers for tracking the location [449]. These facts contribute to the violation of location privacy.

11.3.3 Impact on Identity Privacy

Impending tactile Internet and IoT concepts are expanding the scope of the cyber-space. A method is required to identify the billions of entities and people accommodating it via interfacing UEs [439]. Any knowledge-based (username), possession-based (random number generator—RNG), inherence-based (bio-metric), cyber-address, or physical unclonable function (PUF) in case of UEs are adaptable for proving one identity. Identity is the key to safeguarding private information in the cyber-space. An adversary capable of replicating a user identity could access the entire data cluster mapped to that identity. Attacks such as UE tampering, UE cloning, and masquerading attacks that commit identity theft are exemplifications of privacy violations in the access network. In the core network, a cyber-invader capable of exposing the identity credentials of the users or entities are presented with the opportunity to exploit the MEC system. Thus, preserving the identity of the users with expedient mechanisms are a prime requisite for MEC deployments.

11.3.4 Other Privacy Issues

11.3.4.1 Authorized and Curious Adversaries

An authorized entity could exist that captures disowned data with honest and curious intent. The extracted data could then be used for usage profiling, location tracking, and disclosing credentials [21]. Thus, user privacy is violated despite the motive of the intruder. Such an initiative can be taken by the infrastructure or third-party service providers. Due to the openness of the edge ecosystem, constricting these legitimate practices is a conundrum. Every entity engaging with the ecosystem should be aware of the responsibility inherent to them. Moreover, a mechanism should exist to detect any misbehavior perpetrated by an entity or a user.

11.3.4.2 Computational Offloading

Computational offloading is an exploitable feature of MEC system that could be monitored by the adversaries to determine the locational information [451]. Tracking such intervening attacks is questionable due to the complexity of the offloading process and the non-accountability of the attack origination. Moreover, MEC servers inherit the features to track the usage patterns of the subscribers from the contextual information and channel status parameters. This creates a distinct concern with subscribers for exposing their private information [449]. Thus, computational offloading is a feature that risks the privacy of the MEC system.

11.3.4.3 Service Migration

Migrating services from one MEH to another at the same edge level or at another edge level is possibly dependent on the scope of the subscribed service and the mobility of the UE. Certain services, such as AR, are demanding service migration to satisfy service requirements. Thus, a cyber-eavesdropper capable of reviewing the service statistics of the MEH entities is in a position to track the user from the migrating service patterns [450]. It is leading to violating location and usage pattern privacy aspects. This vulnerability exists with compromised MEC entities that could monitor service statistics of the MEC virtualization infrastructure platform.

11.3.5 GDPR on MEC

The European General Data Protection Regulation is an initiative enforced on the 25th of May 2018; that strengthens the citizens' fundamental rights in the digital age for securing private data and preserving the privacy of individuals subjugated by any IoT application [161]. Accordingly, any hosting service that acquires data from any citizen should ensure their consent before commencing. Handling and extracting information from an unconsented citizen is an obvious offense. Under GDPR, untraceability is a paramount objective proposed to achieve via anonymization and pseudonymization approaches. According to GDPR legislation, data processing should attribute; (i) lawful, fair, and transparent processing (transparency); (ii) legitimacy and legality of the purpose (purpose limitation); (iii) acquiring only the adequate data amount for serving the purpose (data minimization); (iv) ensuring data accuracy, validity and alignment for the purpose; (v) minimal data redundancy and replication; (vi) preserving confidentiality to assure integrity and privacy of data in all aspects of hosting organization including non-digitized materials; (vii) demonstrate compliance for accountability and liability. Any organization or entity disregarding the declared regulations is viable for prosecution under GDPR reforms.

The prime issue for privacy protection in MEC-based deployments is the unawareness of the users regarding the handling of data by the outsourced authority. Even the user consent is drawn, awareness is limited to the inherent knowledge and concentration ability of the users. The declaration of GDPR legislation, however, is a critical juncture of the IT industry that lays the foundation for dissemination of privacy preservation in a lawful context. The presence of a legal authority enhances the engagement of privacy practices and encourages users to be involved in outsourcing IoT applications, which were once considered untrusting. Thus, the GDPR initiative is a timely approach coordinated for the benefit of MEC and other IT-related service platforms.

11.4 Objectives for Privacy Preservation in MEC

The following objectives are proposed as regulatory privacy policies based on the privacy issues and scope for improving privacy-related mechanisms in MEC. The assurance of these objectives is vital for improving the trust of MEC subscribers and potentially improving the quality of the MEC standardization.

11.4.1 Global Compliance for Privacy Policies

Privacy policies should be adhered to as legislation in a global context for mitigating varied standards maintained beyond national or continental borders. Moreover, standardized levels of privacy policies should account for various stakeholders' interests and service domains for the pragmatic assignment of privacy regulations. Such a globalized standard would entail the successful launching of products or services across the globe. Otherwise, certain digital vendors might target less illicit countries or domains to market their products, creating an unfair discrepancy in access for the consumers residing in countries endowing privacy policies. In the MEC context, since the deployment and migration of services are often taking place within a country or a sector, global compliance for local applications or services would not be a critical requirement. The international services, which are far-reaching from their origin, however, might create issues if the privacy policies are not universal. The intricacies of formulating a global standardization are understandable among the current existing standards such as GDPR, California Consumer Privacy Act (CCPA), Personal Information Protection Electronic Documents Act (PIPEDA) in Canada, Health Insurance Portability and Accountability Act (HIPAA), Personal Data Protection Act (PDPA) in Singapore/Malaysia, Australia Privacy Act, and the well-known ISO 27701 [452–454]. On the contrary, these varied standardizations provide the basis for the required framework and enable the discussion at a global audience.

11.4.2 Responsibility of MEC Service Providers and Consumers

MEC service providers should be responsible for any committed privacy violation within its domain, while consumers must report any privacy violation endured by them or other parties. Satisfying the responsibilities enables a preserved privacy eco-system where privacy violations are not unaccounted for. With the decentralized and distributed nature of the MEC service infrastructure, the governing jurisdictions can be altered to apply to the region of interest. However, a general set of rules should be enforced upon all the entities engaging in the MEC system, paving the path for limiting any possible privacy infringements. These responsibilities exceed a typical privacy policy, as a responsible entity should be aware of any ethical, legal, or sensible repercussions accounted for by any irresponsible act. Hence, this responsible scheme creates a trusted domain between the MEC consumers and subscribers.

11.4.3 Privacy Compliance on Integrating Technologies

MEC is integrating technologies such as NFV, SDN, ICN, Network Slicing, IoT, and 5G. These technologies are operated and standardized by diverse institutions and corporations. Thus, it is vital to establish a common code of conduct among such organizations regarding privacy mechanisms and policies for mitigating conflicts of interest. Specifically, implementing privacy policies within the controlling and executing constructs within these technologies should be decided in the MEC service design stages. Privacy-preserving technologies and mechanisms are proposed for such integrating technologies as presented in [364, 455–458]. Each of these mechanisms is formulating different metrics, rules, and policies. Utilizing these constructs in a singular MEC control environment is going to be quite a challenge.

11.4.4 Data Portability

This objective requires disseminating private information among MEC service providers without employing mandated standards [439]. In fact, through these services, customers can share or reuse their own data for different purposes, for different services, without requiring to feed them at every service instigation instance. It ensures the interoperability of the holistic cross-domain MEC system. To enable this within the MEC environment, a unique function should exist to manage the intricacies pointed out under objectives 11.4.1–11.4.3. A model should be contrived to store, fetch, share, and reuse data within the MEC environment that enables the data portability feature [459]. This data model or structure should incorporate the aspects of demanded privacy level and relevant privacy mechanisms embedded in its metadata, in addition to the regular metadata inclusions. Though this data-model formation seems more computationally intensive, it will save time on data reusing and improve the interoperability of user data.

11.4.5 Accountability and Transparency of Data Handling

As diverse stakeholders are acting in a MEC system, each party should declare their intention on data storing, processing, and transferring activities to other parties, including the users, to maintain transparent service obligations. In fact, this objective is quite vital for detecting privacy violations and enabling traceability of the violating actions and violators. Accountability is the tenet for achieving privacy and managing privacy within an intricate setting such as in the MEC system [460, 461]. Accountability should be systematically applied to the entirety of the personal data life cycle, adhering to the general data protection rules. All the aspects of policies, procedures, and practices should be covered under this. At the same time, all the manipulations conducted on the data should be transparent to the authorized entities to enable the analysis and detection of privacy infringements. Blockchain is a promising solution that can be leveraged to achieve this objective in a highly effective manner [462].

11.4.6 Declaring Minimum Specification Requisites of UE for Subscribing MES

Heterogeneous IoT devices subscribing to MEC services inherit variant resources. Depending on the device's specifications, the provisioned level of privacy guarantee is deviating (i.e., a high-end device gets better assurance than a low-end device). Thus, it is vital for MESs to publish the minimum specification requirements to ensure the guaranteed privacy level. With these pre-conditions, only the resource-availing IoT devices will be instigated for service delivery. To deploy this concept realistically, however, a proper privacy quantification framework should be designed and developed to identify levels of privacy metrics [463]. The identified metrics will then be mapped with the required resource level capable of achieving or enabling the corresponding privacy mechanisms. Thus, this objective will guarantee the removal of inconsistencies in privacy-enabled service delivery at the initiation stages.

11.4.7 Optimal Utilization of UE Resources with Embedded Privacy Enhancing Mechanisms

The operating lifetime of a UE is a critical factor for MESs to encourage subscriptions. A MES that performs with least resource consumption would be preferable to the consumers. It is evident that each MES should employ privacy-preserving mechanisms embedded in their service protocols. Thus, achieving optimum resource utilization of the UE with applied privacy-preserving mechanisms is a clear objective for MEC service providers. Objective 4.6 establishes the minimum requisites for launching the desired privacy guarantees with the associated service, while this objective ensures the cost of delivering privacy is optimal in the context of efficient service delivery. In fact, this objective sets the upper limit on the operating resources for the respective IoT device connected to the MES. To achieve this, accountability for energy spent on memory, processing, storage, and transmission is vital. Hence, optimal vantage points on the energy spectrum can be identified through systematic analysis. This objective is ever so vital for the sustainability of the MEC system, especially for enabling privacy in resource-constrained IoT edge applications.

11.5 Privacy Protection Mechanisms for MEC

In this section, we discuss the possible mechanisms to protect the privacy of MEC environments by mapping the state-of-the-art solutions to the objectives specified above.

11.5.1 Regulatory Approaches

Regulation is required to promote the objectives of privacy among different entities. Objective 11.4.1 can be implemented through such regulations conveniently. Regulations can be mainly categorized into three types [464–466].

11.5.1.1 Governmental or Continental Regulations

Governmental or continental-level regulation is the most influential method to regulate privacy since these sectors have the audacity and resources to design, deploy, and maintain privacy policies within their sectorial regions. The governments are responsible bodies for writing and enforcing regulations that are relevant to each country. Government-level representations in multi-nation organizations such as United Nations (UN) and European Union (EU) can effectively extend national-level regulation to the global level [467]. Since MEC is widely promoted within the EU region, and the GDPR initiative is actively adapted to ensure the privacy rights of digital consumers, the EU is the most resourceful to carry the flag forward on privacy implementations.

11.5.1.2 Industry Self-regulation

At the industry level, different industries and industrial groups can develop principles and practices that reflect consensus on the best approach to protect privacy [468]. Moreover, industrial groups can be formed to mandate the regulations. In the telecommunication domain, these groups may include original equipment manufacturers (OEMs), chipset manufacturers and designers, telecommunications operators, service providers, application providers, operating system providers, SMEs, and researchers. European Telecommunication Standards Institute (ETSI), International Telecommunication Union (ITU), Next Generation Mobile Networks (NGMN) Alliance, ONF (Open Networking Foundation), 3rd Generation Partnership Project (3GPP), and 5G Infrastructure Public Private Partnership (5G PPP) are such industrial consortiums that have formalized to drive the future technologies in the telecommunication sector collectively. In fact, these groups are the pioneers in standardizing the MEC concept and ETSI is the leader in its deployment. These groups can influence not only their business partners but also other government-level regulation bodies to meet industry standards on security and privacy [466].

11.5.1.3 Consumer or Market Regulation

Consumers are the actual users of the systems, and they are the ones who need privacy protection. A marketplace is created mainly with consumers, while prosumers, vendors, and service providers are secondary to them. They can enforce the terms to obtain the desired level of privacy. With the decentralized nature and the reach the MEC technology has as an infrastructure service provider across a nation, indirectly, it is contriving a digital marketplace for the entire country. Thus, when a consumer subscribes to the MEC ecosystem to obtain the intrinsic services, all other services run on top of the MEC service infrastructure. It provides the ultimate opportunity to gain the necessary consent from the subscriber, along with all the other metrics required for determining the privacy levels desired by them. They can also influence governments through voting and industries through choices taken in the marketplace engagements [469]. Government regulation can promote many of those targets, at least at the EU level. In particular, traditional telecom players can also contribute to many of the objectives. It will be a challenge to convince new players also to participate.

11.5.2 Privacy Preservation in Mobile Offloading

As described earlier in Section 11.3.4.2, mobile or computational offloading is a prominent aspect of MEC that can violate the privacy of services. He et al. studied the location privacy and usage pattern privacy of the MEC task offloading feature [449]. The paper investigated the balance to restore privacy protection while maintaining maximized operational delay and energy consumption performance. A constrained Markov decision process (CMDP) based scheduling algorithm was proposed as an approach to the task offloading process. As most IoT-based privacy discussions are based on conventional cloud computing deployments, He et al. [451] identify novel vulnerabilities in MEC-enabled IoT deployments emanated by wireless offloading features. An adversary model is formed considering an untrusted/compromised service provider that monitors the UE and MEC server engagements for revealing the location of the UE employing estimation techniques. This offloading problem is formed using Markov decision process (MDP). The privacy-aware solution is introduced as a deep post-decision state (PDS) learning process; formulated by integrating PDS and Deep Q-Network (DQN) techniques that utilize energy harvesting statistics. The deep PDS approach suggests an offloading strategy to the IoT device faster in contrast to the DQN method.

Chi et al. [470] introduced a novel technique called privacy partitioning which is composed of a trusted local partition and an un-trusted remote partition. The proposed method targets privacy preservation of deep learning classification tasks employed in mobile offloading processes. The bipartite topology-based threat modeling and interactive adversarial deep networks are adopted for modeling the threat domain. The intention of the proposed framework is to attenuate the access capacity of the attackers exploiting the internal states of the deep network.

11.5.3 Strong Encryption and Strong Authentication

Encryption is the traditional way of hiding the content of data in many mediums. It may also provide a satisfactory level of individual user privacy protection. The MEC platforms interconnect all types of IoT devices. Therefore, special attention should be made to the battery-powered and constrained IoT devices that cannot perform compute-intensive encryption operations, requiring highly efficient security algorithms and protocols. Moreover, encryption may not comply with data sovereignty legislation [471].

Authentication methods, solely based on username/password, should phase out. Everyone, from well-informed individuals to large corporations, still makes classic password errors like low complexity and reusing the same for several services. Password manager apps, randomly generating passwords and storing them in an encrypted file, should become the standard. In addition, two-factor authentication and even three-factor authentication with the integration of biometrics will significantly increase the protection. However, using biometrics also requires special care since once they are stolen, they are lost forever, as they cannot be changed anymore. Moreover, since biometrics uniquely identify a user, it also facilitates user tracking. Consequently, an additional layer of protection should be built around the biometric data. In the context of privacy, especially within the MEC ecosystem, multi(n_1)-factor multi(n_2)-modal authentication schemes are the way forward [28]. The extent of n_1 and n_2, however, depends entirely on the resource availability of the devices enabled in the service. There are various authentication schemes proposed in current literature focusing on privacy [472–475].

11.5.4 Privacy by Design

Privacy by design is an approach in system engineering that promotes the integration of privacy throughout the whole design process [476]. It was defined in 1995 and recognized in 2010 by regulators from around the world as a fundamental component of fundamental privacy protection. The concept is based on seven foundational principles as indicated in Figure 11.1: proactive and not reactive, privacy as the default setting, privacy embedded into the design, full functionality, end-to-end security, visibility and transparency, and respect

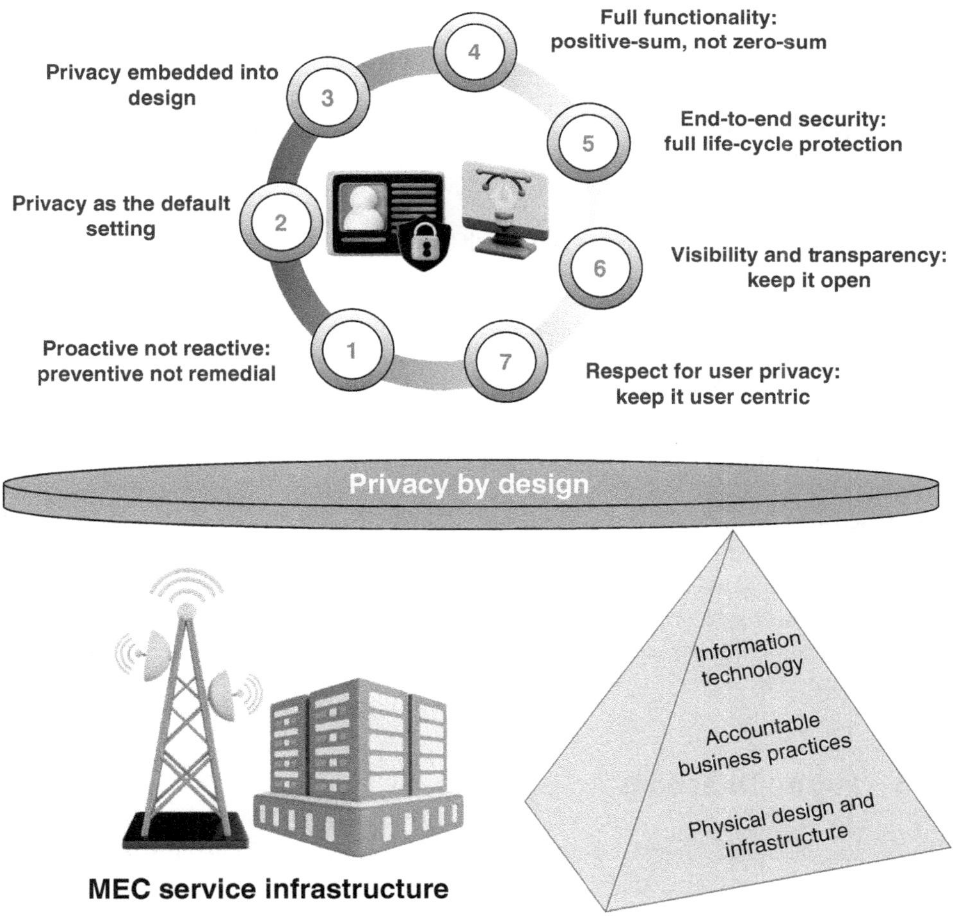

Figure 11.1 Privacy by Design.

for user privacy [477]. Consequently, the main idea of the privacy-by-design approach is that it prevents privacy risks from occurring. It is essential to mention that it does not offer solutions once privacy infractions have materialized [478].

Some guidelines to apply this methodology for the development of web applications are proposed by the OWASP Top 10 Privacy Risk Project.[1] Unfortunately, no concrete details are given on how to establish this approach in the design architecture of IoT. To establish a privacy-by-design approach, the following steps should be realized.

- Definition of a general framework for addressing the requirements of privacy by design policies.
- Development of highly efficient privacy-enhancing technologies (PETs), able to cope with the scalability and interoperability issues within the framework.
- Implementation and evaluation of the solutions.

Existing privacy protection schemes require complex key agreements using highly demanding cryptographic operations like elliptic curve pairing. Also, solutions based on the existence of a trusted third party would cause too much delay in a MEC setting that require 5G level guarantees due to the multiple required inquiries to this remote third party. A solution can be based on secret sharing principles, where the SDN controller is able to choose multiple paths in the network to transmit different parts of the data stream. Only the receiver, who is in possession of the secret shared key, is able to collect and reorganize all these pieces of information.

11.5.5 Software Defined Privacy Approach

"Software defined privacy (SDP)" concept allows easy orchestration of existing tools to enforce privacy requirements of an infrastructure as a service (IaaS) cloud customer [479]. This concept can be further extended to provide privacy protection for MEC deployments. The SDP approach also has the same three-layered model (application, control, and infrastructure layers) as SDN.

The users define the privacy protection level at the application layer by selecting the required privacy policies. Based on the user's selection, "Privacy Officers" define the set of rules which are transmitted to the control layer. This SDP methodology uses an agent-based approach at the control layer. These agents change the underlying infrastructure layer components such as hypervisors, virtual machines, storage systems, switches, and other network components based on application layer privacy rules [479]. This is well-suited for the MEC virtualization infrastructure, as it enables the manipulation of virtualized controlling assets prioritized by privacy.

11.6 Future Directions

This section discusses several future directives that are currently being researched for enhancing privacy measures that can be adapted for MEC use cases and applications.

1 https://www.owasp.org/index.php/OWASP_Top_10_Privacy_Risks_Project.

11.6.1 Hybrid Cloud Approach

A hybrid cloud approach allows the mobile operators to store critical data on the in-house cloud and to process it locally while less sensitive data is stored and processed on the public cloud [480]. In such an approach, operators can have full control of their data and can decide what to share to a public cloud [481]. Some use cases, such as enterprise IoT and health, motivate the use of hybrid cloud architectures consisting of private and public clouds. This concept provides a unique opportunity for leveraging the MEC proximate infrastructure as the private cloud and the distant cloud infrastructures as the public cloud [482]. Moreover, this adaption allows for unburdening the MEC infrastructure of any inessential public data stored in its vicinity, contributing to the efficiency of the overall performance.

11.6.2 Privacy Protection as a Service

Beyond 4G networks have drastically increased the number of new Internet services which are supported over the telecommunication networks: As a result, mobile network operators now compete with a wide range of service providers such as over-the-top (OTT) players, cloud operators, and Internet service providers (ISPs). Thus, mobile operators are now also seeking to provide services for vertical industries such as health care, energy, transport, and other critical infrastructures.

Legacy telecommunication systems have a good reputation for protecting user privacy. Moreover, mobile users are usually trusting the security strength of mobile operators more than the Internet and other IT service providers. Initially, telecom operators are intended to be the MEC service providers since their mobile infrastructures are leveraged for launching the MEC environments. Later on, other service providers can sub-let the infrastructure and become service providers of various flavors. However, in the current context, legacy mobile communications networks are responsible for the privacy protection of users for network access only. MEC service providers are responsible for privacy administering between the access network toward the core network.

The MEC infrastructure provides the ultimate infrastructure required for provisioning privacy protection as a service (PPaaS) due to its attributes of reduced latency with proximate localization, data minimization that can facilitate anonymization/aggregation/encryption, user customizability, context, awareness, local compliance of privacy regulations, and scalability. Though, implementing PPaaS is quite challenging since the launching of this service cannot be done separately from the underlying services. A harmonious nature should be exhibited with the prevailing services to launch such a service directive successfully.

11.6.3 Privacy Preserving Data Aggregation

Privacy-preserving data aggregation (PPDA) refers to a technique that allows sensitive data to be collected and processed while protecting the privacy of individual contributors [483]. It enables organizations and researchers to perform aggregate analysis on large datasets without compromising the confidentiality of personal information. In contrast to traditional data aggregation processes where individual data points are collected and combined to derive meaningful insights or statistics, PPDA employs various privacy-enhancing techniques. One common approach is the use of cryptographic protocols such as secure multi-party computation (MPC) or homomorphic encryption. These techniques allow data to be encrypted and computations to be performed on the encrypted data without revealing the underlying information.

By encrypting the data before aggregation, individual contributors can be assured that their sensitive information remains protected throughout the process. The aggregated results are obtained without compromising privacy, as the decryption keys are only available to authorized parties who have a legitimate need to access the data. Anonymization, which is also employed in PPDA, either removes or randomizes the personally identifiable information from the dataset before aggregation. With this technique, even if the data is received by an intruder, they cannot be traced back to the data owner. PPDA can be embedded into the protocols of the MEC system and therefore assure privacy preservation guarantees to its consumers [484].

11.6.4 Novel Privacy Preserving Techniques

Several novel privacy-preserving techniques have emerged in recent years. Some of the key ones include:

1) **Differential Privacy:** Differential privacy provides a mathematical framework to quantify and control the privacy risks associated with sharing data. It aims to provide strong privacy guarantees by adding carefully calibrated noise to the data or query responses. However, differential privacy application presents a clear trade-off between privacy and data accuracy, which could be vital for machine learning or AI-intensive tasks.
2) **Homomorphic Encryption:** Homomorphic encryption allows computation on encrypted data without decrypting it. This technique ensures that sensitive data remains encrypted during processing, providing privacy guarantees in scenarios where data needs to be outsourced for computation. MEC provides a suitable platform to carry out such holomorphic operations on user data, where user trust is attainable considering its proximate nature. However, there are limitations to the extent of mathematical operations that current holomorphic functions can utilize.
3) **Secure Multi-party Computation (MPC):** MPC allows multiple parties to jointly compute a function on their private inputs without revealing their individual inputs. This technique allows collaborative analysis of private data while preserving privacy. The MEC platform can act as the operational infrastructure for the MPC operations within the local domain.
4) **Federated Learning:** Federated learning enables training machine learning models on decentralized data without the need to share raw data. Instead, models are trained locally on user devices, and only aggregated updates are shared, ensuring the privacy of individual data. In fact, MEC intermediary nodes that are launched as aggregators in IoT applications can act as FL nodes for aggregating local data, while MEC mobile platform can act as the central aggregator for that considered region in the decentralized hierarchy. This will ensure the user privacy and trust for the ML-based MEC services in the future.

5) **Zero-Knowledge Proofs:** Zero-knowledge proofs (ZKPs) allow one party to prove to another party that a statement is true without revealing any other information. This technique can be used to verify the correctness of computations while keeping the underlying data private. ZKPs can be employed in MEC constructs or protocols for privacy-preserving validations, credential verifications, and means of accountability.
6) **Secure Function Evaluation:** Secure function evaluation protocols allow different parties to compute a function jointly while keeping their inputs private. This technique ensures privacy-preserving computation in scenarios where multiple parties need to collaborate.

Besides the examples of the novel privacy-preserving techniques that have been developed, there are ongoing research and advancements in this field, aiming to find more effective and efficient ways to protect privacy in various domains.

12

Telecom and Cloud Integration*

12.1 Introduction

The evolution of mobile networks, by and large, has met the expectations in terms of evolving network services driven by the ever-increasing market demands. Hitherto, we have fully witnessed the first four generations of mobile networks from the first generation (1G) to the fourth generation (4G). The fifth generation (5G) is in the early deployment phase, while, the sixth generation (6G) is in the embryonic stage. Interestingly, the role of software has also evolved with generations of mobile networks; from negligible in 1G to remarkable in 5G. Softwarization is a paradigm where the functionalities, appliances, and services are implemented and managed by software, instead of the traditionally hardware-based solution [407, 485, 486]. Let us have a quick look at the generations of mobile networks and how softwarization gradually advanced with them.

The late 70s marks the advent of 1G of mobile networks with the sole intention of enabling analog voice communications for mobile users. The use of software in 1G was very limited, and it was confined to low-level embedded software [487]. In the early 1990s, the second generation (2G) mobile networks, for the first time, introduced digital systems that offered digital voice as well as text messages. The intermediate generations like 2.5G and 2.75G provided incremental enhancement in data capabilities. In this generation, software was mainly used for control operations including billing [487]. The credit for introducing mobile data to offer fixed wireless and mobile internet services goes to the third generation (3G). This generation arrived roughly in the early 2000s. The intermediate generations often denoted as 3.5G and 3.75G provided improved mobile data services primarily in terms of data rate. 3G made better use of software but it was the fourth generation (4G) that opened doors to programmable networks and open standards. 4G mobile networks surfaced in the late 2000s and in a true sense ushered in the era of fast mobile internet with a plethora of broadband mobile services. Finally, the mobile telecom industries are all set to roll out the fifth generation (5G) of mobile networks. Unlike 4G, 5G is driven by technologies that enable overall softwarization including core networks, edge networks, and transport networks [485].

*This chapter is partly contributed by Anshuman Kalla (CGPIT, Uka Tarsadia University (UTU), Gujarat, India).

Mobile Edge Computing and Communications: Driving Forces, Technology Foundation, and Application Areas, First Edition. Aaron Yi Ding, Chamitha De Alwis, and Madhusanka Liyanage.
© 2025 John Wiley & Sons Ltd. Published 2025 by John Wiley & Sons Ltd.

5G signifies yet another leap in the evolution of mobile networks. What is most fascinating about 5G is that the technology is geared up to provide a kind of never-seen-before connectivity not only to traditional handsets (like cell phones, smartphones, PDAs and mobile broadband modems/adaptors) for enabling human-centric communications, but also to any machine or device for enabling machine-centric communications. 5G is characterized by three broad scenarios which are enhanced mobile broadband (eMBB), ultra-reliable low latency communications (URLLC) and massive machine-type communications (mMTC). Thus, 5G has the potential to connect billions of heterogeneous devices/machines, and cater to a wide range of service requirements for end-users as well as business customers like IoT tenants, Over The Top (OTT) providers, mobile virtual network operators (MVNO), and cloud service providers (CSP). The traditional business model, used by the predecessor generations of 5G, considered only the mobile subscribers as the source of revenue. However, 5G goes beyond this model by taking into account both the end-users and the business customers as sources of revenue [487]. Undoubtedly, 5G as the underlying technology is about to revolutionize the way we communicate and interact with each other and with machines.

From the architecture point-of-view, 5G networks have emerged as softwarized and virtualized networks that allow the existence of generic hardware and commodity servers that can be programmed on-the-fly based on the needs. This allows efficient monitoring and management, provide resiliency and self-reconfigurability, facilitates agile creation of vertical services, offers a flexible level of services and many more advantages.

5G lays the foundation for a massive and ubiquitously networked society of everyone and everything.

12.2 Network Softwarization in 5G Networks

In order to empower 5G to deal with voluminous traffic demands and heterogeneous use cases, numerous key enabling technologies are integrated in a profitable manner. Some of these novel technologies are multi-access edge computing (MEC), software defined networking (SDN), network function virtualization (NFV), network slicing (NS) and cloud computing. Apparently, together these technologies aim to transform mobile networks to softwarized, virtualized, cloudified and edgified networks. Softwarization turns out to be a new paradigm shift in the realm of cellular networks, and it promotes the programmability of networks, the agility of network services and the reusability of expensive infrastructure. The moment networks become programable then, for example, the job of introducing (or improving) any specific network service becomes as easy as developing a new software (or upgrading an existing one) without incurring any change in available hardware.

The rationale of SDN is to decouple the control plane and data plane in a way that controllability and intelligence are logically centralized at the network control (plane) elements (called SDN controllers), whereas, processing and forwarding of packets are confined to network data (plane) elements (called as SDN switches). NFV ushers in the virtualization of network functionalities within the premises of 5G. It solves the problem of proprietary hardware which can only perform hardwired tasks, by substituting them with virtualized, dynamic, and softwarized instances of functionalities that are more responsive, cost-effective, and can be regulated [488]. Cloud computing and MEC will provide

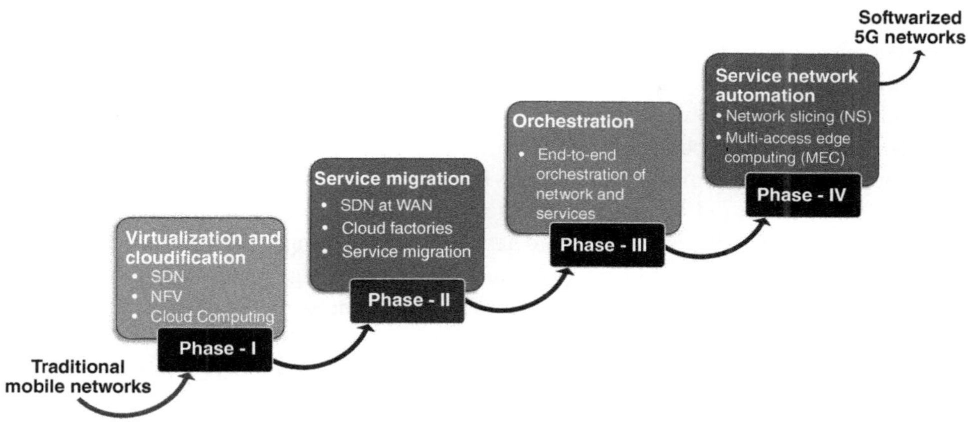

Figure 12.1 Four Phases Toward Softwarization of Mobile Networks. Source: Khan et al. [379]/IEEE.

the on-demand scalability for the network. NS is a way to customize end-to-end services offered by mobile networks, keeping the underlying physical infrastructure unaltered, with the aim to satisfy the differential and stringent requirements of diversified customers of 5G networks.

Figure 12.1, redrawn from [379], illustrates the four phases of network softwarization which make the path for 5G. It illustrates how above technologies will enable the deployment of a softwarized 5G network, spanning from inflexible fixed-mobile architecture to more and more dynamic software-based architecture. These architectural changes in 5G are expected to fuel the digital transformation that all the industry is witnessing. This will also result in generating new service models and new value chains which will lead to significant socio-economic impact.

12.3 MEC Integration in 5G Backhaul

This section explains the integration of MEC in 5G networks while giving answers to questions like; (i) what is MEC?, (ii) what are the benefits of integrating MEC in 5G networks, and (iii) what are the challenges related in MEC in 5G?

12.3.1 Why MEC Needed in 5G

MEC is considered an evolution of cloud computing since it allows applications, previously being served exclusively by the centralized data centers, to be served by the edge of the mobile network [15]. The fundamental idea behind MEC is to extend the capabilities of the centralized cloud to the edge, i.e., within the premises of mobile radio access network (RAN). This results in the establishment of a service platform closer to the mobile (end) users. Realization of MEC is made possible, first, by strengthening the RAN with a computing and storing facility, and second, by leveraging IT virtualization technologies to optimally utilize this facility at RAN (aka MEC server). Thus, by virtue of MEC, the required applications can be dynamically migrated, instantiated, and run at the network's edge. With the emergence of MEC, 5G networks can be visualized to have a two-tier infrastructure model; the first level of infrastructure is provided by MEC at the edge and the second is provided by the cloud at the core. The use of MEC in 5G offers a multitude of benefits including ultra-low latency, high bandwidth, proximity, real-time access, location awareness, context awareness and network cognizance [25].

European Telecommunications Standards Institute (ETSI) has been working toward standardization of MEC and has been continuously putting concentrated efforts toward the integration of MEC with other technologies which are fueling the growth of 5G. As per the initial proposition of ETSI [25], MEC stood for mobile edge computing. The aim was to standardize edge computing in the realm of 5G mobile networks after thoroughly studying its benefits and use case scenarios. Later in the year 2017, ETSI renamed its Mobile Edge Computing Industry Specification Group (MEC ISG) to multi-access edge computing with the intent to expand the scope of the edge computing paradigm, beyond the premises of typical mobile access environments, for more generic environments (like fixed access and WiFi) [489].

Unlike 4G, where MEC was deployed as an incremental patch, it has been considered as key technology behind the evolution of 5G mobile networks [15]. In particular, MEC turns out to be a key enabler for URLLC (aka mission-critical communications) services since it is characterized by close proximity, ultra-low latency, and real-time and high-speed data connectivity. For instance, sectors like automated industrial manufacturing, autonomous vehicles, and critical e-health systems, can be immensely benefited since real-time high-speed data collection, processing, and decision-making can be performed almost locally. This results in a significant reduction in the traffic that traverses the backhaul network and core network to reach cloud servers.

MEC also paves the way for mMTC scenarios since all sorts of log maintenance, data aggregation, analytics, and decision-making can be carried out right away at the closest set of MEC servers located at the edge of the network. This kind of paradigm has immense potential to improve the efficiency and performance of IoT networks as well as increase the battery life of massive IoT devices [9, 490].

With the use of MEC, a variety of new services that can be realized are personalized services as well as localized and content-aware services. *Personalized services* are based on the real-time awareness and edge analytics computed at MEC servers. *Location-based and context-aware* services are offered based on the user's location, network's conditions and contextual information. The first two pieces of information are easily captured by MEC nodes by leveraging the readily available radio signals [234] and contextual information is collected using standardized APIs [15]. In particular, localized services can be selectively offered based on the real-time location of users, for example, suggesting services from nearby hospitals, traffic management, augmented navigation, security services, etc. This kind of service potentially improves QoE several folds for end users [490].

Another advantage of integrating MEC in 5G networks comes in the form of new business opportunities for mobile network operators (MNOs). They can open their (MEC augmented) RAN edge networks to third parties for hosting innovative applications and services [389]. Third parties as a new set of customers would increase revenues for MNOs. Moreover, the MEC platform can be considered as an immediate solution to conduct the pilot testing of innovative new services in virtual retail space to verify the correct functioning, test the return on investment (RoI), and get the feel of market response [491].

In a nutshell, it can be said that MEC ushers a new paradigm shift in the way delay-sensitive and lightweight processing (or pre-processing) is carried out in the 5G networks. Numerous new players (like third-party application/service providers) and innovative business models will get attracted with a progressive adaptation of MEC.

12.3.2 Challenges Related to MEC in 5G Backhaul Networks

General challenges associated with edge computing, as listed in [21], are (i) infrastructure management and interoperability among various edge-infrastructure providers [492], (ii) life cycle management of virtual machines (VM), (iii) resource optimization leveraging scheduling and offloading to the cloud server, (iv) mobility handling, etc. More specific challenges pertaining to MEC 5G networks are presented in [234] which include network integration as well as security and privacy. Let us briefly discuss some of these challenges, next.

To harness the full potential of MEC in 5G networks, dynamic and optimal allocation of edge resources to MEC applications/services is required. However, this requirement is challenging since MEC is characterized by finite infrastructure as well as distributed deployment [492]. Use of a centralized entity for optimization may result in increased backhaul overhead (required for reporting network status), computational complexity, and increased latency [234].

Next, since 5G is going to be deployed as denser and smaller cells, even small-scale mobility will need handover. Handling such frequent mobility is going to challenging in MEC 5G networks. For instance, while MEC is processing a computational-related offload request from a user, s/he may move different cells [234]. A more difficult scenario would be MEC-enabled IoT systems where an IoT device may traverse multiple small cells during the offload [9].

The security and privacy of MEC 5G networks is yet another challenge to be addressed. On the one hand, keeping the communication between mobile users and MEC servers secure and privacy protected would require strong encryption techniques. On the other hand, implementing such encryption techniques might not be feasible under all scenarios. For instance, resource-constrained IoT devices would find it difficult to run usual cryptographic algorithms. This calls for dedicated efforts to design a dedicated communication mechanism that uses lightweight cryptographic techniques.

12.4 Key Technologies for MEC 5G Integration

MEC has immense potential to realize the 5G vision (IMT-2020) as defined by ITU-R in the year 2015 [493]. This section discusses the key technologies that help to expedite the smooth integration of MEC in 5G and to ensure the robustness of the overall architecture. The considered technologies are NFV, SDN, ICN, and NS. The way each technology is discussed is as follows; first, the fundamental concepts and ideas behind the technology are presented, second, the mutual benefits of integrating the technology with MEC are discussed, and third, the challenges that need to be addressed are pinpointed.

12.4.1 Network Function Virtualization

NFV is a networking concept that opposes the use of closed and specific-purpose-built network equipment comprising of a rigidly embedded set of functionalities that cannot be altered or changed for its entire lifetime. Use of such network equipment leads to a myriad of issues like tedious and time-consuming deployment, technical complexities in integrating them with existing network infrastructure, ossification since such functionalities cannot be moved once deployed at a particular location, and increase in CAPEX and OPEX [494]. The increase in the CAPEX happens when new network functionalities are required to meet the ever-increasing demands, whereas, the increase in OPEX is ascribed to continuous monitoring, fault diagnosis, and rectification, configuration updates, etc. [494]. NFV offers an intriguing solution to these issues.

The principal idea of NFV is to decouple network functions from the network equipment such that the former can be developed as a software and deployed on the general-purpose commercial-off-the-self (COTS) network equipment. This results in the independent existence of network functions in the form of software and generic processing in the form of hardware. Such softwarized network functions are named virtual network functions (VNFs) based on the fact that they do not exist physically as a part of any network equipment, and they can be instantiated on the fly when required over existing generic network hardware. A running instance of VNFs help in the realization of virtual appliances like Dynamic Host Configuration Protocol (DHCP) server, firewall, intrusion detection system (IDS), network address translator (NAT), deep packet inspection (DPI), packet data network gateway (PGW), and virtual private network (VPN) [359]. Ordered chaining of multiple VNFs represents a unique service and the concept is referred as service function chain (SFC) [486]. NFV is realized by leveraging software virtualization technologies that come from the IT world.

Many organizations (like ETSI, IRTF, IETF, and ATIS) have been working toward the standardization of NFV [495]. The most popular and adopted NFV architecture is the one proposed by ETSI [496]. As per the reference architecture proposed by ETSI [497] there are various functional blocks and some main references between these blocks. Interestingly, these functional blocks can be broadly divided into two different parts: network service and management and orchestration (MANO) [498]. The network service part includes NFV Infrastructure (NFVI), VNFs plus element managers (EMs), and operation and business support systems (OSS/BSS), whereas, management and orchestration comprises of NFV orchestrator (NFVO), VNF managers (VNFMs), and virtualized infrastructure managers (VIMs).

VNFs run over virtualized resources (computing, storage and network) that are offered by NFVI. It is the responsibility of NFVI to virtualize heterogeneous physical resources using a virtualization layer and create a pool of virtualized resources [496]. VIMs manage NFVI's resources and govern the interaction between VNFs and resources. VNFM is in charge of lifecycle management of VNF from instantiation to removal as well as update. NFVO is responsible for resource orchestration in coordination with VIMs and network service orchestration in coordination with VNFMs.

Most of the advantages of NFV come from the fact that it makes networks programmable. Some noteworthy advantages of NFV are:

- The development and advancement in network functions can happen independently at their own pace, without bothering about the hardware on which they will run,
- NFV allows real-time agility in network service provisioning as a chain of VNFs can be migrated (usually from the cloud) and instantiated on the fly at any location in the network,
- It enables resource optimization by dynamically scaling up or down the allocated resources to run VNFs,
- It expedites the process of creation and deployment of new services [359],
- Reduces the cost of establishing or upgradation network infrastructure,
- It creates new business opportunities for example third-party VNF vendors who can develop and offer VNFs to mobile network operators (MNOs) [499].

12.4.1.1 Mutual Benefits of NFV MEC Integration

Principally, the plane of action of the two technologies are different; MEC is deployed at the edge of the 5G networks (i.e., within or close to the access network), whereas, NFV is deployed for 5G core network. Initially, different frameworks were proposed by ETSI for MEC and NFV. However, virtualization lies at the root of both technologies. Moreover, there are many similarities between the MEC architecture and NFV architecture. For instance, both architectures have a virtualization infrastructure manager, (VNF/MEC) manager, and (NFV/MEC) orchestrator as functional blocks. Further, the underlying infrastructure that hosts MEC applications and VNFs is fairly similar [25]. These similarities pave the way for the integration of the two technologies. In this direction, ETSI has proposed to implement MEC as part of the NFV environment where MEC applications are considered just like VNFs which can be migrated from the cloud to the edge of the network and instantiated there. Figure 12.2 shows the MEC in the NFV environment as proposed by ETSI in [500]. There are numerous benefits that can be harnessed with the integration of MEC and NFV.

- First and the foremost advantage is reusing NFV's infrastructure and its management functionalities for hosting both network functions and MEC applications as simply VNFs [25]. In other words, when MEC is deployed in an NFV environment, MEC applications are treated as VNFs and NFVO acts as a proxy to handle much of the work of mobile edge orchestration [351, 500]. This promises a reduction in cost and a higher return on Investment for 5G mobile network operators.
- Second, using the MEC philosophy for NFV makes the deployment of VNFs possible much closer to the end-users. That equally holds for the entire service which is merely an ordered chaining of VNFs. This results in a reduction in the latency of the network's functioning.
- Third, implementing MEC in the NFV environment improves the scalability of MEC applications. NFV environment provides dynamic allocation of network resources based on the real-time need of the MEC applications.

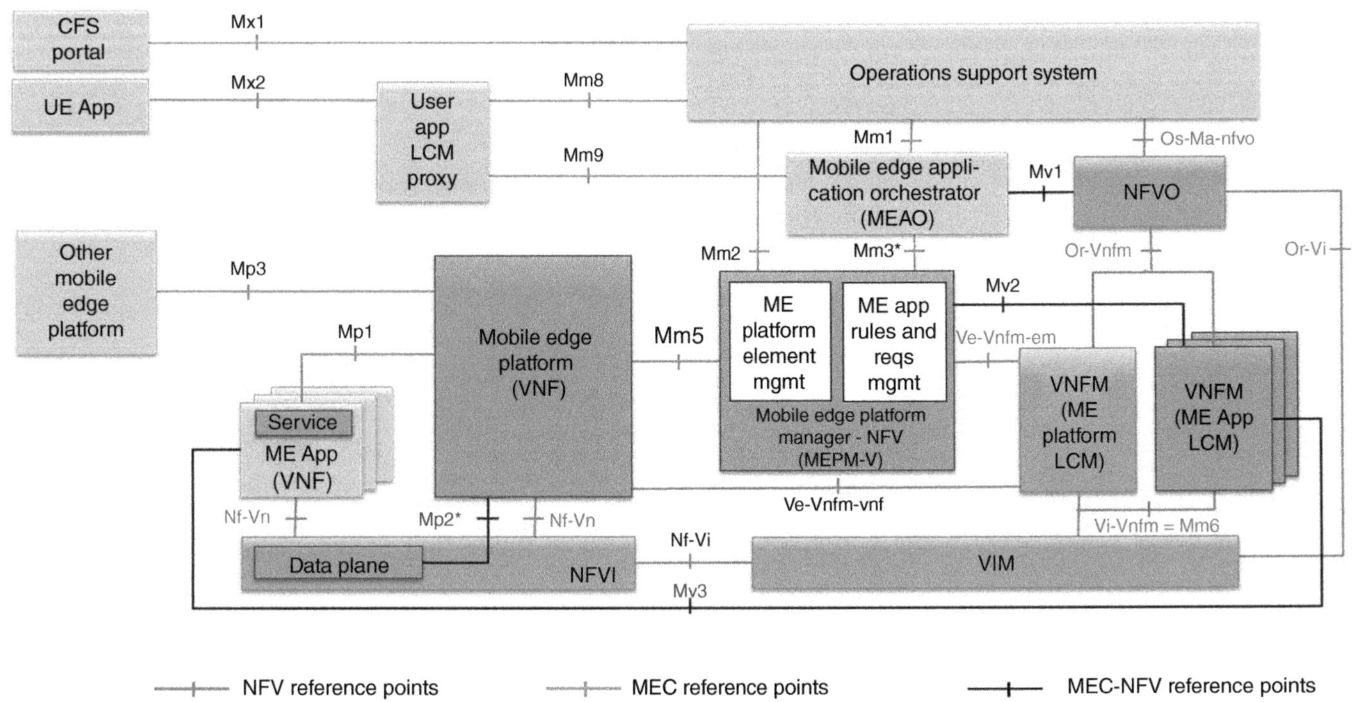

Figure 12.2 MEC in NFV Architecture. Source: ETSI [500]/with permission of ETSI.

Thus blending MEC and NFV offers mutual benefits and enhances the capability of 5G mobile networks.

12.4.1.2 Challenges and Obstacles of NFV MEC Integration

The main research challenge is the absence of full-fledged industry standards for NFV-MEC integration. The ETSI report [500] presents a reference architecture to integrate MEC in the NFV environment. The report also lists various issues and recommends solutions to mitigate them. To harness the maximum benefits of MEC NFV integration, there is a need to finalize the industry standard with collaborative efforts and clearly define architectural components [351]. An initial comparison carried out by ETSI (in [500]) highlights the similarities as well as the differences between VNF Descriptor (VNFD) used in NFV to instantiate a VNF, and application descriptor (AppD) used in MEC to instantiate a MEC application. AppD lays more emphasis on the requirements in terms of infrastructure and service availability than VNFD. Moreover, the attributes of AppD define templates for a single virtual resource required for a MEC application, whereas, attributes of VNFD can define templates for multiple virtual resources for every VNF. Hence, thorough mapping (much more than simple field matching) between AppD and VNFD needs to be in place for smooth integration [496]. Such kind of clear specifications will allow wide adaption of the two technologies.

To gain maximum benefit from MEC in 5G networks, the resources at the edge must be optimally utilized by efficiently deploying MEC services [351]. However, the same is difficult to achieve when MEC has implemented the NFV environment. This is because of the fact that both the MEC platform and MEC applications are deployed as VNFs which makes the placement and overall life-cycle management of MEC applications challenging [496].

Another important set of challenges comes from the security perspective. Deploying MEC in an NFV environment extends the security vulnerabilities from NFV to the MEC domain. For instance, attacks like VNF manipulation, VNF location shift, DDoS, etc. on a targeted VNF can hamper other VNFs running on the same virtual machine [351]. The situation is aggravated by the resource pooling feature of the NFV environment.

12.4.2 Software Defined Networking

The traditional network equipment comes with tightly bundled proprietary software that prohibits customization as per the needs of network administrators. Moreover, such network equipment has established the coexistence of a control plane and data plane as the most popular way to design them. Nevertheless, with a year of experience, the network community realized that such closed and vendor-specific network equipment has restricted the design flexibility, rapid deployability, service agility, and network manageability of the networks.

In this direction, SDN has emerged as an innovative networking concept. SDN profitably decouples the control plane from the data plane by leveraging network-level programmability with more emphasis on the programmable control plane. Conceptually, SDN comprises SDN controllers and SDN switches. The former manages control plane operations, and the latter performs data plane operations. SDN controller is implemented as software that runs on commodity servers and maintains a consistent global view of the network. The job of the SDN controller is to determine the rules (i.e., actions to be taken) for a particular flow of data packets and communicate those to the SDN switches. This communication happens over a secure channel using a standard interface called as southbound interface. On the other hand, the job of the SDN switch is to process and forward the data packets based on the rules provided by the SDN controller. To do so, SDN switches maintain flow tables which are populated and updated based on the instructions received from the SDN controller.

From the other end, services at the application layer communicate with the SDN controller using the northbound interface. Thus, SDN allows the introduction of newer services at the pace of programmability.

12.4.2.1 Mutual Benefits of SDN MEC Integration

Leveraging SDN for 5G networks would imply shifting network control functionalities to logically centralized SDN controllers. However, to make this approach effective, the SDN controller should be placed as close as possible to the data plane to support delay-sensitive use cases and meet up the requirement of 1 ms user plane latency. To achieve this, MEC turns out to be a promising solution and can host SDN controllers. Hence, from the perspective of SDN incorporation in 5G, the role of MEC is highly beneficial.

Now, let us look at the important role that SDN can play in the realization of MEC in 5G. From the above discussion, we know that SDN empowers network operators with the capability to orchestrate their networks and reconfigure their resources to provide a multitude of network services to users transparently. Consequently, the SDN can potentially be used for the smooth implementation of the multi-tier MEC paradigm. With the presence of intelligence coupled with the global view of the network at the SDN controllers, SDN can optimally route the traffic flowing between MEC servers and the cloud servers. This helps in shifting efficiently the required functionalities, as and when required, from the cloud to MEC servers. Moreover, the use of SDN allows the existence of light MEC servers since it takes care of the operations like network orchestration, service discovery, traffic flow management, flow-based anomaly detection, deep packet inspection, etc. [501]. The possible use of SDN for MEC deployment is shown in Figure 12.3.

12.4.2.2 Challenges for SDN MEC Integration

Though the coexistence of SDN and MEC ushers in the obvious benefits, nevertheless, there are some research challenges that need due attention to harness the full potential of the two technologies.

As the resources available at the MEC servers are limited compared to the cloud servers thus the level of security offered by MEC is proportionately lower [356]. Further, with the use of SDN for MEC systems, security concerns are even aggravated. This is because of two reasons. First, SDN brings along with it not just its benefits but also its integral security issues. Some of the concerns with SDN are DoS attacks because of logical centralization, flow poisoning, hijacking of SDN controller, and overall malicious programming of network functions [358]. Second, the intrinsic security issues of MEC like DoS, rouge gateways, privacy leakage, service manipulation, etc. [21, 501], get exacerbated with the use of SDN. The reason is SDN employs software-based programmable backhaul devices in place of legacy closed black-box devices.

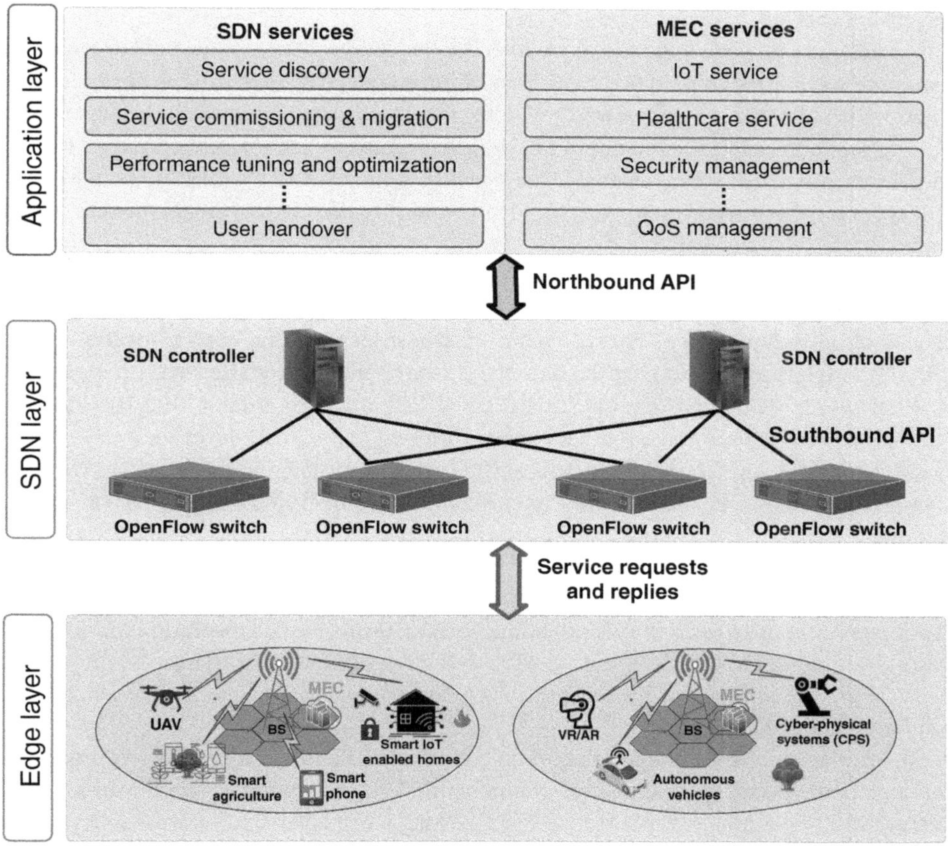

Figure 12.3 The Use of SDN for MEC. Source: Liyanage et al. [351]/with permission of arXiv.

12.4.3 Information Centric Networking

Over the last few decades, numerous issues like mobility, security, availability, named hosts (tight coupling of content with its location), TCP connection overhead, and many more have raised the difficulty of TCP/IP-based networks [502]. Hitherto, these issues have been handled by supplementary solutions that have transformed the TCP/IP into a frail network architecture. In this context, the networking community has lately proposed ICN as a new clean slate networking paradigm to mitigate the issues in a comprehensive manner.

ICN aims to retire the host-centric model that lies at the heart of prevailing TCP/IP architecture and replaces it with the content-centric model. The salient feature of ICN are:

Named Content

CCN advocates named content in such a way that a name identifies the associated content and not the location of the host that serves the content. Thus ICN completely decouples the name from its location and establishes the name as an identifier. *Name* in place of *IP address* introduces intelligence within the network which is in opposition to the end-to-end principle. These names can be made content and/or context-dependent.

In-Network Caching

ICN establishes network-wide caching which implies every node is empowered to cache the content flowing through it. In other words, caching is conceived as one of the network layer functions besides routing and forwarding. This proposition becomes possible because of the dipping cost of memory. Three different types of in-network caching are on-path caching, off-path caching, and hybrid caching. The benefits of in-network caching are reduced cost of data retrieval, reduced latency, efficient load handling, effective retransmission, and high availability.

Data-Level Security

Instead of putting enormous effort to secure the host location and communication channel, ICN emphasizes on making the data itself secure and binds names with the content. In other words, ICN envisions the existence of self-verifiable data so that the location from where the data is retrieved becomes indifferent. Thus, data-level security enables location-independent retrieval of content. Further, cryptographically signed routing (control) messages will make routing secure, and cases like prefix hijacking will not be possible.

Efficient Routing

Three types of routing used in ICN are name-based routing, named resolution routing, and hybrid routing. In general, this routing allows efficient utilization of resources

Pull-Based Communications

There is a one-to-one relationship between content and its request, i.e., one request pulls only one corresponding content from the network. The pull-based communication helps to exercise hop-by-hop flow control at layer 3. Note, since layer 3 is a universal layer this means that flow control is done at all the points between sender and receiver. This is in contrast with TCP/IP-based networks where end-to-end flow control as performed.

Other interesting features of ICN are intrinsic mobility support, multipath routing, multicasting, and multi-homing. Some of the popular ICN projects are Named Data Networking (NDN) [503], Scalable and Adaptive Internet Solutions (SAIL) [504], PSIRP [505], PURSUIT [506], COMNET [507], etc. With this overview, it is easily conceivable that ICN can well nurture the applications that are characterized by bursty and heavy traffic and demands low latency.

12.4.3.1 Mutual Benefits of ICN MEC Integration

MEC and ICN are mutually complementary technologies, thus, in this section we intend to explore and understand their reciprocal benefits.

The underpinning of ICN for MEC manifests captivating benefits for a wide range of use cases that are promised by 5G. The named-based rationale of ICN allows the creation of named services which leads to service-centric ICN architectures [508, 509]. Such service-aware ICN offers advantages like service-level multicast, dynamic service multipath transmission and dynamic rollout & management of services in addition to above mentioned salient features of ICN. Thus, a service-centric extension of ICN better supports the MEC-enabled URLLC scenario.

ICN-coupled MEC network has the ability to sustain massive traffic both real-time and non-real-time [510]. Such a network, being conscious of content and its requests, can very well aggregate content requests at the edge of the network. At any given, one request per unique content time is pushed upstream via the core network. Further, in-network caching facilitates caching of retrieved contents (at all the nodes or the strategic nodes) and helps efficient dissemination of data. For instance, ICN MEC combination can drastically improve the dissemination of live video content without hogging the core network and congesting the communication channels between the core and access network [490].

The use of name-based ICN cooperative forwarding strategies permits the exploration of nearby MEC nodes. When a request for named content arrives at an ingress MEC node, then availability of that content can be cooperatively checked in the neighboring MEC nodes before forwarding the request upstream via core network. This is potentially enabled with the use of Mp3 point between mobile edge platforms, which was originally established by ETSI to exchange control information [511]. The use of such a cooperative approach can reduce latency, off-loads the core network, and reduce the cost of retrieval, if any.

12.4.3.2 Challenges and Obstacles of ICN MEC Integration

To harness the full benefits of the complementary technologies, collaborative efforts are required to define the MEC-ICN interface and to design easy-to-use secure APIs for communication between systems.

In general, ICN has capabilities to boost the real-time context-aware applications running over the MEC platform. However, simultaneous tracking of a variety of contextual information is still a challenging job.

Though ICN, by design, supports consumer (i.e., receivers') mobility, however, producer mobility is challenging. Producer mobility implies the content's point of the source is mobile. The Producer mobility is typically important for MEC since the services running at MEC servers are migrated (when required) from the cloud, instantiated, and then removed. However, recently numerous works have shown ways to achieve it [512, 513].

Access control is yet another challenge arising with using ICN for IoT applications supported by MEC. In ICN, named-based routing is used for forwarding content requests, and the retrieval of content is allowed from any cache-enabled ICN node. Thus, traditional ways of exercising authorization and access control techniques are infeasible in ICN. For instance, maintaining a network-wide access control list (ACL) per named content at all the dynamic point-of-sources (i.e., ICN routers), which keeps updating with time, will lead to exorbitant overheads [514].

12.4.4 Network Slicing

The types of services offered by the predecessors of 5G (i.e., 4G, 3G, and the intermediate generations) were mostly human-(user)-centric like voice, text, video, etc. Thus, before 5G, the use of a "one-fit-for-all" approach worked fine, where a common network was used to support all the prevalent services. 5G, under three broad scenarios, targets to deliver distinctive use cases, services, and applications, all of which have stringent requirements regarding latency, mobility, bandwidth, reliability, scalability, etc. The conventional way of using one network is not apt to meet these targeted use cases. This has given rise to the concept of network slicing, which is being conceptualized by Next Generation Mobile Networks (NGMN) and is under standardization by Service and System Aspects (SA) Working Group 2 (SA2) under 3GPP (3rd Generation Partnership Project) [334].

Network slicing implies building multiple end-to-end virtual networks over common physical infrastructure that includes a core network (along with its core cloud), transport network, access network, and if required, user equipment as well [515]. Each of these virtual networks offers a unique set of services that are tailored to a specific scenario. Virtualization, lying at the heart of the NS, is leveraged for slicing virtually the physical network such that each slice comprises of a set of virtual resources which can be dynamically scaled (up or down). Hence, NS is a way to flexibly realize heterogeneous networks while keeping the underlying physical infrastructure the same. With the help of NS, network providers can lease their unused resources by employing network-as-a-service (NaaS) or infrastructure-as-a-service (IaaS) models [516]. Moreover, NS expedites the roll-out of network operator services and third-party services [516]. Allowing flexible participation of third parties to control virtual networks is another advantage of network slicing [517].

12.4.4.1 Mutual Benefits of Slicing MEC Integration

The smooth amalgamation of NS and MEC has indeed immense potential to help achieve 5G vision. The end-to-end specialized service environment NS creates gets reinforced with the edification offered by MEC. More specifically, MEC boosts the differentiated services (offered by different network slices) in terms of latency, operational cost (OPEX), scalability, edge-based analytics, and QoE. Thus, MEC NS integration benefits the use cases under mMTC and URLLC scenarios. Further, as per 3GPP, MEC is considered a key driver that, along with NS will help attain full multi-tenancy.

As shown in Figure 12.4, different slices (virtual networks) can exist serving different use cases. The first network slice is dedicated to human-centric communications using mobile

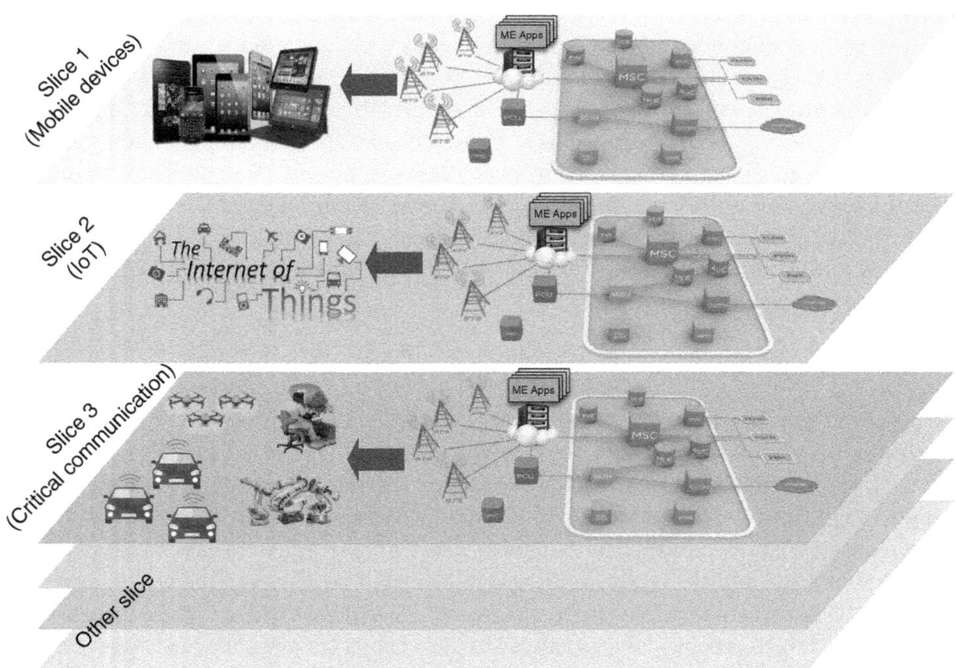

Figure 12.4 Use of Network Slicing and MEC in Different 5G Applications.

handheld devices and comes under the eMBB scenario. The second network slice represents the Industrial Internet of Things (IIoT) use case which comes under mMTC scenario. From one end, MEC can help offload IoT devices by performing caching, data processing and analytics at MEC nodes. From the other end, MEC helps in damping traffic load on the core network as well as load on cloud servers.

Yet another network slice in Figure 12.4 is to handle use cases like autonomous vehicles, time-critical cyber-physical systems (CPSs), UAV/drones and tactile internet. These use case comes under URLLC scenario and essentially needs extremely low latency and traffic prioritization. The former requirement is met using MEC, whereas the later is the job NS.

Recently, 3GPP has put forward three Slice/Service Type (SST) values: 1 for eMBB, 2 for URLLC, and 3 for MIoT [518]. SST specifies features and services anticipated by a network slice. During the device registration process, user equipment (UE) indicates the preferred SST for receiving differentiated services. As per 3GPP, a UE can simultaneously establish connection and avail services from eight network slices [518].

12.4.4.2 Challenges and Obstacles of Slicing MEC Integration

There are clear challenges that are to be addressed in order to make the best use of the two technologies, which have different genesis; ETSI is contributing toward standardization of MEC, whereas 3GPP is putting efforts for standardization of NS. The group recently created by ETSI pin-pointedly aims to analyze the core concepts of NS, identify the prominent use cases put forward by NS, and evaluate the gaps that need to be abridged [518].

Proper working of NS demands the creation of traffic silos per network slice and secure orchestration are essential requirements. This implies unauthorized access and traffic leakage from one slice to another must not happen at any cost. With MEC NS integration, the security and privacy requirements become even more stringent and challenging. For instance, a given MEC service, part of one network slice, should not be availed by another network slice [519].

As discussed above, a network slice encompasses radio access network (RAN); hence instances of virtualized RAN are expected to exist. In this context, clear isolation and efficient management of radio resources become vital. The use of software-defined RAN controllers at MEC nodes can play an alleviating role [351].

Thus, the vertical coordination between the two technologies should be clearly defined by efficient handling of resources, extending or creating new interfaces, secure API-based management, adaptation of MEC Application Orchestrator (MEAO) as per Network Slice Instance (NSI), etc. [351, 518].

12.5 Conclusion

5G marks a step change in the evolution of mobile communications. It comes out from its conventional human-centric communication use case and offers enhanced capabilities to support massive machine-centric communications as well. 5G has radically changed how mobile networks are configured, deployed, managed, and maintained to achieve this vision. More specifically, it has brought network softwarization and virtualization at its core.

MEC being one of the key enabling technologies for 5G, uses virtualized resources and softwarized applications/services. Numerous advantages of MEC in the realm of 5G have been presented. The chapter presents how the MEC can be profitably integrated with other technologies like NFV, SDN, ICN, and NS. Further, it highlights some of the pertinent challenges that need to be addressed for the seamless integration of MEC with each of the discussed technologies.

Part IV

Outlook

The future of MEC^2 consists of opportunities and challenges. To inspire future research and development, we conclude this book by highlighting open challenges and sharing our outlook for future directions.

Mobile Edge Computing and Communications: Driving Forces, Technology Foundation, and Application Areas, First Edition. Aaron Yi Ding, Chamitha De Alwis, and Madhusanka Liyanage.
© 2025 John Wiley & Sons Ltd. Published 2025 by John Wiley & Sons Ltd.

13

Conclusion and Outlook for *MEC*^{2}

Driven by the vision for a hyper-connected world and cloud-edge continuum, combining the virtually limitless computing power of the cloud with distributed devices at the edge can help create a foundation for building innovative services. *MEC*^{2} can be a strong drive and enabler for these services.

On the outlook, inline with the rapid advancement of AI technologies, integrating Foundation Models such as large language models (LLMs) [520, 521] into the edge services and 5G-beyond networks can be an exciting direction for future research. Since LLMs have been explored largely in the cloud due to their high computational demand, running LLMs on edge devices requires a paradigm shift, which generates ample research opportunities for *MEC*^{2}. In terms of hardware, energy-efficient accelerators have already found their ways on edge devices. For example, edge-TPUs from Google are becoming available as sub-modules for embedded devices, which can be utilized for running "optimized" LLMs. Meanwhile, given the constraints of *MEC*^{2} including limited processing power and energy/battery life on edge devices, we need to explore learning and inference techniques such as neural architecture search, quantization and sparsification. The exploration insights can lead to innovations in future AI application and system design.

From system perspective, as edge applications become more prevalent, interoperability and scalability are crucial for the success of *MEC*^{2} in the long run. To tackle the growing heterogeneity, we need to develop standardized protocols and orchestration frameworks that enable seamless integration of diverse edge devices and corresponding services. One important aspect is to explore universal data formats for system interactions. For scalability in the context of computing continuum, we need effective protocols for discovering edge devices and services under dynamic settings (e.g., devices/services become online or offline spontaneously).

The technology foundation and driving forces for *MEC*^{2} are still in the their development stage. Our book offers a "Top-Down" lens to illustrate the ongoing and future research treads, use cases, and key technologies, which shall offer insights for active actors and stakeholders by serving as a reference material and guide on mobile edge computing and communications.

Mobile Edge Computing and Communications: Driving Forces, Technology Foundation, and Application Areas, First Edition. Aaron Yi Ding, Chamitha De Alwis, and Madhusanka Liyanage.
© 2025 John Wiley & Sons Ltd. Published 2025 by John Wiley & Sons Ltd.

References

1 D. D. Clark, C. Partridge, R. T. Braden, B. Davie, S. Floyd, V. Jacobson, D. Katabi, G. Minshall, K. K. Ramakrishnan, T. Roscoe, I. Stoica, J. Wroclawski, and L. Zhang, "Making the world (of communications) a different place," *ACM SIGCOMM Computer Communication Review*, vol. 35, no. 3, pp. 91–96, Jul. 2005.

2 Gartner, "Top 10 Trends in Data and Analytics, 2021," 2021.

3 B. Varghese, E. de Lara, A. Y. Ding, C.-H. Hong, F. Bonomi, S. Dustdar, P. Harvey, P. Hewkin, W. Shi, M. Thiele, and P. Willis, "Revisiting the arguments for edge computing research," *IEEE Internet Computing*, vol. 25, no. 5, pp. 36–42, 2021.

4 L. Lockhart, P. Harvey, P. Imai, P. Willis, and B. Varghese, "Scission: Performance-driven and context-aware cloud-edge distribution of deep neural networks," in *2020 IEEE/ACM 13th International Conference on Utility and Cloud Computing (UCC)*, 2020, pp. 257–268.

5 N. Mohan, L. Corneo, A. Zavodovski, S. Bayhan, W. Wong, and J. Kangasharju, "Pruning edge research with latency shears," in *Proceedings of the 19th ACM Workshop on Hot Topics in Networks*, ser. HotNets '20. New York, NY, USA: Association for Computing Machinery, 2020, pp. 182–189. [Online]. Available: https://doi.org/10.1145/3422604.3425943.

6 CRN, "AWS CEO Admits Some Workloads Will 'Never Move' To Cloud," 2021. [Online]. Available: https://www.crn.com/news/data-center/aws-ceo-admits-some-workloads-will-never-move-to-cloud.

7 M. Satyanarayanan, P. Bahl, R. Caceres, and N. Davies, "The case for VM-based cloudlets in mobile computing," *IEEE Pervasive Computing*, vol. 8, no. 4, pp. 14–23, 2009.

8 F. Bonomi, R. Milito, J. Zhu, and S. Addepalli, "Fog computing and its role in the Internet of Things," in *Proceedings of the 1st Edition of the MCC Workshop on Mobile Cloud Computing*, ser. MCC '12. New York, NY, USA: Association for Computing Machinery, 2012, pp. 13–16. [Online]. Available: https://doi.org/10.1145/2342509.2342513.

9 P. Porambage, J. Okwuibe, M. Liyanage, M. Ylianttila, and T. Taleb, "Survey on multi-access edge computing for Internet of Things Realization," *IEEE Communications Surveys & Tutorials*, vol. 20, no. 4, pp. 2961–2991, 2018.

Mobile Edge Computing and Communications: Driving Forces, Technology Foundation, and Application Areas, First Edition. Aaron Yi Ding, Chamitha De Alwis, and Madhusanka Liyanage.
© 2025 John Wiley & Sons Ltd. Published 2025 by John Wiley & Sons Ltd.

10 P. Ranaweera, A. D. Jurcut, and M. Liyanage, "Survey on multi-access edge computing security and privacy," *IEEE Communications Surveys & Tutorials*, vol. 23, no. 2, pp. 1078–1124, 2021.

11 A. Reznik, Y. Fang, and S. Ullah, "MEC in an enterprise setting: A solution outline," *ETSI White Paper #30*, vol. 2, no. 30, 2018, last accessed 16 May 2019. [Online]. Available: https://www.etsi.org/.

12 P. Ranaweera, M. Liyanage, and A. D. Jurcut, "Novel MEC based approaches for smart hospitals to combat COVID-19 pandemic," *IEEE Consumer Electronics Magazine*, vol. 10, no. 2, pp. 80–91, 2020.

13 T. Taleb, K. Samdanis, B. Mada, H. Flinck, S. Dutta, and D. Sabella, "On multi-access edge computing: A survey of the emerging 5G network edge cloud architecture and orchestration," *IEEE Communications Surveys & Tutorials*, vol. 19, no. 3, pp. 1657–1681, 2017.

14 ETSI, "Mobile edge computing (MEC); Framework and reference architecture," *ETSI White Paper #3*, 2016, last accessed 16 May 2019. [Online]. Available: https://www.etsi.org/?gs/MEC/001?099/003/01.01.01?60/gs?MEC003v010101p.pdf.

15 S. Kekki, W. Featherstone, Y. Fang, P. Kuure, A. Li, A. Ranjan, D. Purkayastha, F. Jiangping, D. Frydman, G. Verin, K.-W. Wen, K. Kim, R. Arora, A. Odgers, L. M. Contreras, and S. Scarpina, "MEC in 5G networks," *ETSI White Paper #28*, vol. 28, no. 28, pp. 1–28, 2018, last accessed 16 May 2019. [Online]. Available: https://www.etsi.org/.

16 C. Tranoris, "OpenSlice: An opensource OSS for delivering network slice as a service," *arXiv preprint arXiv:2102.03290*, 2021.

17 P. Ranaweera, V. N. Imrith, M. Liyanag, and A. D. Jurcut, "Security as a service platform leveraging multi-access edge computing infrastructure provisions," in *ICC 2020-2020 IEEE International Conference on Communications (ICC)*. IEEE, 2020, pp. 1–6.

18 P. Ranaweera, A. D. Jurcut, and M. Liyanage, "Realizing multi-access edge computing feasibility: Security perspective," in *2019 IEEE Conference on Standards for Communications and Networking (CSCN)*. IEEE, 2019, pp. 1–7.

19 ETSI-GS, "Multi-access edge computing (MEC); Framework and reference architecture," *ETSI White Paper*, vol. 3, 2019, last accessed 16 May 2019. [Online]. Available: https://www.etsi.org/?gs/MEC/001?099/003/02.01.01?60/gs?MEC003v020101p.pdf.

20 ETSI, "Multi-access Edge Computing (MEC); UE application interface," *ETSI Group Specification*, 2019, last accessed 02 September 2021. [Online]. Available: https://www.etsi.org/?gs/mec/001?099/016/02.01.01?60/gs?mec016v020101p.pdf.

21 R. Roman, J. Lopez, and M. Mambo, "Mobile edge computing, Fog et al.: A survey and analysis of security threats and challenges," *Future Generation Computer Systems*, vol. 78, pp. 680–698, 2018.

22 ETSI, "Mobile-edge computing–introductory technical White Paper," *ETSI White Paper #1*, 2014, last accessed 16 May 2019. [Online]. Available: https://portal.etsi.org/.

23 N. Abbas, Y. Zhang, A. Taherkordi, and T. Skeie, "Mobile edge computing: A survey," *IEEE Internet of Things Journal*, vol. 5, no. 1, pp. 450–465, 2018.

24 ETSI-GS-MEC, "Mobile-edge computing (MEC); Proof of concept framework," *ETSI White Paper #5*, vol. 1, 2015, last accessed 16 May 2019. [Online]. Available: https://www.etsi.org/.

25 Y. C. Hu, M. Patel, D. Sabella, N. Sprecher, and V. Young, "Mobile edge computing—A key technology towards 5G," *ETSI White Paper #11*, vol. 11, no. 11, pp. 1–16, 2015, last accessed 16 May 2019. [Online]. Available: https://www.etsi.org/.

26 ETSI, "Multi-access edge computing (MEC); Phase 2: Use cases and requirements," *ETSI White Paper*, vol. 2, 2018, last accessed 16 May 2019. [Online]. Available: https://www.etsi.org/?gs/MEC/001?099/002/02.01.01?60/gs?MEC002v020101p.pdf.

27 D. E. Kouicem, A. Bouabdallah, and H. Lakhlef, "Internet of Things security: A top-down survey," *Computer Networks*, vol. 141, pp. 199–221, 2018.

28 A. D. Jurcut, P. Ranaweera, and L. Xu, "Introduction to IoT security," *IoT security: Advances in authentication*, pp. 27–64, 2020.

29 F. A. Alaba, M. Othman, I. A. T. Hashem, and F. Alotaibi, "Internet of Things security: A survey," *Journal of Network and Computer Applications*, vol. 88, pp. 10–28, 2017.

30 S.-S. Yang, J.-W. Pang, X.-M. Jin, Z.-Y. Wu, X.-Y. Yang, W.-Q. Guo, Z.-Q. Zhao, and N.-Q. Ren, "Let a sewage plant running smart," in *IOP Conference Series: Earth and Environmental Science*, vol. 127, no. 1. IOP Publishing, 2018, p. 012013.

31 A. Molina, "Water governance in the smart city," *WIT Transactions on The Built Environment*, vol. 179, no. 1, pp. 13–22, 2018.

32 R. F. Hussain, M. A. Salehi, A. Kovalenko, S. Salehi, and O. Semiari, "Robust resource allocation using edge computing for smart oil fields," in *Proceedings of the International Conference on Parallel and Distributed Processing Techniques and Applications (PDPTA).* The Steering Committee of The World Congress in Computer Science, Computer …, 2018, pp. 204–210.

33 E. Mohyeldin, "Minimum technical performance requirements for imt-2020 radio interface (s)," in *ITU-R Workshop on IMT-2020 Terrestrial Radio Interfaces*, 2016, pp. 1–12.

34 Y. Mao, C. You, J. Zhang, K. Huang, and K. B. Letaief, "A survey on mobile edge computing: The communication perspective," *IEEE Communications Surveys & Tutorials*, vol. 19, no. 4, pp. 2322–2358, 2017.

35 O. Mäkinen, "Streaming at the edge: Local service concepts utilizing mobile edge computing," in *2015 9th International Conference on Next Generation Mobile Applications, Services and Technologies*. IEEE, 2015, pp. 1–6.

36 K. Bilal and A. Erbad, "Edge computing for interactive media and video streaming," in *2017 2nd International Conference on Fog and Mobile Edge Computing (FMEC)*. IEEE, 2017, pp. 68–73.

37 S. Chen, R. Ma, H.-H. Chen, H. Zhang, W. Meng, and J. Liu, "Machine-to-machine communications in ultra-dense networks—A survey," *IEEE Communications Surveys & Tutorials*, vol. 19, no. 3, pp. 1478–1503, 2017.

38 M. Usman, M. R. Asghar, I. S. Ansari, and M. Qaraqe, "Security in wireless body area networks: From in-body to off-body communications," *IEEE Access*, vol. 6, pp. 58064–58074, 2018.

39 M. Li, R. Yu, P. Si, and Y. Zhang, "Energy-efficient machine-to-machine (M2M) communications in virtualized cellular networks with mobile edge computing (MEC)," *IEEE Transactions on Mobile Computing*, vol. 18, no. 7, pp. 1541–1555, 2018.
40 L. Dong and G. Wang, "Information centric approach in achieving anycast service in machine type communications," in *2017 IEEE Conference on Standards for Communications and Networking (CSCN)*. IEEE, 2017, pp. 157–162.
41 J. Petit and S. E. Shladover, "Potential cyberattacks on automated vehicles," *IEEE Transactions on Intelligent Transportation Systems*, vol. 16, no. 2, pp. 546–556, 2015.
42 L. Zhu, F. R. Yu, Y. Wang, B. Ning, and T. Tang, "Big data analytics in intelligent transportation systems: A survey," *IEEE Transactions on Intelligent Transportation Systems*, vol. 20, no. 1, pp. 383–398, 2018.
43 I. Tal and G.-M. Muntean, "Towards reasoning vehicles: A survey of fuzzy logic-based solutions in vehicular networks," *ACM Computing Surveys (CSUR)*, vol. 50, no. 6, pp. 1–37, 2017.
44 S. Garg, A. Singh, S. Batra, N. Kumar, and L. T. Yang, "UAV-empowered edge computing environment for cyber-threat detection in smart vehicles," *IEEE Network*, vol. 32, no. 3, pp. 42–51, 2018.
45 W. Xu, C. Yan, W. Jia, X. Ji, and J. Liu, "Analyzing and enhancing the security of ultrasonic sensors for autonomous vehicles," *IEEE Internet of Things Journal*, vol. 5, no. 6, pp. 5015–5029, 2018.
46 J.-H. Lee and H. Kim, "Security and privacy challenges in the Internet of Things [security and privacy matters]," *IEEE Consumer Electronics Magazine*, vol. 6, no. 3, pp. 134–136, 2017.
47 E. Uhlemann, "The US and Europe advances V2V deployment [connected vehicles]," *IEEE Vehicular Technology Magazine*, vol. 12, no. 2, pp. 18–22, 2017.
48 M. S. Elbamby, C. Perfecto, M. Bennis, and K. Doppler, "Toward low-latency and ultra-reliable virtual reality," *IEEE Network*, vol. 32, no. 2, pp. 78–84, 2018.
49 J. Mitchell, D. Soldani, and M. Shore, "The Path to 5G in Australia: Architecture Evolution from 4G to 5G," 2018, last accessed 16 May 2019. [Online]. Available: http://huaweihub.com.au/wp-content/uploads/2018/07/The-path-to-5G-in-Australia-03-August-2018-2.pdf.
50 M. Langfinger, M. Schneider, D. Stricker, and H. D. Schotten, "Addressing security challenges in industrial augmented reality systems," in *15th International Conference on Industrial Informatics (INDIN)*. IEEE, 2017, pp. 299–304.
51 P. Mach and Z. Becvar, "Mobile edge computing: A survey on architecture and computation offloading," *IEEE Communications Surveys & Tutorials*, vol. 19, no. 3, pp. 1628–1656, 2017.
52 X. Qiao, P. Ren, S. Dustdar, and J. Chen, "A new era for web AR with mobile edge computing," *IEEE Internet Computing*, vol. 22, no. 4, pp. 46–55, 2018.
53 D. He, S. Chan, and M. Guizani, "Security in the Internet of Things supported by mobile edge computing," *IEEE Communications Magazine*, vol. 56, no. 8, pp. 56–61, 2018.
54 F. A. Forecast, *Fiscal Years 2019–2039*. Federal Aviation Administration: Washington, DC, USA, 2019.

55 B. Li, Z. Fei, and Y. Zhang, "UAV communications for 5G and beyond: Recent advances and future trends," *IEEE Internet of Things Journal*, vol. 6, no. 2, pp. 2241–2263, 2018.

56 Z. Sanaei, S. Abolfazli, A. Gani, and R. Buyya, "Heterogeneity in mobile cloud computing: Taxonomy and open challenges," *IEEE Communications Surveys & Tutorials*, vol. 16, no. 1, pp. 369–392, 2014.

57 A. N. Khan, M. M. Kiah, S. U. Khan, and S. A. Madani, "Towards secure mobile cloud computing: A survey," *Future Generation Computer Systems*, vol. 29, no. 5, pp. 1278–1299, 2013.

58 M. Mukherjee, L. Shu, and D. Wang, "Survey of fog computing: Fundamental, network applications, and research challenges," *IEEE Communications Surveys & Tutorials*, vol. 20, no. 3, pp. 1826–1857, 2018.

59 C. Mouradian, D. Naboulsi, S. Yangui, R. H. Glitho, M. J. Morrow, and P. A. Polakos, "A Comprehensive survey on fog computing: state-of-the-art and research challenges," *IEEE Communications Surveys & Tutorials*, vol. 20, no. 1, pp. 416–464, 2018.

60 A. Yousefpour, C. Fung, T. Nguyen, K. Kadiyala, F. Jalali, A. Niakanlahiji, J. Kong, and J. P. Jue, "All one needs to know about fog computing and related edge computing paradigms: A complete survey," *Journal of Systems Architecture*, vol. 98, pp. 289–330, 2019.

61 H. T. Dinh, C. Lee, D. Niyato, and P. Wang, "A survey of mobile cloud computing: Architecture, applications, and approaches," *Wireless Communications and Mobile Computing*, vol. 13, no. 18, pp. 1587–1611, 2013.

62 N. Antonopoulos and L. Gillam, *Cloud computing*. Springer, 2010.

63 K. Ha, P. Pillai, W. Richter, Y. Abe, and M. Satyanarayanan, "Just-in-time provisioning for cyber foraging," in *Proceeding of the 11th Annual International Conference on Mobile Systems, Applications, and Services*, 2013, pp. 153–166.

64 N. Dimitrova, H.-J. Zhang, B. Shahraray, I. Sezan, T. Huang, and A. Zakhor, "Applications of video-content analysis and retrieval," *IEEE MultiMedia*, vol. 9, no. 3, pp. 42–55, 2002.

65 V. Kastrinaki, M. Zervakis, and K. Kalaitzakis, "A survey of video processing techniques for traffic applications," *Image and Vision Computing*, vol. 21, no. 4, pp. 359–381, 2003. [Online]. Available: https://www.sciencedirect.com/science/article/pii/S0262885603000040.

66 H. Liu, S. Chen, and N. Kubota, "Intelligent video systems and analytics: A survey," *IEEE Transactions on Industrial Informatics*, vol. 9, no. 3, pp. 1222–1233, 2013.

67 R. Xu, S. Razavi, and R. Zheng, "Edge video analytics: A survey on applications, systems and enabling techniques," *IEEE Communications Surveys & Tutorials*, vol. 25, no. 4, pp. 2951–2982, Oct. 2023. [Online]. Available: https://doi.org/10.1109/COMST.2023.3323091.

68 Q. Zhang, H. Sun, X. Wu, and H. Zhong, "Edge video analytics for public safety: A review," *Proceedings of the IEEE*, vol. 107, no. 8, pp. 1675–1696, 2019.

69 R. F. H. Schulzrinne, S. Casner and V. Jacobson, "RTP: A transport protocol for real-time applications," Internet Requests for Comments, RFC Editor, RFC 3550, 2003.

70 H. Alvestrand, “Overview: Real-time protocols for browser-based applications,” Internet Requests for Comments, RFC Editor, RFC 8825, 2021.

71 ISO, “Information technology – dynamic adaptive streaming over http (dash) – Part 1: Media presentation description and segment formats,” ISO/IEC 23009-1:2014, ISO/IEC, Tech. Rep., 2014.

72 B. Bross, Y.-K. Wang, Y. Ye, S. Liu, J. Chen, G. J. Sullivan, and J.-R. Ohm, “Overview of the versatile video coding (VVC) standard and its applications,” *IEEE Transactions on Circuits and Systems for Video Technology*, vol. 31, no. 10, pp. 3736–3764, 2021.

73 K. He, X. Zhang, S. Ren, and J. Sun, “Deep residual learning for image recognition,” in *2016 IEEE Conference on Computer Vision and Pattern Recognition (CVPR)*, 2016, pp. 770–778.

74 H. Zhang, G. Ananthanarayanan, P. Bodik, M. Philipose, P. Bahl, and M. J. Freedman, “Live video analytics at scale with approximation and delay-tolerance,” in *Proceedings of the 14th USENIX Conference on Networked Systems Design and Implementation*, ser. NSDI’17. USA: USENIX Association, 2017, pp. 377–392.

75 M. Hu, Z. Luo, A. Pasdar, Y. C. Lee, Y. Zhou, and D. Wu, “Edge-based video analytics: A survey,” *arXiv preprint arXiv:2303.14329*, 2023.

76 R. Long, T. Tuna, and J. Subhlok, “Lecture video analytics as an instructional resource,” in *2018 IEEE Frontiers in Education Conference (FIE)*, 2018, pp. 1–7.

77 G. Ananthanarayanan, P. Bahl, P. Bodík, K. Chintalapudi, M. Philipose, L. Ravindranath, and S. Sinha, “Real-time video analytics: The killer app for edge computing,” *Computer*, vol. 50, no. 10, pp. 58–67, 2017.

78 R. Morabito, V. Cozzolino, A. Y. Ding, N. Beijar, and J. Ott, “Consolidate IoT edge computing with lightweight virtualization,” *IEEE Network*, vol. 32, no. 1, pp. 102–111, 2018.

79 KubeEdge, 2023. [Online]. Available: https://kubeedge.io/.

80 X. Zhang, A. Zhang, J. Sun, X. Zhu, Y. E. Guo, F. Qian, and Z. M. Mao, “EMP: Edge-assisted multi-vehicle perception,” in *Proceedings of the 27th Annual International Conference on Mobile Computing and Networking*, ser. MobiCom ’21. New York, NY, USA: Association for Computing Machinery, 2021, pp. 545–558. [Online]. Available: https://doi.org/10.1145/3447993.3483242.

81 L. Liu, X. Zhang, M. Qiao, and W. Shi, “SafeShareRide: Edge-based attack detection in ridesharing services,” in *2018 IEEE/ACM Symposium on Edge Computing (SEC)*, 2018, pp. 17–29.

82 C. Westphal, “Challenges in networking to support augmented reality and virtual reality,” *IEEE ICNC*, 2017.

83 “Multi-access Edge Computing (MEC); Phase 2: Use Cases and Requirements.” [Online]. Available: https://www.etsi.org/?gs/MEC/001?099/002/02.01.01?60/gs?MEC002v020101p.pdf.

84 R. Azuma, Y. Baillot, R. Behringer, S. Feiner, S. Julier, and B. MacIntyre, “Recent advances in augmented reality,” *IEEE Computer Graphics and Applications*, vol. 21, no. 6, pp. 34–47, 2001.

85 P. Milgram and F. Kishino, “A taxonomy of mixed reality visual displays,” *IEICE Transactions on Information and Systems*, vol. 77, no. 12, pp. 1321–1329, 1994.

86 D. Chatzopoulos, C. Bermejo, Z. Huang, and P. Hui, "Mobile augmented reality survey: From where we are to where we go," *IEEE Access*, vol. 5, pp. 6917–6950, 2017.

87 F. Zhou, H. B.-L. Duh, and M. Billinghurst, "Trends in augmented reality tracking, interaction and display: A review of ten years of ISMAR," in *Proceedings of the 7th IEEE/ACM International Symposium on Mixed and Augmented Reality. IEEE Computer Society*, 2008, pp. 193–202.

88 I. Rabbi and S. Ullah, "A survey on augmented reality challenges and tracking," *Acta graphica: znanstveni časopis za tiskarstvo i grafičke komunikacije*, vol. 24, no. 1–2, pp. 29–46, 2013.

89 C. Bermejo and P. Hui, "A survey on haptic technologies for mobile augmented reality," *arXiv preprint arXiv:1709.00698*, 2017.

90 G. Papagiannakis, G. Singh, and N. Magnenat-Thalmann, "A survey of mobile and wireless technologies for augmented reality systems," *Computer Animation and Virtual Worlds*, vol. 19, no. 1, pp. 3–22, 2008.

91 D. Van Krevelen and R. Poelman, "Augmented Reality: Technologies, Applications, and Limitations," *Vrije Univ*. Amsterdam, Dep. Comput. Sci., 2007.

92 P. Haynes, S. Hehl-Lange, and E. Lange, "Mobile augmented reality for flood visualisation," *Environmental Modelling & Software*, vol. 109, pp. 380–389, 2018.

93 M. H. Kurniawan and G. Witjaksono, "Human anatomy learning systems using augmented reality on mobile application," *Procedia Computer Science*, vol. 135, pp. 80–88, 2018.

94 P. Dangkham, "Mobile augmented reality on web-based for the tourism using HTML5," in *2018 International Conference on Information Networking (ICOIN)*. IEEE, 2018, pp. 482–485.

95 S. Larabi, "Augmented reality for mobile devices: Textual annotation of outdoor locations," in *Augmented Reality and Virtual Reality*. Springer, 2018, pp. 353–362.

96 A. B. Craig, *Understanding Augmented Reality: Concepts and Applications*. Newnes, 2013.

97 Google Glass, [Online]. Available: https://www.google.com/glass/start/.

98 O. J. Muensterer, M. Lacher, C. Zoeller, M. Bronstein, and J. Kübler, "Google glass in pediatric surgery: An exploratory study," *International Journal of Surgery*, vol. 12, no. 4, pp. 281–289, 2014.

99 M. C. Leue, T. Jung, and D. tom Dieck, "Google glass augmented reality: Generic learning outcomes for art galleries," in *Information and Communication Technologies in Tourism 2015*. Springer, 2015, pp. 463–476.

100 Pokémon Go, [Online]. Available: https://www.pokemongo.com/en-us/.

101 Google Lens, [Online]. Available: https://lens.google.com/.

102 Microsoft HoloLens, [Online]. Available: https://www.microsoft.com/en-us/hololens/.

103 B. Shi, J. Yang, Z. Huang, and P. Hui, "Offloading guidelines for augmented reality applications on wearable devices," in *Proceedings of the 23rd ACM International Conference on Multimedia*. ACM, 2015, pp. 1271–1274.

104 R. Shea, A. Sun, S. Fu, and J. Liu, "Towards fully offloaded cloud-based AR: Design, implementation and experience," in *Proceedings of the 8th ACM on Multimedia Systems Conference*. ACM, 2017, pp. 321–330.

105 W. Zhang, B. Han, and P. Hui, "On the networking challenges of mobile augmented reality," in *Proceedings of the Workshop on Virtual Reality and Augmented Reality Network*, 2017, pp. 24–29.

106 A. Younis, T. X. Tran, B. Qiu, and D. Pompili, "Demo abstract: Mobile augmented reality leveraging cloud radio access networks," in *2019 IEEE 20th International Symposium on" A World of Wireless, Mobile and Multimedia Networks" (WoWMoM)*. IEEE, 2019, pp. 1–3.

107 3GPP, "Requirements for Evolved UTRA (E-UTRA) and Evolved UTRAN (E-UTRAN)," Technical Report, Dec. 2009. [Online]. Available: https://www.3gpp.org/.

108 Ericsson, "LTE – An Introduction," 2009.

109 3GPP, "Study on Communication Services for Critical Medical Applications," Technical Report, Sept. 2019. [Online]. Available: https://www.3gpp.org/.

110 J. G. Andrews, S. Buzzi, W. Choi, S. V. Hanly, A. Lozano, A. C. Soong, and J. C. Zhang, "What will 5G be?" *IEEE Journal on Selected Areas in Communications*, vol. 32, no. 6, pp. 1065–1082, 2014.

111 A. Osseiran, F. Boccardi, V. Braun, K. Kusume, P. Marsch, M. Maternia, O. Queseth, M. Schellmann, H. Schotten, H. Taoka, and H. Tullberg, "Scenarios for 5G mobile and wireless communications: The vision of the METIS project," *IEEE communications magazine*, vol. 52, no. 5, pp. 26–35, 2014.

112 ABIresearch, "Augmented and Virtual Reality: The First Wave of 5G Killer Apps," 2017. [Online]. Available: https://www.qualcomm.com/media/documents/files/augmented-and-virtual-reality-the-first-wave-of-5g-killer-apps.pdf.

113 ITU, "IMT Vision – Framework and Overall Objectives of the Future Development of IMT for 2020 and beyond," Rec. ITU-R M.2083, Sept. 2015.

114 M. Shafi, A. F. Molisch, P. J. Smith, T. Haustein, P. Zhu, P. Silva, F. Tufvesson, A. Benjebbour, and G. Wunder, "5G: A tutorial overview of standards, trials, challenges, deployment, and practice," *IEEE Journal on Selected Areas in Communications*, vol. 35, no. 6, pp. 1201–1221, 2017.

115 P. K. Agyapong, M. Iwamura, D. Staehle, W. Kiess, and A. Benjebbour, "Design considerations for a 5G network architecture," *IEEE Communications Magazine*, vol. 52, no. 11, pp. 65–75, 2014.

116 E. Vlachos and A. S. Lalos, "Enabling extended reality applications over 5G mobile networks," *ERCIM NEWS*, p. 30, 2019.

117 M. Erol-Kantarci and S. Sukhmani, "Caching and computing at the edge for mobile augmented reality and virtual reality (AR/VR) in 5G," in *Ad Hoc Networks*. Springer, 2018, pp. 169–177.

118 Z. Huang, W. Li, P. Hui, and C. Peylo, "CloudRidAR: A cloud-based architecture for mobile augmented reality," in *Proceedings of the 2014 Workshop on Mobile Augmented Reality and Robotic Technology-based Systems*. ACM, 2014, pp. 29–34.

119 B.-R. Huang, C. H. Lin, and C.-H. Lee, "Mobile augmented reality based on cloud computing," in *Anti-counterfeiting, Security, and Identification*. IEEE, 2012, pp. 1–5.

120 B.-S. P. Lin, W.-H. Tsai, C. Wu, P. Hsu, J. Huang, and T.-H. Liu, "The design of cloud-based 4G/LTE for mobile augmented reality with smart mobile devices," in

2013 IEEE 7th International Symposium on Service-Oriented System Engineering. IEEE, 2013, pp. 561–566.

121 K. Ab Aziz, N. A. Ab Aziz, A. M. Yusof, and A. Paul, "Potential for providing augmented reality elements in special education via cloud computing," *Procedia Engineering*, vol. 41, pp. 333–339, 2012.

122 K. Chen, T. Li, H.-S. Kim, D. E. Culler, and R. H. Katz, "MARVEL: Enabling mobile augmented reality with low energy and low latency," in *Proceedings of the 16th ACM Conference on Embedded Networked Sensor Systems*, ser. SenSys '18. New York, NY, USA: ACM, 2018, pp. 292–304. [Online]. Available: http://doi.acm.org/10.1145/3274783.3274834.

123 M. Noreikis, Y. Xiao, and A. Ylä-Jääski, "SeeNav: Seamless and energy-efficient indoor navigation using augmented reality," in *Proceedings of the on Thematic Workshops of ACM Multimedia 2017*. ACM, 2017, pp. 186–193.

124 J. A. Juanes, D. Hernández, P. Ruisoto, E. García, G. Villarrubia, and A. Prats, "Augmented reality techniques, using mobile devices, for learning human anatomy," in *Proceedings of the 2nd International Conference on Technological Ecosystems for Enhancing Multiculturality*. ACM, 2014, pp. 7–11.

125 P. Jain, J. Manweiler, and R. Roy Choudhury, "Low bandwidth offload for mobile AR," in *Proceedings of the 12th International on Conference on emerging Networking EXperiments and Technologies*. ACM, 2016, pp. 237–251.

126 GSMA, "Cloud AR/VR Whitepaper," GSMA Future Networks, Apr. 2019. [Online]. Available: https://www.gsma.com/.

127 M. Schneider, J. Rambach, and D. Stricker, "Augmented reality based on edge computing using the example of remote live support," in *2017 IEEE International Conference on Industrial Technology (ICIT)*. IEEE, 2017, pp. 1277–1282.

128 P. Zhou, W. Zhang, T. Braud, P. Hui, and J. Kangasharju, "Enhanced augmented reality applications in vehicle-to-edge networks," in *2019 22nd Conference on Innovation in Clouds, Internet and Networks and Workshops (ICIN)*. IEEE, 2019, pp. 167–174.

129 W. Liu, J. Ren, G. Huang, Y. He, and G. Yu, "Data offloading and sharing for latency minimization in augmented reality based on mobile-edge computing," in *2018 IEEE 88th Vehicular Technology Conference (VTC-Fall)*. IEEE, 2018, pp. 1–5.

130 P. Zhou, W. Zhang, T. Braud, P. Hui, and J. Kangasharju, "ARVE: Augmented reality applications in vehicle to edge networks," in *Proceedings of the 2018 Workshop on Mobile Edge Communications*, 2018, pp. 25–30.

131 P. Renukdas, R. Ghundiyal, H. Gadhil, and V. Pathare, "Markerless augmented reality android app for interior decoration," *International Journal of Engineering Research & Technology (IJERT)*, vol. 2, no. 4, pp. 1367–1373, 2013.

132 N. I. Adhani and R. D. R. Awang, "A survey of mobile augmented reality applications," in *1st International Conference on Future Trends in Computing and Communication Technologies*, 2012, pp. 89–96.

133 M. Gervautz and D. Schmalstieg, "Anywhere interfaces using handheld augmented reality," *Computer*, vol. 45, no. 7, pp. 26–31, 2012.

134 A. Mulloni, H. Seichter, and D. Schmalstieg, "Handheld augmented reality indoor navigation with activity-based instructions," in *Proceedings of the 13th International*

Conference on Human Computer Interaction with Mobile Devices and Services. ACM, 2011, pp. 211–220.

135 J. Ren, Y. He, G. Huang, G. Yu, Y. Cai, and Z. Zhang, "An edge-computing based architecture for mobile augmented reality," *IEEE Network*, vol. 33, no. 4, pp. 162–169, 2019.

136 P. Fraga-Lamas, T. M. Fernández-Caramés, Ó. Blanco-Novoa, and M. A. Vilar-Montesinos, "A review on industrial augmented reality systems for the industry 4.0 shipyard," *IEEE Access*, vol. 6, pp. 13358–13375, 2018.

137 T. Fernández-Caramés, P. Fraga-Lamas, M. Suárez-Albela, and M. Vilar-Montesinos, "A fog computing and cloudlet based augmented reality system for the industry 4.0 shipyard," *Sensors*, vol. 18, no. 6, pp. 1798, 2018.

138 O. Blanco-Novoa, T. M. Fernández-Caramés, P. Fraga-Lamas, and M. A. Vilar-Montesinos, "A practical evaluation of commercial industrial augmented reality systems in an industry 4.0 shipyard," *IEEE Access*, vol. 6, pp. 8201–8218, 2018.

139 P. Ren, X. Qiao, J. Chen, and S. Dustdar, "Mobile edge computing–a booster for the practical provisioning approach of web-based augmented reality," in *2018 IEEE/ACM Symposium on Edge Computing (SEC).* IEEE, 2018, pp. 349–350.

140 A. Braeken, P. Porambage, A. Puvaneswaran, and M. Liyanage, "ESSMAR: Edge supportive secure mobile augmented reality architecture for healthcare," in *5th International Conference on Cloud Computing and Artificial Intelligence: Technologies and Applications (CloudTech'20).* IEEE, 2020.

141 F. Roesner, T. Kohno, and D. Molnar, "Security and privacy for augmented reality systems," *Communications of the ACM*, vol. 57, no. 4, pp. 88–96, 2014.

142 S. Baldassi, T. Kohno, F. Roesner, and M. Tian, "Challenges and new directions in augmented reality, computer security, and neuroscience–Part 1: Risks to sensation and perception," *arXiv preprint arXiv:1806.10557*, 2018.

143 J. A. De Guzman, K. Thilakarathna, and A. Seneviratne, "Security and privacy approaches in mixed reality: A literature survey," *ACM Computing Surveys (CSUR)*, vol. 52, no. 6, pp. 110, 2019.

144 3GPP, "Study on Communication for Automation in Vertical Domains (CAV)," Technical Report, Dec. 2018. [Online]. Available: https://portal.3gpp.org/.

145 T. Blajić, D. Nogulić, and M. Družijanić, "Latency improvements in 3G long term evolution," in *Proceedings of the International Convention on Information and Communication Technology, Electronics and Microelectronics*, 2006, pp. 1–6.

146 H. Zhu, C. K. Wu, C. H. Koo, Y. T. Tsang, Y. Liu, H. R. Chi, and K.-F. Tsang, "Smart healthcare in the era of Internet-of-Things," *IEEE Consumer Electronics Magazine*, vol. 8, no. 5, pp. 26–30, 2019.

147 A. J. Ferrer, J. M. Marquès, and J. Jorba, "Towards the decentralised cloud: Survey on approaches and challenges for mobile, ad hoc, and edge computing," *ACM Computing Surveys (CSUR)*, vol. 51, no. 6, pp. 1–36, 2019.

148 P. P. Ray, D. Dash, and D. De, "Edge computing for internet of things: A survey, e-healthcare case study and future direction," *Journal of Network and Computer Applications*, vol. 140, pp. 1–22, 2019.

149 S. Oueida, Y. Kotb, M. Aloqaily, Y. Jararweh, and T. Baker, "An edge computing based smart healthcare framework for resource management," *Sensors*, vol. 18, no. 12, pp. 4307, 2018.

150 S. P. Mohanty, "Healthcare cyber-physical system is more important than before," *IEEE Consumer Electronics Magazine*, vol. 9, no. 5, pp. 6–7, 2020.

151 A. K. Tripathy, A. G. Mohapatra, S. P. Mohanty, E. Kougianos, A. M. Joshi, and G. Das, "EasyBand: A wearable for safety-aware mobility during pandemic outbreak," *IEEE Consumer Electronics Magazine*, vol. 9, no. 5, pp. 57–61, 2020.

152 Y. Siriwardhana, C. De Alwis, G. Gur, M. Ylianttila, and M. Liyanage, "The fight against COVID-19 pandemic with 5G technologies," *IEEE Engineering Management Review*, vol. 48, no. 3, pp. 72–84, 2020.

153 Y. Siriwardhana, G. Gur, M. Ylianttila, and M. Liyanage, "The role of 5G for digital healthcare against COVID-19 pandemic: Opportunities and challenges," *Elsevier ICT Express*, vol. 7, no. 2, pp. 244–252, 2020.

154 G.-Z. Yang, B. J. Nelson, R. R. Murphy, H. Choset, H. Christensen, S. H. Collins, P. Dario, K. Goldberg, K. Ikuta, N. Jacobstein, and Kragic, D., "Combating COVID-19—The role of robotics in managing public health and infectious diseases," *Science Robotics*, vol. 5, no. 40, p. eabb5589, 2020.

155 L. Manring, J. Pederson, D. Potts, B. Boardman, D. Mascarenas, T. Harden, and A. Cattaneo, "Augmented reality for interactive robot control," in *Special Topics in Structural Dynamics & Experimental Techniques, Volume* 5. Springer, 2020, pp. 11–18.

156 WHO, "Clinical management of COVID-19: Interim guidance," May 2020. [Online]. Available: https://www.who.int/.

157 R. Kimmig, R. H. Verheijen, and M. Rudnicki, "Robot assisted surgery during the COVID-19 pandemic, specially for gynecological cancer: A statement of the society of European Robotic Gynaecological Surgery (SERGS)," *Journal of Gynecologic Oncology*, vol. 31, no. 3, p. e59, 2020.

158 D. Ni, A. Yew, S. Ong, and A. Nee, "Haptic and visual augmented reality interface for programming welding robots," *Advances in Manufacturing*, vol. 5, no. 3, pp. 191–198, 2017.

159 S. Tanwar, K. Parekh, and R. Evans, "Blockchain-based electronic healthcare record system for healthcare 4.0 applications," *Journal of Information Security and Applications*, vol. 50, p. 102407, 2020.

160 B. Deng, "Machine ethics: The robot's dilemma," *Nature News*, vol. 523, no. 7558, p. 24, 2015.

161 EUGDPR, "European General Data Protection Regulation," 2018, last accessed May 16, 2019. [Online]. Available: https://gdpr.eu/.

162 A. Y. Ding, E. Peltonen, T. Meuser, A. Aral, C. Becker, S. Dustdar, T. Hiessl, D. Kranzlmüller, M. Liyanage, S. Maghsudi, and N. Mohan, "Roadmap for edge AI: A dagstuhl perspective," *ACM SIGCOMM Computer Communication Review*, vol. 52, no. 1, pp. 28–33, 2022.

163 I. Goodfellow, Y. Bengio, and A. Courville, *Deep Learning*. MIT Press, 2016. [Online]. Available: http://www.deeplearningbook.org.

164 Z. Zhou, X. Chen, E. Li, L. Zeng, K. Luo, and J. Zhang, "Edge intelligence: Paving the last mile of artificial intelligence with edge computing," *Proceedings of the IEEE*, vol. 107, no. 8, pp. 1738–1762, 2019.

165 E. Peltonen, I. Ahmad, A. Aral, M. Capobianco, A. Y. Ding, F. Gil-Castiñeira, E. Gilman, E. Harjula, M. Jurmu, T. Karvonen, M. Kelanti, T. Leppänen, L. Lovén, T. Mikkonen, N. Mohan, P. Nurmi, S. Pirttikangas, P. Sroka, S. Tarkoma, and T. Yang, "The many faces of edge intelligence," *IEEE Access*, vol. 10, pp. 104769–104782, 2022.

166 IBM, "What is edge AI?" 2024, last accessed 29 February 2024. [Online]. Available: https://www.ibm.com/topics/edge-ai.

167 T. Rausch and S. Dustdar, "Edge intelligence: The convergence of humans, things, and AI," in *2019 IEEE International Conference on Cloud Engineering (IC2E)*, 2019, pp. 86–96.

168 S. Deng, H. Zhao, W. Fang, J. Yin, S. Dustdar, and A. Y. Zomaya, "Edge intelligence: The confluence of edge computing and artificial intelligence," *IEEE Internet of Things Journal*, vol. 7, no. 8, pp. 7457–7469, 2020.

169 S. Dustdar, V. C. Pujol, and P. K. Donta, "On distributed computing continuum systems," *IEEE Transactions on Knowledge and Data Engineering*, vol. 35, no. 4, pp. 4092–4105, 2023.

170 D. Xu, T. Li, Y. Li, X. Su, S. Tarkoma, T. Jiang, J. Crowcroft, and P. Hui, "Edge intelligence: Empowering intelligence to the edge of network," *Proceedings of the IEEE*, vol. 109, no. 11, pp. 1778–1837, 2021.

171 J. Chen and X. Ran, "Deep learning with edge computing: A review," *Proceedings of the IEEE*, vol. 107, no. 8, pp. 1655–1674, 2019.

172 D. Katare, D. Perino, J. Nurmi, M. Warnier, M. Janssen, and A. Y. Ding, "A survey on approximate edge AI for energy efficient autonomous driving services," *IEEE Communications Surveys & Tutorials*, vol. 25, no. 4, pp. 2714–2754, 2023.

173 P. Kairouz, H. B. McMahan, B. Avent, A. Bellet, M. Bennis, A. N. Bhagoji, K. Bonawitz, Z. Charles, G. Cormode, R. Cummings, et al., "Advances and open problems in federated learning," *Foundations and Trends in Machine Learning*, vol. 14, no. 1–2, pp. 1–210, 2021. [Online]. Available: http://dx.doi.org/10.1561/2200000083.

174 Y. Kang, J. Hauswald, C. Gao, A. Rovinski, T. Mudge, J. Mars, and L. Tang, "Neurosurgeon: Collaborative intelligence between the cloud and mobile edge," in *Proceedings of the 22nd International Conference on Architectural Support for Programming Languages and Operating Systems*, ser. ASPLOS '17. New York, NY, USA: Association for Computing Machinery, 2017, pp. 615–629. [Online]. Available: https://doi.org/10.1145/3037697.3037698.

175 V. Vaikuntanathan, "Computing blindfolded: New developments in fully homomorphic encryption," in *2011 IEEE 52nd Annual Symposium on Foundations of Computer Science*, 2011, pp. 5–16.

176 S. Teerapittayanon, B. McDanel, and H.-T. Kung, "BranchyNet: Fast inference via early exiting from deep neural networks," in *2016 23rd International Conference on Pattern Recognition (ICPR)*. IEEE, 2016, pp. 2464–2469.

177 R. Taheri, S. Dustdar, A. Zomaya, and S. Deng, *Edge Intelligence From Theory to Practice*. Springer Cham, 2023. [Online]. Available: https://doi.org/10.1007/978-3-031-22155-2.

178 T. Elsken, J. H. Metzen, and F. Hutter, "Neural architecture search: A survey," *Journal of Machine Learning Research*, vol. 20, no. 1, pp. 1997–2017, Jan. 2019.

179 I. Stoica, D. Song, R. A. Popa, D. A. Patterson, M. W. Mahoney, R. H. Katz, A. D. Joseph, M. Jordan, J. M. Hellerstein, J. Gonzalez, K. Goldberg, A. Ghodsi, D. E. Culler, and P. Abbeel, "A Berkeley View of Systems Challenges for AI," Technical Report, EECS Department, University of California, Berkeley, Tech. Rep., Oct. 2017. [Online]. Available: https://www2.eecs.berkeley.edu/Pubs/TechRpts/2017/EECS-2017-159.html.

180 A. Y. Ding, M. Janssen, and J. Crowcroft, "Trustworthy and sustainable edge AI: A research agenda," in *2021 3rd IEEE International Conference on Trust, Privacy and Security in Intelligent Systems and Applications (TPS-ISA)*, 2021, pp. 164–172.

181 WiFi Alliance, "*WiFi-Direct*," last accessed 26 October 2020. [Online]. Available: https://www.wi-fi.org/discover-wi-fi/wi-fi-direct.

182 S. Tarkoma, M. Siekkinen, E. Lagerspetz, and Y. Xiao, *Smartphone energy consumption: Modeling and optimization*. Cambridge University Press, 2014.

183 Y. Ding, "Collaborative traffic offloading for mobile systems," Ph.D. dissertation, University of Helsinki, Nov. 2015.

184 R. Draves and D. Thaler, "Default router preferences and more-specific routes," Internet Requests for Comments, RFC Editor, RFC 4191, 2005.

185 I. F. Akyildiz, X. Jiang, and S. Mohanty, "A survey of mobility management in next-generation all-IP-based wireless systems," *IEEE Wireless Communications*, vol. 11, no. 4, pp. 16–28, 2004.

186 M. Stemm and R. H. Katz, "Vertical handoffs in wireless overlay networks," *Mobile Networks and Applications*, vol. 3, pp. 335–350, 1998.

187 F. Rebecchi, M. Dias de Amorim, V. Conan, A. Passarella, R. Bruno, and M. Conti, "Data offloading techniques in cellular networks: A survey," *IEEE Communications Surveys & Tutorials*, vol. 17, no. 2, pp. 580–603, 2015.

188 A. Y. Ding, B. Han, Y. Xiao, P. Hui, A. Srinivasan, M. Kojo, and S. Tarkoma, "Enabling energy-aware collaborative mobile data offloading for smartphones," in *IEEE International Conference on Sensing, Communications and Networking (SECON)*, 2013, pp. 487–495.

189 B. Han, P. Hui, V. S. A. Kumar, M. V. Marathe, J. Shao, and A. Srinivasan, "Mobile data offloading through opportunistic communications and social participation," *IEEE Transactions on Mobile Computing*, vol. 11, no. 5, pp. 821–834, 2012.

190 E. Cuervo, A. Balasubramanian, D.-k. Cho, A. Wolman, S. Saroiu, R. Chandra, and P. Bahl, "MAUI: Making smartphones last longer with code offload," in *Proceedings of the 8th International Conference on Mobile Systems, Applications, and Services*, 2010, pp. 49–62. [Online]. Available: https://doi.org/10.1145/1814433.1814441.

191 K. Kumar and Y.-H. Lu, "Cloud computing for mobile users: Can offloading computation save energy?" *Computer*, vol. 43, no. 4, pp. 51–56, 2010.

192 B. Cao, L. Zhang, Y. Li, D. Feng, and W. Cao, "Intelligent offloading in multi-access edge computing: A state-of-the-art review and framework," *IEEE Communications Magazine*, vol. 57, no. 3, pp. 56–62, 2019.

193 F. A. Salaht, F. Desprez, and A. Lebre, "An overview of service placement problem in fog and edge computing," *ACM Computing Surveys (CSUR)*, vol. 53, no. 3, pp. 1–35, 2020.

194 C. Clark, K. Fraser, S. Hand, J. G. Hansen, E. Jul, C. Limpach, I. Pratt, and A. Warfield, "Live migration of virtual machines," in *Proceedings of the 2nd conference on Symposium on Networked Systems Design & Implementation-Volume 2*, 2005, pp. 273–286.

195 B. Chun, S. Ihm, P. Maniatis, M. Naik, and A. Patti, "CloneCloud: Elastic execution between mobile device and cloud," in *Proceedings of ACM EuroSys*, ser. EuroSys '11. Association for Computing Machinery, 2011, pp. 301–314. [Online]. Available: https://doi.org/10.1145/1966445.1966473.

196 S. Kosta, A. Aucinas, P. Hui, R. Mortier, and X. Zhang, "ThinkAir: Dynamic resource allocation and parallel execution in the cloud for mobile code offloading," in *Proceedings of IEEE INFOCOM*, 2012, pp. 945–953.

197 H. Flores, X. Su, V. Kostakos, A. Y. Ding, P. Nurmi, S. Tarkoma, P. Hui, and Y. Li, "Large-scale offloading in the Internet of Things," in *International Conference on Pervasive Computing and Communications Workshops (PerCom Workshops)*. IEEE, 2017, pp. 479–484.

198 J. Ren, G. Yu, Y. Cai, and Y. He, "Latency optimization for resource allocation in mobile-edge computation offloading," *IEEE Transactions on Wireless Communications*, vol. 17, no. 8, pp. 5506–5519, 2018.

199 J. Xu and J. Yao, "Exploiting physical-layer security for multiuser multicarrier computation offloading," *IEEE Wireless Communications Letters*, vol. 8, no. 1, pp. 9–12, 2018.

200 Automotive Edge Computing Consortium (AECC), "Driving Data to the Edge: The Challenge of Traffic Distribution," 2020, last accessed 04 February 2024. [Online]. Available: https://aecc.org/wp-content/uploads/2020/07/AECC_WG2_TR_v2.0_July2020.pdf.

201 GOV.UK, "Data protection," 2018, last accessed 10 June 2024. [Online]. Available: https://www.gov.uk/data-protection.

202 State of California Department of Justice, "California Consumer Privacy Act (CCPA)," 2018, last accessed 10 June 2024. [Online]. Available: https://oag.ca.gov/privacy/ccpa.

203 Justice Laws Website, "Personal Information Protection and Electronic Documents Act," 2019, last accessed 10 June 2024. [Online]. Available: https://laws-lois.justice.gc.ca/eng/acts/p-8.6/.

204 B. Zolfaghari, G. Srivastava, S. Roy, H. R. Nemati, F. Afghah, T. Koshiba, A. Razi, K. Bibak, P. Mitra, and B. K. Rai, "Content delivery networks: State of the art, trends, and future roadmap," *ACM Computing Surveys (CSUR)*, vol. 53, no. 2, Apr. 2020. [Online]. Available: https://doi.org/10.1145/3380613.

205 A. Vakali and G. Pallis, "Content delivery networks: Status and trends," *IEEE Internet Computing*, vol. 7, no. 6, pp. 68–74, 2003.

206 V. Cozzolino, J. Ott, A. Y. Ding, and R. Mortier, "ECCO: Edge-cloud chaining and orchestration framework for road context assessment," in *2020 IEEE/ACM 5th International Conference on Internet-of-Things Design and Implementation (IoTDI)*. IEEE, 2020, pp. 223–230.

207 P. Guo, B. Hu, R. Li, and W. Hu, "FoggyCache: Cross-device approximate computation reuse," in *Proceedings of the 24th Annual International Conference on Mobile Computing and Networking*, ser. MobiCom '18. New York, NY, USA:

Association for Computing Machinery, 2018, pp. 19–34. [Online]. Available: https://doi.org/10.1145/3241539.3241557.

208 T. Braud, P. Zhou, J. Kangasharju, and P. Hui, "Multipath computation offloading for mobile augmented reality," in *In Proceedings of the IEEE International Conference on Pervasive Computing and Communications (PerCom 2020),* Austin USA, 2020.

209 G. Bartolomeo, M. Yosofie, S. Bäurle, O. Haluszczynski, N. Mohan, and J. Ott, "Oakestra: A lightweight hierarchical orchestration framework for edge computing," in *2023 USENIX Annual Technical Conference (USENIX ATC 23)*. Boston, MA, USA: USENIX Association, Jul. 2023, pp. 215–231. [Online]. Available: https://www.usenix.org/conference/atc23/presentation/bartolomeo.

210 Bluetooth SIG, "*Bluetooth Core Specification Version 5.0*," last accessed 26 October 2020. [Online]. Available: https://www.bluetooth.com/specifications/bluetooth-core-specification/.

211 WiFi Alliance, "WiFi Alliance," last accessed 24 February 2024. [Online]. Available: https://www.wi-fi.org/.

212 R. Tafazolli, C. Wang, and P. Chatzimisios, *Wiley 5G Ref: The essential 5G reference online*. John Wiley & Sons, Limited, 2019. [Online]. Available: https://books.google.nl/books?id=tZlPzQEACAAJ.

213 J. Iyengar and M. Thomson, "QUIC: A UDP-based multiplexed and secure transport," Internet Requests for Comments, RFC Editor, RFC 9000, 2021.

214 A. Ford, C. Raiciu, M. Handley, O. Bonaventure, and C. Paasch, "TCP extensions for multipath operation with multiple addresses," Internet Requests for Comments, RFC Editor, RFC 8684, 2020.

215 B. Ahlgren, C. Dannewitz, C. Imbrenda, D. Kutscher, and B. Ohlman, "A survey of information-centric networking," *IEEE Communications Magazine*, vol. 50, no. 7, pp. 26–36, 2012.

216 S. K. Fayazbakhsh, Y. Lin, A. Tootoonchian, A. Ghodsi, T. Koponen, B. Maggs, K. Ng, V. Sekar, and S. Shenker, "Less pain, most of the gain: Incrementally deployable ICN," in *Proceedings of the ACM SIGCOMM 2013 Conference on SIGCOMM*, ser. SIGCOMM '13. New York, NY, USA: Association for Computing Machinery, 2013, pp. 147–158. [Online]. Available: https://doi.org/10.1145/2486001.2486023.

217 Docker Inc., "Docker," 2024, last accessed 04 February 2024. [Online]. Available: https://www.docker.com/.

218 A. Madhavapeddy and D. J. Scott, "Unikernels: The rise of the virtual library operating system," *Communications of the ACM*, vol. 57, no. 1, pp. 61–69, 2014.

219 J. Ren, D. Zhang, S. He, Y. Zhang, and T. Li, "A survey on end-edge-cloud orchestrated network computing paradigms: Transparent computing, mobile edge computing, fog computing, and cloudlet," *ACM Computing Surveys (CSUR)*, vol. 52, no. 6, pp. 1–36, 2019.

220 S. A. Huda and S. Moh, "Survey on computation offloading in UAV-enabled mobile edge computing," *Journal of Network and Computer Applications*, vol. 201, p. 103341, 2022. [Online]. Available: https://www.sciencedirect.com/science/article/pii/S1084804522000108.

221 J. Meng, Z. J. Kong, Y. C. Hu, M. G. Choi, and D. Lal, "Do we need sophisticated system design for edge-assisted augmented reality?" in *Proceedings of the 5th International*

Workshop on Edge Systems, Analytics and Networking, ser. EdgeSys '22. New York, NY, USA: Association for Computing Machinery, 2022, pp. 7–12. [Online]. Available: https://doi.org/10.1145/3517206.3526267.

222 N. H. Motlagh, P. Kortoçi, X. Su, L. Lovén, H. K. Hoel, S. Bjerkestrand Haugsvær, V. Srivastava, C. F. Gulbrandsen, P. Nurmi, and S. Tarkoma, "Unmanned aerial vehicles for air pollution monitoring: A survey," *IEEE Internet of Things Journal*, vol. 10, no. 24, pp. 21687–21704, 2023.

223 S. Yu, X. Wang, and R. Langar, "Computation offloading for mobile edge computing: A deep learning approach," in *2017 IEEE 28th Annual International Symposium on Personal, Indoor, and Mobile Radio Communications (PIMRC)*, 2017, pp. 1–6.

224 B. Chen, Z. Yan, H. Guo, Z. Yang, A. Ali-Eldin, P. Shenoy, and K. Nahrstedt, "Deep contextualized compressive offloading for images," in *Proceedings of the 19th ACM Conference on Embedded Networked Sensor Systems*, ser. SenSys '21. New York, NY, USA: Association for Computing Machinery, 2021, pp. 467–473. [Online]. Available: https://doi.org/10.1145/3485730.3493452.

225 T. Murad, A. Nguyen, and Z. Yan, "DAO: Dynamic adaptive offloading for video analytics," in *Proceedings of the 30th ACM International Conference on Multimedia*, ser. MM '22. New York, NY, USA: Association for Computing Machinery, 2022, pp. 3017–3025. [Online]. Available: https://doi.org/10.1145/3503161.3548249.

226 L. Kaelbling, M. Littman, and A. Moore, "Reinforcement learning: A survey," *Journal of Artificial Intelligence Research*, vol. 4, pp. 237–285, 1996.

227 Z. Zabihi, A. M. Eftekhari Moghadam, and M. H. Rezvani, "Reinforcement learning methods for computation offloading: A systematic review," *ACM Computing Surveys*, vol. 56, no. 1, Aug. 2023. [Online]. Available: https://doi.org/10.1145/3603703.

228 M. Tang and V. W. Wong, "Deep reinforcement learning for task offloading in mobile edge computing systems," *IEEE Transactions on Mobile Computing*, vol. 21, no. 6, pp. 1985–1997, 2022.

229 X. Wang, G. Gao, X. Wu, Y. Lyu, and W. Wu, "Dynamic DNN model selection and inference off loading for video analytics with edge-cloud collaboration," ser. NOSSDAV '22. New York, NY, USA: Association for Computing Machinery, 2022, pp. 64–70. [Online]. Available: https://doi.org/10.1145/3534088.3534352.

230 B. Trinh and G.-M. Muntean, "A deep reinforcement learning-based offloading scheme for multi-access edge computing-supported extended reality systems," *IEEE Transactions on Vehicular Technology*, vol. 72, no. 1, pp. 1254–1264, 2023.

231 C. De Alwis, A. Kalla, Q.-V. Pham, P. Kumar, K. Dev, W.-J. Hwang, and M. Liyanage, "Survey on 6G frontiers: Trends, applications, requirements, technologies and future research," *IEEE Open Journal of the Communications Society*, vol. 2, pp. 836–886, 2021.

232 I. Lujic, V. De Maio, and I. Brandic, "Efficient edge storage management based on near real-time forecasts," in *2017 IEEE 1st International Conference on Fog and Edge Computing (ICFEC)*. IEEE, 2017, pp. 21–30.

233 H. Chen, J. Yu, H. Zhou, T. Zhou, F. Liu, and Z. Cai, "SmartStore: A blockchain and clustering based intelligent edge storage system with fairness and resilience," *International Journal of Intelligent Systems*, vol. 36, no. 9, pp. 5184–5209, 2021.

234 Q.-V. Pham, F. Fang, V. N. Ha, M. J. Piran, M. Le, L. B. Le, W.-J. Hwang, and Z. Ding, "A survey of multi-access edge computing in 5G and beyond: Fundamentals, technology integration, and state-of-the-art," *IEEE Access*, vol. 8, pp. 116974–117017, 2020.

235 A.-C. Nicolaescu, S. Mastorakis, and I. Psaras, "Store edge networked data (send): A data and performance driven edge storage framework," in *IEEE INFOCOM 2021-IEEE Conference on Computer Communications*. IEEE, 2021, pp. 1–10.

236 D. Liu, B. Chen, C. Yang, and A. F. Molisch, "Caching at the wireless edge: Design aspects, challenges, and future directions," *IEEE Communications Magazine*, vol. 54, no. 9, pp. 22–28, 2016.

237 S. Nayak, R. Patgiri, L. Waikhom, and A. Ahmed, "A review on edge analytics: Issues, challenges, opportunities, promises, future directions, and applications," *Digital Communications and Networks*, vol. 10, no. 3, pp. 783–804, 2022.

238 J. Xia, G. Cheng, S. Gu, and D. Guo, "Secure and trust-oriented edge storage for internet of things," *IEEE Internet of Things Journal*, vol. 7, no. 5, pp. 4049–4060, 2019.

239 L. Yuan, Q. He, F. Chen, J. Zhang, L. Qi, X. Xu, Y. Xiang, and Y. Yang, "CSEdge: Enabling collaborative edge storage for multi-access edge computing based on blockchain," *IEEE Transactions on Parallel and Distributed Systems*, vol. 33, no. 8, pp. 1873–1887, 2021.

240 K. Sonbol, Ö. Özkasap, I. Al-Oqily, and M. Aloqaily, "EdgeKV: Decentralized, scalable, and consistent storage for the edge," *Journal of Parallel and Distributed Computing*, vol. 144, pp. 28–40, 2020.

241 C. de Alwis, H. K. Arachchi, V. De Silva, A. Fernando, and A. Kondoz, "Robust video communication using random linear network coding with pre-coding and interleaving," in *2012 19th IEEE International Conference on Image Processing*. IEEE, 2012, pp. 2269–2272.

242 F. Qiao, J. Wu, J. Li, A. K. Bashir, S. Mumtaz, and U. Tariq, "Trustworthy edge storage orchestration in intelligent transportation systems using reinforcement learning," *IEEE Transactions on Intelligent Transportation Systems*, vol. 22, no. 7, pp. 4443–4456, 2020.

243 aws, "Amazon Caching Overview," last accessed 20 June 2023. [Online]. Available: https://aws.amazon.com/caching/.

244 J. Yao, T. Han, and N. Ansari, "On mobile edge caching," *IEEE Communications Surveys & Tutorials*, vol. 21, no. 3, pp. 2525–2553, 2019.

245 G. Gür, A. Kalla, C. De Alwis, Q.-V. Pham, K.-H. Ngo, M. Liyanage, and P. Porambage, "Integration of ICN and MEC in 5G and beyond networks: Mutual benefits, use cases, challenges, standardization, and future research," *IEEE Open Journal of the Communications Society*, vol. 3, pp. 1382–1412, 2022.

246 K. Yu, S. Eum, T. Kurita, Q. Hua, T. Sato, H. Nakazato, T. Asami, and V. P. Kafle, "Information-centric networking: Research and standardization status," *IEEE Access*, vol. 7, pp. 126164–126176, 2019.

247 L. Zhang, D. Estrin, J. Burke, V. Jacobson, J. D. Thornton, D. K. Smetters, B. Zhang, G. Tsudik, D. Massey, C. Papadopoulos, T. Abdelzaher, L. Wang, P. Crowley, and E. Yeh, "Named data networking (NDN) project," *Relatório Técnico NDN-0001, Xerox Palo Alto Research Center-PARC*, vol. 157, p. 158, 2010.

248 C. Han, Y. Wu, Z. Chen, and X. Wang, "Terahertz Communications (TeraCom): Challenges and Impact on 6G Wireless Systems," *arXiv preprint arXiv:1912.06040*, 2019.

249 W. Saad, M. Bennis, and M. Chen, "A vision of 6G wireless systems: Applications, trends, technologies, and open research problems," *IEEE Network*, vol. 34, no. 3, pp. 134–142, 2019.

250 ETSI, "Zero-touch network and service management (ZSM); Terminology for concepts in ZSM," Aug. 2019, last accessed 29 March 2021. [Online]. Available: https://www.etsi.org/?gs/ZSM/001?099/007/01.01.01?60/gs?ZSM007v010101p.pdf.

251 ETSI, "Zero-touch network and Service Management (ZSM); Requirements based on documented scenarios," Oct. 2019, last accessed 29 March 2021. [Online]. Available: https://www.etsi.org/?gs/ZSM/001?099/001/01.01.01?60/gs?ZSM001v010101p.pdf.

252 Y. Lu, "The blockchain: State-of-the-art and research challenges," *Journal of Industrial Information Integration*, vol. 15, pp. 80–90, 2019.

253 T. M. Hewa, A. Kalla, A. Nag, M. E. Ylianttila, and M. Liyanage, "Blockchain for 5G and IoT: Opportunities and challenges," in *2020 IEEE 8th International Conference on Communications and Networking (ComNet)*. IEEE, 2020, pp. 1–8.

254 Q. Yang, Y. Liu, T. Chen, and Y. Tong, "Federated machine learning: Concept and applications," *ACM Transactions on Intelligent Systems and Technology*, vol. 10, no. 2, pp. 1–19, 2019.

255 M. Z. Chowdhury, M. Shahjalal, M. K. Hasan, Y. M. Jang, "The role of optical wireless communication technologies in 5G/6G and IoT solutions: Prospects, directions, and challenges," *Applied Sciences*, vol. 9, no. 20, p. 4367, 2019.

256 D. Karunatilaka, F. Zafar, V. Kalavally, and R. Parthiban, "Led based indoor visible light communications: State of the art," *IEEE Communications Surveys & Tutorials*, vol. 17, no. 3, pp. 1649–1678, 2015.

257 N. Gisin and R. Thew, "Quantum communication," *Nature Photonics*, vol. 1, no. 3, pp. 165–171, 2007.

258 Z. Na, Y. Liu, J. Shi, C. Liu, and Z. Gao, "UAV-supported clustered NOMA for 6G-enabled Internet of Things: Trajectory planning and resource allocation," *IEEE Internet of Things Journal*, vol. 8, no. 20, pp. 15041–15048, 2020.

259 M. Alsenwi, N. H. Tran, M. Bennis, S. R. Pandey, A. K. Bairagi, and C. S. Hong, "Intelligent resource slicing for eMBB and URLLC coexistence in 5G and beyond: A deep reinforcement learning based approach," *IEEE Transactions on Wireless Communications*, vol. 20, no. 7, pp. 4585–4600, 2021.

260 Y. Xing and T. S. Rappaport, "Propagation measurement system and approach at 140 GHz-moving to 6G and above 100 GHz," in *2018 IEEE Global Communications Conference (GLOBECOM)*. IEEE, 2018, pp. 1–6.

261 G. Gür, "Expansive networks: Exploiting spectrum sharing for capacity boost and 6G vision," *Journal of Communications and Networks*, vol. 22, no. 6, pp. 444–454, 2020.

262 M. A. Siddiqi, H. Yu, and J. Joung, "5G ultra-reliable low-latency communication implementation challenges and operational issues with IoT devices," *Electronics*, vol. 8, no. 9, p. 981, 2019.

263 F. Fang, Y. Xu, Q.-V. Pham, and Z. Ding, "Energy-efficient design of IRS-NOMA networks," *IEEE Transactions on Vehicular Technology*, vol. 69, no. 11, pp. 14088–14092, 2020.

264 C. Huang, A. Zappone, G. C. Alexandropoulos, M. Debbah, and C. Yuen, "Reconfigurable intelligent surfaces for energy efficiency in Wireless Communication," *IEEE Transactions on Wireless Communications*, vol. 18, no. 8, pp. 4157–4170, 2019.

265 S. P. Rout, "6G wireless communication: Its vision, viability, application, requirement, technologies, encounters and research," in *2020 11th International Conference on Computing, Communication and Networking Technologies (ICCCNT)*. IEEE, 2020, pp. 1–8.

266 J. Navarro-Ortiz, P. Romero-Diaz, S. Sendra, P. Ameigeiras, J. J. Ramos-Munoz, and J. M. Lopez-Soler, "A survey on 5G usage scenarios and traffic models," *IEEE Communications Surveys & Tutorials*, vol. 22, no. 2, pp. 905–929, 2020.

267 C.-X. Wang, J. Huang, H. Wang, X. Gao, X. You, and Y. Hao, "6G wireless channel measurements and models: Trends and challenges," *IEEE Vehicular Technology Magazine*, vol. 15, no. 4, pp. 22–32, 2020.

268 W. Li and M. Liewig, "A survey of AI accelerators for edge environment," in *Trends and Innovations in Information Systems and Technologies, Volume* 2 8. Springer, 2020, pp. 35–44.

269 L. Pomšár, A. Brecko, and I. Zolotová, "Brief overview of edge AI accelerators for energy-constrained edge," in *2022 IEEE 20th Jubilee World Symposium on Applied Machine Intelligence and Informatics (SAMI)*, 2022, pp. 000461–000466.

270 A. Mishra, P. Yadav, and S. Kim, "Artificial intelligence accelerators," in *Artificial Intelligence and Hardware Accelerators*. Springer, 2023, pp. 1–52.

271 A. Reuther, P. Michaleas, M. Jones, V. Gadepally, S. Samsi, and J. Kepner, "Survey of machine learning accelerators," in *2020 IEEE High Performance Extreme Computing Conference (HPEC)*. IEEE, 2020, pp. 1–12.

272 A. Reuther, P. Michaleas, M. Jones, V. Gadepally, S. Samsi, and J. Kepner, "AI and ML accelerator survey and trends," in *2022 IEEE High Performance Extreme Computing Conference (HPEC)*. IEEE, 2022, pp. 1–10.

273 D. Liu, H. Kong, X. Luo, W. Liu, and R. Subramaniam, "Bringing AI to edge: From deep learning's perspective," *Neurocomputing*, vol. 485, pp. 297–320, 2022.

274 H. J. Damsgaard, A. Ometov, and J. Nurmi, "Approximation opportunities in edge computing hardware: A systematic literature review," *ACM Computing Surveys*, vol. 55, no. 12, pp. 1–49, 2023.

275 D. Pathak, R. N. Sarangapani, D. Katare, A. S. Gaikwad, and M. El-Sharkawy, "IoT based solution for level detection using CNN and OpenCV," in *Proceedings on the International Conference on Internet Computing (ICOMP)*, 2018, pp. 83–87.

276 A. Raha, S. K. Kim, D. A. Mathaikutty, G. Venkataramanan, D. Mohapatra, R. Sung, C. Brick, and G. N. Chinya, "Design considerations for edge neural network accelerators: An industry perspective," in *2021 34th International Conference on VLSI Design and 2021 20th International Conference on Embedded Systems (VLSID)*, 2021, pp. 328–333.

277 E. Manor and S. Greenberg, "Custom hardware inference accelerator for tensorflow lite for microcontrollers," *IEEE Access*, vol. 10, pp. 73484–73493, 2022.

278 E. Asanghanwa, "The enclave device blueprint for confidential computing at the edge - Azure blog," last accessed 26 September 2023. [Online]. Available: https://azure.microsoft.com/en-us/blog/the-enclave-device-blueprint-for-confidential-computing-at-the-edge/.

279 Amazon, "Edge networking and content delivery - AWS," last accessed 28 September 2023. [Online]. Available: https://aws.amazon.com/products/networking/edge-networking/.

280 Amazon, "AI Chip - Amazon Inferentia - AWS — aws.amazon.com," last accessed 28 September 2023. [Online]. Available: https://aws.amazon.com/machine-learning/inferentia/.

281 S. Iftikhar, S. S. Gill, C. Song, M. Xu, M. S. Aslanpour, A. N. Toosi, J. Du, H. Wu, S. Ghosh, D. Chowdhury, and M. Golec, "AI-based fog and edge computing: A systematic review, taxonomy and future directions," *Internet of Things*, vol. 21, p. 100674, 2022.

282 R. Machupalli, M. Hossain, and M. Mandal, "Review of ASIC accelerators for deep neural network," *Microprocessors and Microsystems*, vol. 89, p. 104441, 2022.

283 N. Perryman, C. Wilson, and A. George, "Evaluation of xilinx versal architecture for next-gen edge computing in space," in *2023 IEEE Aerospace Conference*. IEEE, 2023, pp. 1–11.

284 I. K. Ganusov, M. A. Iyer, N. Cheng, and A. Meisler, "Agilex™ generation of intel® fpgas," in *2020 IEEE Hot Chips 32 Symposium (HCS)*. IEEE Computer Society, 2020, pp. 1–26.

285 A. Durand and J. Pasquier, "Physical unclonable functions for IoT security using free software," in *Proceedings of the 11th International Conference on the Internet of Things*, 2021, pp. 173–176.

286 N. Jouppi, "Google supercharges machine learning tasks with TPU custom chip," 2016, last accessed 22 February 2024. [Online]. Available: https://cloud.google.com/blog/products/ai-machine-learning/google-supercharges-machine-learning-tasks-with-custom-chip.

287 U. H. Jeff Dean, "Build and train machine learning models on our new google cloud TPU," 2017, last accessed 22 February 2024. [Online]. Available: https://blog.google/products/google-cloud/google-cloud-offer-tpus-machine-learning/.

288 K.-C. Hsu and H.-W. Tseng, "Accelerating applications using edge tensor processing units," in *Proceedings of the International Conference for High Performance Computing, Networking, Storage and Analysis*, 2021, pp. 1–14.

289 K. Seshadri, B. Akin, J. Laudon, R. Narayanaswami, and A. Yazdanbakhsh, "An evaluation of edge TPU accelerators for convolutional neural networks," in *2022 IEEE International Symposium on Workload Characterization (IISWC)*. IEEE, 2022, pp. 79–91.

290 R. Prokscha, M. Schneider, and A. Höß, "Efficient edge deployment demonstrated on YOLOv5 and coral edge TPU," in *Industrial Artificial Intelligence Technologies and Applications*. River Publishers, 2023, pp. 141–155.

291 V. Taran, Y. Gordienko, O. Rokovyi, O. Alienin, Y. Kochura, and S. Stirenko, "Edge intelligence for medical applications under field conditions," in *The International Conference on Artificial Intelligence and Logistics Engineering*. Springer, 2022, pp. 71–80.

292 S. Hosseininoorbin, S. Layeghy, M. Sarhan, R. Jurdak, and M. Portmann, "Exploring edge TPU for network intrusion detection in IoT," *Journal of Parallel and Distributed Computing*, vol. 179, p. 104712, 2023.

293 K. Karras, E. Pallis, G. Mastorakis, Y. Nikoloudakis, J. M. Batalla, C. X. Mavromoustakis, and E. Markakis, "A hardware acceleration platform for AI-based inference at the edge," *Circuits, Systems, and Signal Processing*, vol. 39, pp. 1059–1070, 2020.

294 C. Marantos, N. Karavalakis, V. Leon, V. Tsoutsouras, K. Pekmestzi, and D. Soudris, "Efficient support vector machines implementation on Intel/Movidius Myriad 2," in *2018 7th International Conference on Modern Circuits and Systems Technologies (MOCAST)*. IEEE, 2018, pp. 1–4.

295 L. Deng, G. Li, S. Han, L. Shi, and Y. Xie, "Model compression and hardware acceleration for neural networks: A comprehensive survey," *Proceedings of the IEEE*, vol. 108, no. 4, pp. 485–532, 2020.

296 H. Rahimi, Y. Picaud, K. D. Singh, G. Madhusudan, S. Costanzo, and O. Boissier, "Design and simulation of a hybrid architecture for edge computing in 5G and beyond," *IEEE Transactions on Computers*, vol. 70, no. 8, pp. 1213–1224, 2021.

297 R. Singh and S. S. Gill, "Edge AI: A survey," *Internet of Things and Cyber-Physical Systems*, vol. 3, pp. 71–92, 2023.

298 Y. Xiong, Y. Sun, L. Xing, and Y. Huang, "Extend cloud to edge with KubeEdge," in *2018 IEEE/ACM Symposium On Edge Computing (SEC)*. IEEE, 2018, pp. 373–377.

299 E. Li, L. Zeng, Z. Zhou, and X. Chen, "Edge AI: On-demand accelerating deep neural network inference via edge computing," *IEEE Transactions on Wireless Communications*, vol. 19, no. 1, pp. 447–457, 2019.

300 H. Ning, Y. Li, F. Shi, and L. T. Yang, "Heterogeneous edge computing open platforms and tools for Internet of Things," *Future Generation Computer Systems*, vol. 106, pp. 67–76, 2020.

301 "Azure IoT," 2023. [Online]. Available: https://azure.microsoft.com/.

302 "Extending cloud to the edge," 2023. [Online]. Available: https://www.dell.com/en-uk/dt/learn/cloud/edge-cloud.htm.

303 X. Wang, Y. Han, V. C. Leung, D. Niyato, X. Yan, and X. Chen, *Edge AI: Convergence of edge computing and artificial intelligence*. Springer, 2020.

304 Transforma Insights, "Global IoT Market will Grow to 24.1 Billion Devices in 2030, Generating $1.5 Trillion Annual Revenue," Transforma Insights Research, May 2020, last accessed 29 March 2021. [Online]. Available: https://transformainsights.com/news/iot-market-24-billion-usd15-trillion-revenue-2030.

305 M. H. Miraz, M. Ali, P. S. Excell, and R. Picking, "A review on Internet of Things (IoT), Internet of Everything (IoE) and Internet of Nano Things (IoNT)," in *2015 Internet Technologies and Applications (ITA)*. IEEE, 2015, pp. 219–224.

306 N. Janbi, I. Katib, A. Albeshri, and R. Mehmood, "Distributed artificial intelligence-as-a-service (DAIaaS) for smarter IoE and 6G environments," *Sensors*, vol. 20, no. 20, p. 5796, 2020.

307 S. B. Azmy, R. A. Sneineh, N. Zorba, and H. S. Hassanein, "Small Data in IoT: An MCS Perspective," in *Performability in Internet of Things*. Springer, 2019, pp. 209–229.

308 S. Hu, F. Rusek, and O. Edfors, "Beyond massive MIMO: The potential of data transmission with large intelligent surfaces," *IEEE Transactions on Signal Processing*, vol. 66, no. 10, pp. 2746–2758, 2018.

309 A. Taha, M. Alrabeiah, and A. Alkhateeb, "Enabling large intelligent surfaces with compressive sensing and deep learning," *IEEE Access*, vol. 9, pp. 44304–44321, 2021.

310 S. Nahavandi, "Industry 5.0—A human-centric solution," *Sustainability*, vol. 11, no. 16, p. 4371, 2019.

311 V. Özdemir and N. Hekim, "Birth of industry 5.0: Making sense of big data with artificial intelligence, "The Internet of Things" and next-generation technology policy," *Omics: A Journal of Integrative Biology*, vol. 22, no. 1, pp. 65–76, 2018.

312 S. Nayak and R. Patgiri, "6G communication technology: A vision on intelligent healthcare," *arXiv preprint arXiv:2005.07532*, 2020.

313 L. Mucchi, S. Jayousi, S. Caputo, E. Paoletti, P. Zoppi, S. Geli, and P. Dioniso, "How 6G technology can change the future wireless healthcare," in *2020 2nd 6G Wireless Summit (6G SUMMIT)*. IEEE, 2020, pp. 1–6.

314 P. Mehta, R. Gupta, and S. Tanwar, "Blockchain envisioned UAV networks: Challenges, solutions, and comparisons," *Computer Communications*, vol. 151, pp. 518–538, 2020.

315 L. Gupta, R. Jain, and G. Vaszkun, "Survey of important issues in UAV communication networks," *IEEE Communications Surveys & Tutorials*, vol. 18, no. 2, pp. 1123–1152, 2015.

316 P. K. R. Maddikunta, S. Hakak, M. Alazab, S. Bhattacharya, T. R. Gadekallu, W. Z. Khan, and Q.-V. Pham, "Unmanned aerial vehicles in smart agriculture: Applications, requirements, and challenges," *IEEE Sensors Journal*, vol. 21, no. 16, pp. 17608–17619, 2021.

317 R. Gupta, A. Kumari, S. Tanwar, and N. Kumar, "Blockchain-envisioned softwarized multi-swarming UAVs to tackle COVID-I9 situations," *IEEE Network*, vol. 35, no. 2, pp. 160–167, 2021.

318 M. Giordani, M. Polese, M. Mezzavilla, S. Rangan, and M. Zorzi, "Toward 6G networks: Use cases and technologies," *IEEE Communications Magazine*, vol. 58, no. 3, pp. 55–61, 2020.

319 F. Yang, S. Wang, J. Li, Z. Liu, and Q. Sun, "An overview of internet of vehicles," *China Communications*, vol. 11, no. 10, pp. 1–15, 2014.

320 G. Raja, Y. Manaswini, G. D. Vivekanandan, H. Sampath, K. Dev, and A. K. Bashir, "AI-powered blockchain-a decentralized secure multiparty computation protocol for IoV," in *IEEE INFOCOM 2020-IEEE Conference on Computer Communications Workshops (INFOCOM WKSHPS)*. IEEE, 2020, pp. 865–870.

321 A. P. Dani, I. Salehi, G. Rotithor, D. Trombetta, and H. Ravichandar, "Human-in-the-loop robot control for human-robot collaboration: Human intention estimation and safe trajectory tracking control for collaborative tasks," *IEEE Control Systems Magazine*, vol. 40, no. 6, pp. 29–56, 2020.

322 European Telecommunications Standards Institute, 2018, last accessed 18 March 2018. [Online]. Available: https://www.etsi.org/.

323 The Next Generation Mobile Networks, last accessed 29 March 2021. [Online]. Available: https://www.ngmn.org/.

324 Alliance for Telecommunications Industry Solutions, last accessed 29 March 2021. [Online]. Available: https://www.atis.org/.

325 Next G Alliance, last accessed 29 March 2021. [Online]. Available: https://nextgalliance.org/.

326 R. Von Solms and J. Van Niekerk, "From information security to cyber security," *Computers & Security*, vol. 38, pp. 97–102, 2013.

327 D. Schatz, R. Bashroush, and J. Wall, "Towards a more representative definition of cyber security," *Journal of Digital Forensics, Security and Law*, vol. 12, no. 2, p. 8, 2017.

328 M. Ehrlich, L. Wisniewski, H. Trsek, and J. Jasperneite, "Modelling and automatic mapping of cyber security requirements for industrial applications: Survey, problem exposition, and research focus," in *2018 14th IEEE International Workshop on Factory Communication Systems (WFCS)*. IEEE, 2018, pp. 1–9.

329 R. Leszczyna, "A review of standards with cybersecurity requirements for smart grid," *Computers & Security*, vol. 77, pp. 262–276, 2018.

330 I. Afolabi, T. Taleb, K. Samdanis, A. Ksentini, and H. Flinck, "Network slicing and softwarization: A survey on principles, enabling technologies, and solutions," *IEEE Communications Surveys & Tutorials*, vol. 20, no. 3, pp. 2429–2453, 2018.

331 F.-Y. Wang, L. Yang, X. Cheng, S. Han, and J. Yang, "Network softwarization and parallel networks: Beyond software-defined networks," *IEEE Network*, vol. 30, no. 4, pp. 60–65, 2016.

332 M. A. Salahuddin, A. Al-Fuqaha, M. Guizani, K. Shuaib, and F. Sallabi, "Softwarization of Internet of Things infrastructure for secure and smart healthcare," *arXiv preprint arXiv:1805.11011*, 2018.

333 P. Leitao, S. Karnouskos, L. Ribeiro, J. Lee, T. Strasser, and A. W. Colombo, "Smart agents in industrial cyber-physical systems," *Proceedings of the IEEE*, vol. 104, no. 5, pp. 1086–1101, 2016.

334 J. Kim, D. Kim, and S. Choi, "3GPP SA2 architecture and functions for 5G mobile communication system," *ICT Express*, vol. 3, no. 1, pp. 1–8, 2017.

335 X. Lu, D. Niyato, N. Privault, H. Jiang, and P. Wang, "Managing physical layer security in wireless cellular networks: A cyber insurance approach," *IEEE Journal on Selected Areas in Communications*, vol. 36, no. 7, pp. 1648–1661, 2018.

336 S. Wang, Y. Zhao, J. Xu, J. Yuan, and C.-H. Hsu, "Edge server placement in mobile edge computing," *Journal of Parallel and Distributed Computing*, vol. 127, pp. 160–168, 2018.

337 O. N. Hamoud, T. Kenaza, and Y. Challal, "Security in device-to-device communications: A survey," *IET Networks*, vol. 7, no. 1, pp. 14–22, 2017.

338 T. Kumar, P. Porambage, I. Ahmad, M. Liyanage, E. Harjula, and M. Ylianttila, "Securing gadget-free digital services," *Computer*, vol. 51, no. 11, pp. 66–77, 2018.

339 V. Marttin, U. Yüzgeç, C. Bayılmış, and K. Küçük, "D2D communication and energy efficiency on LTE for public safety networks," in *2018 3rd International Conference on Computer Science and Engineering (UBMK)*. IEEE, 2018, pp. 547–551.

340 I. Ahmad, M. Liyanage, S. Namal, M. Ylianttila, A. Gurtov, M. Eckert, T. Bauschert, Z. Faigl, L. Bokor, E. Saygun, and H. A. Akyıldız, "New concepts for traffic, resource and mobility management in software-defined mobile networks," in *2016 12th Annual*

Conference on Wireless On-demand Network Systems and Services (WONS). IEEE, 2016, pp. 1–8.

341 K. Peng, V. Leung, X. Xu, L. Zheng, J. Wang, and Q. Huang, "A survey on mobile edge computing: Focusing on service adoption and provision," *Wireless Communications and Mobile Computing*, vol. 2018, no. 1, p. 8267838, 2018.

342 T. Choi, S. Yoon, and S. Song, "Information fusion based agile streaming telemetry for intelligent traffic analytics of softwarized network," in *2017 19th Asia-Pacific Network Operations and Management Symposium (APNOMS)*. IEEE, 2017, pp. 399–402.

343 S. Lal, T. Taleb, and A. Dutta, "NFV: Security threats and best practices," *IEEE Communications Magazine*, vol. 55, no. 8, pp. 211–217, 2017.

344 Y. Ai, M. Peng, and K. Zhang, "Edge computing technologies for Internet of Things: A primer," *Digital Communications and Networks*, vol. 4, no. 2, pp. 77–86, 2018.

345 D. Kumar, N. R. Prakash, and S. Singh, "Techniques and concepts to mitigate and neutralize electro-magnetic weapon attacks," in *2015 13th International Conference on Electromagnetic Interference and Compatibility (INCEMIC)*. IEEE, 2015, pp. 69–73.

346 A. El Oualkadi and J. Zbitou, *Handbook of research on advanced trends in microwave and communication engineering, IGI Global*, 2016.

347 S. Dalal and S. Devi, "Security framework against denial of service attacks in wireless mesh network networks," *International Journal of Computer Networks and Communications Security*, vol. 4, no. 8, p. 237, 2016.

348 K. Guan, A. M. Tulino, P. J. Winzer, and E. Soljanin, "Secrecy capacities in space-division multiplexed fiber optic communication systems," *IEEE Transactions on Information Forensics and Security*, vol. 10, no. 7, pp. 1325–1335, 2015.

349 H. Ko, K. Lim, J. Oh, and J.-K. K. Rhee, "Informatic analysis for hidden pulse attack exploiting spectral characteristics of optics in plug-and-play quantum key distribution system," *Quantum Information Processing*, vol. 15, no. 10, pp. 4265–4282, 2016.

350 E. Gündüzhan and K. D. Brown, "Narrowband satellite communications: Challenges and emerging solutions," *John Hopkins APL Technical Digest*, vol. 33, no. 1, pp. 52–56, 2015.

351 M. Liyanage, P. Porambage, and A. Y. Ding, "Five driving forces of multi-access edge computing," *arXiv preprint arXiv:1810.00827*, 2018.

352 R. Solozabal, A. Sanchoyerto, E. Atxutegi, B. Blanco, J. O. Fajardo, and F. Liberal, "Exploitation of mobile edge computing in 5G distributed mission-critical push-to-talk service deployment," *IEEE Access*, vol. 6, pp. 37665–37675, 2018.

353 E. Piri, P. Ruuska, T. Kanstrén, J. Mäkelä, J. Korva, A. Hekkala, A. Pouttu, O. Liinamaa, M. Latva-Aho, K. Vierimaa, and H. Valasma, "5GTN: A test network for 5G application development and testing," in *(EuCNC), 2016 European Conference on Networks and Communications*. IEEE, 2016, pp. 313–318.

354 X. Costa-Perez, A. Garcia-Saavedra, X. Li, T. Deiss, T. De La Oliva, A. Di Giglio, P. Iovanna, and A. Moored, "5G-crosshaul: An SDN/NFV integrated fronthaul/backhaul transport network architecture," *IEEE Wireless Communications*, vol. 24, no. 1, pp. 38–45, 2017.

355 T. X. Tran, A. Hajisami, P. Pandealjhuaniy, and D. Pompili, "Collaborative mobile edge computing in 5G networks: New paradigms, scenarios, and challenges," *IEEE Communications Magazine*, vol. 55, no. 4, pp. 54–61, Apr. 2017.

356 I. Ahmad, T. Kumar, M. Liyanage, J. Okwuibe, M. Ylianttila, and A. Gurtov, "5G security: Analysis of threats and solutions," in *2017 IEEE Conference on Standards for Communications and Networking (CSCN)*. IEEE, 2017, pp. 193–199.

357 M. Liyanage, I. Ahmad, M. Ylianttila, A. Gurtov, A. B. Abro, and E. M. de Oca, "Leveraging LTE security with SDN and NFV," in *2015 IEEE 10th International Conference on Industrial and Information Systems (ICIIS)*. IEEE, 2015, pp. 220–225.

358 M. Liyanage, A. B. Abro, M. Ylianttila, and A. Gurtov, "Opportunities and challenges of software-defined mobile networks in network security," *IEEE Security & Privacy*, vol. 14, no. 4, pp. 34–44, 2016.

359 B. Han, V. Gopalakrishnan, L. Ji, and S. Lee, "Network function virtualization: Challenges and opportunities for innovations," *IEEE Communications Magazine*, vol. 53, no. 2, pp. 90–97, 2015.

360 R. Wojtczuk, *Poacher turned gamekeeper: Lessons learned from eight years of breaking hypervisors. Black Hat USA*, 2014.

361 A. Aljuhani and T. Alharbi, "Virtualized network functions security attacks and vulnerabilities," in *7th Annual Computing and Communication Workshop and Conference (CCWC). IEEE*, 2017, pp. 1–4.

362 F. Reynaud, F.-X. Aguessy, O. Bettan, M. Bouet, and V. Conan, "Attacks against network functions virtualization and software-defined networking: State-of-the-art," in *NetSoft Conference and Workshops (NetSoft), 2016 IEEE*. IEEE, 2016, pp. 471–476.

363 K. Leach, F. Zhang, and W. Weimer, "Scotch: Combining software guard extensions and system management mode to monitor cloud resource usage," in *International Symposium on Research in Attacks, Intrusions, and Defenses*. Springer, 2017, pp. 403–424.

364 J. Ni, X. Lin, and X. S. Shen, "Efficient and secure service-oriented authentication supporting network slicing for 5G-enabled IoT," *IEEE Journal on Selected Areas in Communications*, vol. 36, no. 3, pp. 644–657, 2018.

365 R. Ravindran, A. Chakraborti, S. O. Amin, A. Azgin, and G. Wang, "5G-ICN: Delivering ICN services over 5G using network slicing," *IEEE Communications Magazine*, vol. 55, no. 5, pp. 101–107, 2017.

366 S. Zhang, "An overview of network slicing for 5G," *IEEE Wireless Communications*, vol. 26, no. 3, pp. 111–117, 2019.

367 R. Harel and S. Babbage, "5G Security Recommendations Package 2: Network Slicing," 2016, last accessed 16 May 2019. [Online]. Available: https://www.ngmn.org/.

368 R. Wen, G. Feng, J. Tang, T. Q. Quek, G. Wang, W. Tan, and S. Qin, "On robustness of network slicing for next-generation mobile networks," *IEEE Transactions on Communications*, vol. 67, no. 1, pp. 430–444, 2019.

369 H. Hantouti, N. Benamar, T. Taleb, and A. Laghrissi, "Traffic steering for service function chaining," *IEEE Communications Surveys & Tutorials*, vol. 21, no. 1, pp. 487–507, 2018.

370 K. Fysarakis, N. E. Petroulakis, A. Roos, K. Abbasi, P. Vizarreta, G. Petropoulos, E. Sakic, G. Spanoudakis, and I. Askoxylakis, "A reactive security framework for operational wind parks using service function chaining," in *Computers and Communications (ISCC), 2017 IEEE Symposium on* IEEE, 2017, pp. 663–668.

371 S. Wang, J. Xu, N. Zhang, and L. Yujiong, “A survey on service migration in mobile edge computing,” *IEEE Access*, vol. 6, pp. 23511–23528, 2018.

372 F. Zhang, G. Liu, X. Fu, and R. Yahyapour, “A survey on virtual machine migration: Challenges, techniques, and open issues,” *IEEE Communications Surveys & Tutorials*, vol. 20, no. 2, pp. 1206–1243, 2018.

373 Y. Siriwardhana, P. Porambage, M. Liyanage, J. S. Walia, M. Matinmikko-Blue, and M. Ylianttila, “Micro-operator driven local 5G network architecture for industrial internet,” in *IEEE Wireless Communications and Networking Conference (WCNC) 2019*. IEEE, 2019, pp. 1–8.

374 P. Ahokangas, S. Moqaddamerad, M. Matinmikko, A. Abouzeid, I. Atkova, J. F. Gomes, and M. Iivari, “Future micro operators business models in 5G,” *The Business & Management Review*, vol. 7, no. 5, p. 143, 2016.

375 A. Prasad, Z. Li, S. Holtmanns, and M. A. Uusitalo, “5G micro-operator networks—A key enabler for new verticals and markets,” in *Telecommunication Forum (TELFOR), 2017 25th*. IEEE, 2017, pp. 1–4.

376 A. Braeken, M. Liyanage, P. Kumar, and J. Murphy, “Novel 5G authentication protocol to improve the resistance against active attacks and malicious serving networks,” *IEEE Access*, vol. 7, pp. 64040–64052, 2019.

377 U. Meyer and S. Wetzel, “On the impact of GSM encryption and man-in-the-middle attacks on the security of interoperating GSM/UMTS networks,” in *2004 IEEE 15th International Symposium on Personal, Indoor and Mobile Radio Communications (IEEE Cat. No. 04TH8754)*, volume 4. IEEE, 2004, pp. 2876–2883.

378 M. Frustaci, P. Pace, and G. Aloi, “Securing the IoT world: Issues and perspectives,” in *2017 IEEE Conference on Standards for Communications and Networking (CSCN)*. IEEE, 2017, pp. 246–251.

379 R. Khan, P. Kumar, D. N. K. Jayakody, and M. Liyanage, “A survey on security and privacy of 5G technologies: Potential solutions, recent advancements, and future directions,” *IEEE Communications Surveys & Tutorials*, vol. 22, no. 1, pp. 196–248, 2019.

380 M. H. Eiza, Q. Ni, and Q. Shi, “Secure and privacy-aware cloud-assisted video reporting service in 5G-enabled vehicular networks,” *IEEE Transactions on Vehicular Technology*, vol. 65, no. 10, pp. 7868–7881, 2016.

381 Z. Chen, F. Zhang, P. Zhang, J. K. Liu, J. Huang, H. Zhao, and J. Shen, “Verifiable keyword search for secure big data-based mobile healthcare networks with fine-grained authorization control,” *Future Generation Computer Systems*, vol. 87, pp. 712–724, 2018.

382 5GPPP-Security-WG, “5G PPP phase 1 security landscape,” *5GPP White Paper*, 2017, last accessed 11 November 2019. [Online]. Available: https://5g-ppp.eu/wp-content/uploads/2014/02/5G-PPP_White-Paper_Phase-1-Security-Landscape_June-2017.pdf.

383 N. Raad, T. Hasan, A. Chalak, and J. Waleed, “Secure data in LoRaWAN network by adaptive method of elliptic-curve cryptography,” in *2019 International Conference on Computing and Information Science and Technology and Their Applications (ICCISTA)*. IEEE, 2019, pp. 1–6.

384 Y. Jeon, H.-I. Ju, and S. Yoon, "Design of an LPWAN communication module based on secure element for smart parking application," in *2018 IEEE International Conference on Consumer Electronics (ICCE)*. IEEE, 2018, pp. 1–2.

385 L. Chen, L. Chen, S. Jordan, Y.-K. Liu, D. Moody, R. Peralta, R. Perlner, and D. Smith-Tone, *Report on post-quantum cryptography*. US Department of Commerce, National Institute of Standards and Technology, 2016.

386 S. Gupta, B. L. Parne, and N. S. Chaudhari, "Security vulnerabilities in handover authentication mechanism of 5G network," in *2018 1st International Conference on Secure Cyber Computing and Communication (ICSCCC)*. IEEE, 2018, pp. 369–374.

387 P. Hao and X. Wang, "Integrating PHY security into NDN-IoT networks by exploiting MEC: Authentication efficiency, robustness, and accuracy enhancement," *arXiv preprint arXiv:1904.03283*, 2019.

388 G. Choudhary, J. Kim, and V. Sharma, "Security of 5G-mobile backhaul networks: A survey," *arXiv preprint arXiv:1906.11427*, 2019.

389 D. Sabella, A. Vaillant, P. Kuure, U. Rauschenbach, and F. Giust, "Mobile-edge computing architecture: The role of MEC in the Internet of Things," *IEEE Consumer Electronics Magazine*, vol. 5, no. 4, pp. 84–91, 2016.

390 X. Zhang, A. Kunz, and S. Schröder, "Overview of 5G security in 3GPP," in *2017 IEEE Conference on Standards for Communications and Networking (CSCN)*. IEEE, 2017, pp. 181–186.

391 H. Zhu and C. Huang, "Availability-aware mobile edge application placement in 5G networks," in *GLOBECOM 2017-2017 IEEE Global Communications Conference*. IEEE, 2017, pp. 1–6.

392 M. Liyanage, A. Braeken, P. Kumar, and M. Ylianttila, *IoT security: Advances in authentication*. John Wiley & Sons, 2020.

393 M. H. Mahalat, S. Saha, A. Mondal, and B. Sen, "A PUF based light weight protocol for secure WiFi authentication of IoT devices," in *2018 8th International Symposium on Embedded Computing and System Design (ISED)*. IEEE, 2018, pp. 183–187.

394 G. Li and P. Bours, "Studying WiFi and accelerometer data based authentication method on mobile phones," in *Proceedings of the 2018 2nd International Conference on Biometric Engineering and Applications*. ACM, 2018, pp. 18–23.

395 Z. Zhao, G. Min, Y. Pang, W. Gao, and J. Lv, "Towards fast and reliable WiFi authentication by utilizing visible light diversity," in *2019 16th Annual IEEE International Conference on Sensing, Communication, and Networking (SECON)*. IEEE, 2019, pp. 1–9.

396 S. Kambou and A. Bouabdallah, "Improving strong mobile authentication with structural diversity and user-friendliness," in *2019 10th IFIP International Conference on New Technologies, Mobility and Security (NTMS)*. IEEE, 2019, pp. 1–5.

397 S. Kambou and A. Bouabdallah, "Using structural diversity to enforce strong authentication of mobiles to the cloud," in *2019 IEEE Conference on Communications and Network Security (CNS)*. IEEE, 2019, pp. 1–9.

398 Y. Zhang, F. Ren, A. Wu, T. Zhang, J. Cao, and D. Zheng, "Certificateless multi-party authenticated encryption for NB-IoT terminals in 5G networks," *IEEE Access*, vol. 7, pp. 114721–114730, 2019.

399 J. Cao, P. Yu, M. Ma, and W. Gao, "Fast authentication and data transfer scheme for massive NB-IoT devices in 3GPP 5G network," *IEEE Internet of Things Journal*, vol. 6, no. 2, pp. 1561–1575, 2018.

400 J. Cao, P. Yu, X. Xiang, M. Ma, and H. Li, "Anti-quantum fast authentication and data transmission scheme for massive devices in 5G NB-IoT system," *IEEE Internet of Things Journal*, vol. 6, no. 6, pp. 9794–9805, 2019.

401 P. Gope, R. Amin, S. H. Islam, N. Kumar, and V. K. Bhalla, "Lightweight and privacy-preserving RFID authentication scheme for distributed IoT infrastructure with secure localization services for smart city environment," *Future Generation Computer Systems*, vol. 83, pp. 629–637, 2018.

402 Z. Li, H. Zhao, X. Su, and C. Wan, "Asymmetric cryptography based unidirectional authentication method for RFID," in *2018 International Conference on Cyber-Enabled Distributed Computing and Knowledge Discovery (CyberC)*. IEEE, 2018, pp. 374–3743.

403 S.-C. Cha, M.-S. Chuang, K.-H. Yeh, Z.-J. Huang, and C. Su, "A user-friendly privacy framework for users to achieve consents with nearby BLE devices," *IEEE Access*, vol. 6, pp. 20779–20787, 2018.

404 T. Gu and P. Mohapatra, "BF-IoT: Securing the IoT networks via fingerprinting-based device authentication," in *2018 IEEE 15th International Conference on Mobile Ad Hoc and Sensor Systems (MASS)*. IEEE, 2018, pp. 254–262.

405 P. Yu, J. Cao, M. Ma, H. Li, B. Niu, and F. Li, "Quantum-resistance authentication and data transmission scheme for NB-IoT in 3GPP 5G networks," in *2019 IEEE Wireless Communications and Networking Conference (WCNC)*. IEEE, 2019, pp. 1–7.

406 Y. Khettab, M. Bagaa, D. L. C. Dutra, T. Taleb, and N. Toumi, "Virtual security as a service for 5G verticals," in *2018 IEEE Wireless Communications and Networking Conference (WCNC)*. IEEE, 2018, pp. 1–6.

407 M. Liyanage, A. Gurtov, and M. Ylianttila, *Software defined mobile networks (SDMN): Beyond LTE network architecture*. John Wiley & Sons, 2015.

408 M. Hussain and H. Abdulsalam, "SECaaS: Security as a service for cloud-based applications," in *Proceedings of the Second Kuwait Conference on e-Services and e-Systems*, 2011, pp. 1–4.

409 V. N. Imrith, P. Ranaweera, R. A. Jugurnauth, and M. Liyanage, "Dynamic orchestration of security services at fog nodes for 5G IoT," in *ICC 2020-2020 IEEE International Conference on Communications (ICC)*. IEEE, 2020, pp. 1–6.

410 S. Jindal, N. Marriwala, A. Sharma, and R. Bhatia, "Methodological analysis with informative science in bioinformatics," in *Soft Computing for Intelligent Systems*. Springer, 2021, pp. 49–57.

411 M. Ylianttila, R. Kantola, A. Gurtov, L. Mucchi, I. Oppermann, Z. Yan, T. H. Nguyen, F. Liu, T. Hewa, M. Liyanage, and A. Ijaz, "6G white paper: Research challenges for trust, security and privacy," *arXiv preprint arXiv:2004.11665*, 2020.

412 P. Porambage, T. Kumar, M. Liyanage, J. Partala, L. Lovén, M. Ylianttila, and T. Seppänen, "Sec-EdgeAI: AI for Edge Security Vs Security for Edge AI," 2019, last accessed May 16, 2019. [Online]. Available: https://www.researchgate.net/publication/330838792_Sec-EdgeAI_AI_for_Edge_Security_Vs_Security_for_Edge_AI.

413 X. Lin, J. Li, J. Wu, H. Liang, and W. Yang, "Making knowledge tradable in edge-AI enabled IoT: A consortium blockchain-based efficient and incentive approach," *IEEE Transactions on Industrial Informatics*, vol. 15, no. 12, pp. 6367–6378, 2019.

414 Oulu, "Multi-Access Edge Computing (MEC) Artificial Intelligence (AI)," Feb. 2018, last accessed May 16, 2019. [Online]. Available: http://www.edgeai.info/project/mec-ai/.

415 Y. Zhang, Y. Yang, and X. Wang, "A novel android malware detection approach based on convolutional neural network," in *Proceedings of the 2nd International Conference on Cryptography, Security and Privacy*, 2018, pp. 144–149.

416 D. Kong, "Science driven innovations powering mobile product: Cloud AI vs. device AI solutions on smart device," *arXiv preprint arXiv:1711.07580*, 2017.

417 P. Porambage, T. Kumar, M. Liyanage, J. Partala, L. Lovén, M. Ylianttila, and T. Seppänen, "Sec-EdgeAI: A Vision for using Artificial Intelligence for Securing the Edge," 2019, last accessed May 16, 2019. [Online]. Available: https://www.researchgate.net/publication/330620630_EdgeAI_A_Vision_for_Privacy-preserving_Machine_Learning_on_the_Edge.

418 M. A. Kumara and C. Jaidhar, "Leveraging virtual machine introspection with memory forensics to detect and characterize unknown malware using machine learning techniques at hypervisor," *Digital Investigation*, vol. 23, pp. 99–123, 2017.

419 P. K. Das, D. Ghosh, P. Jagtap, A. Joshi, and T. Finin, "Preserving user privacy and security in context-aware mobile platforms," in *Censorship, Surveillance, and Privacy: Concepts, Methodologies, Tools, and Applications*. IGI Global, 2019, pp. 1203–1230.

420 H. Lin, Z. Yan, and Y. Fu, "Adaptive security-related data collection with context awareness," *Journal of Network and Computer Applications*, vol. 126, pp. 88–103, 2019.

421 J. Zhang, X. Tan, X. Wang, A. Yan, and Z. Qin, "T2FA: Transparent two-factor authentication," *IEEE Access*, vol. 6, pp. 32677–32686, 2018.

422 P. Hao, X. Wang, and W. Shen, "A collaborative PHY-aided technique for end-to-end IoT device authentication," *IEEE Access*, vol. 6, pp. 42279–42293, 2018.

423 A. Jurcut, T. Niculcea, P. Ranaweera, and N.-A. Le-Khac, "Security considerations for internet of things: A survey," *SN Computer Science*, vol. 1, pp. 1–19, 2020.

424 C. Yapa, C. de Alwis, M. Liyanage, and J. Ekanayake, "Survey on blockchain for future smart grids: Technical aspects, applications, integration challenges and future research," *Energy Reports*, vol. 7, pp. 6530–6564, 2021.

425 E. H. Jayatunga, P. S. Ranaweera, and I. A. M. Balapuwaduge, "Blockchain advances and security practices in WSN, CRN, SDN, opportunistic mobile networks, delay tolerant networks," in *Revolutionary Applications of Blockchain-Enabled Privacy and Access Control*. IGI Global, 2021, pp. 1–34.

426 M. A. Rahman, M. S. Hossain, G. Loukas, E. Hassanain, S. S. Rahman, M. F. Alhamid, and M. Guizani, "Blockchain-based mobile edge computing framework for secure therapy applications," *IEEE Access*, vol. 6, pp. 72469–72478, 2018.

427 M. Liu, F. R. Yu, Y. Teng, V. C. Leung, and M. Song, "Distributed resource allocation in blockchain-based video streaming systems with mobile edge computing," *IEEE Transactions on Wireless Communications*, vol. 18, no. 1, pp. 695–708, 2018.

428 M. Liu, Y. Teng, F. R. Yu, V. C. Leung, and M. Song, "A mobile edge computing (MEC)-enabled transcoding framework for blockchain-based video streaming," *IEEE Wireless Communications*, vol. 27, no. 2, pp. 81–87, 2020.

429 M. Liu, F. R. Yu, Y. Teng, V. C. Leung, and M. Song, "Computation offloading and content caching in wireless blockchain networks with mobile edge computing," *IEEE Transactions on Vehicular Technology*, vol. 67, no. 11, pp. 11008–11021, 2018.

430 A. Braeken and M. Liyanage, "Highly efficient key agreement for remote patient monitoring in MEC-enabled 5G networks," *The Journal of Supercomputing*, vol. 77, no. 6, pp. 5562–5585, 2021.

431 T. Dangl, B. Taubmann, and H. P. Reiser, "RapidVMI: Fast and multi-core aware active virtual machine introspection," in *The 16th International Conference on Availability, Reliability and Security*, 2021, pp. 1–10.

432 T. Garfinkel and M. Rosenblum, "A virtual machine introspection based architecture for intrusion detection," in *Network and Distributed System Security Symposium*, vol. 3, no. 2003, 2003, pp. 191–206.

433 P. Mishra, I. Verma, S. Gupta, V. S. Rana, and K. Kadarla, "vProVal: Introspection based process validation for detecting malware in KVM-based cloud environment," in *2019 4th International Conference on Fog and Mobile Edge Computing (FMEC)*. IEEE, 2019, pp. 271–277.

434 M. Liyanage, M. Ylianttila, and A. Gurtov, "Fast transmission mechanism for secure VPLS architectures," in *2017 IEEE International Conference on Computer and Information Technology (CIT)*. IEEE, 2017, pp. 192–196.

435 M. Liyanage, M. Ylianttila, and A. Gurtov, "Improving the tunnel management performance of secure VPLS architectures with SDN," in *2016 13th IEEE Annual Consumer Communications & Networking Conference (CCNC). IEEE*, 2016, pp. 530–536.

436 D. J. Solove, *Understanding privacy*, Harvard University Press, 2008.

437 M. Al-Mutawa and S. Mishra, "Data partitioning: An approach to preserving data privacy in computation offload in pervasive computing systems," in *Proceedings of the 10th ACM Symposium on QoS and Security for Wireless and Mobile Networks*, 2014, pp. 51–60.

438 M. Satyanarayanan, "The emergence of edge computing," *Computer*, vol. 50, no. 1, pp. 30–39, 2017.

439 M. Liyanage, J. Salo, A. Braeken, T. Kumar, S. Seneviratne, and M. Ylianttila, "5G privacy: Scenarios and solutions," in *2018 IEEE 5G World Forum (5GWF)*. IEEE, 2018, pp. 197–203.

440 T. Kumar, M. Liyanage, I. Ahmad, A. Braeken, and M. Ylianttila, "User privacy, identity and trust in 5G," *A Comprehensive Guide to 5G Security*, p. 267, 2018.

441 M. Liyanage, I. Ahmad, A. B. Abro, A. Gurtov, and M. Ylianttila, *A comprehensive guide to 5G security*. Wiley Online Library, 2018.

442 R. Di Taranto, S. Muppirisetty, R. Raulefs, D. Slock, T. Svensson, and H. Wymeersch, "Location-aware communications for 5G networks: How location information can improve scalability, latency, and robustness of 5G," *IEEE Signal Processing Magazine*, vol. 31, no. 6, pp. 102–112, 2014.

443 M. Koivisto, A. Hakkarainen, M. Costa, P. Kela, K. Leppanen, and M. Valkama, "High-efficiency device positioning and location-aware communications in dense 5G networks," *IEEE Communications Magazine*, vol. 55, no. 8, pp. 188–195, 2017.

444 P. Schneider and G. Horn, "Towards 5G Security," in *Trustcom/BigDataSE/ISPA*, volume 1. IEEE, 2015, pp. 1165–1170.

445 K. Norrman, M. Näslund, and E. Dubrova, "Protecting IMSI and user privacy in 5G networks," in *Proceedings of the 9th EAI International Conference on Mobile Multimedia Communications*. ICST (Institute for Computer Sciences, Social-Informatics and Telecommunications Engineering), 2016, pp. 159–166.

446 S. Khajuria and K. E. Skouby, "Privacy and economics in a 5G environment," *Wireless Personal Communications*, vol. 95, no. 1, pp. 145–154, Jul. 2017. [Online]. Available: https://doi.org/10.1007/s11277-017-4421-y.

447 P. Zhang, M. Durresi, and A. Durresi, "Mobile privacy protection enhanced with multi-access edge computing," in *2018 IEEE 32nd International Conference on Advanced Information Networking and Applications (AINA)*. IEEE, 2018, pp. 724–731.

448 K. Gai, Y. Wu, L. Zhu, L. Xu, and Y. Zhang, "Permissioned blockchain and edge computing empowered privacy-preserving smart grid networks," *IEEE Internet of Things Journal*, vol. 6, no. 5, pp. 7992–8004, 2019.

449 X. He, J. Liu, R. Jin, and H. Dai, "Privacy-aware offloading in mobile-edge computing," in *GLOBECOM 2017-2017 IEEE Global Communications Conference*. IEEE, 2017, pp. 1–6.

450 T. He, E. N. Ciftcioglu, S. Wang, and K. S. Chan, "Location privacy in mobile edge clouds: A chaff-based approach," *IEEE Journal on Selected Areas in Communications*, vol. 35, no. 11, pp. 2625–2636, 2017.

451 X. He, R. Jin, and H. Dai, "Deep PDS-learning for privacy-aware offloading in MEC-enabled IoT," *IEEE Internet of Things Journal*, vol. 6, no. 3, pp. 4547–4555, 2018.

452 M. P. Goodyear, "A rising tide lifts all consumers: Penumbras of foreign data protection laws in the United States," *Rich. JL & Tech.*, vol. 27, p. 1, 2020.

453 S. Raab, "Protection of children's personal data and where reforms are needed," *Journal of Data Protection & Privacy*, vol. 4, no. 4, pp. 346–361, 2021.

454 Y. Şahin and İ. Dogru, "An enterprise data privacy governance model: Security-centric multi-model data anonymization," *International Journal of Engineering Research and Development*, vol. 15, no. 2, pp. 574–583.

455 Q. Chen, S. Shi, X. Li, C. Qian, and S. Zhong, "SDN-based privacy preserving cross domain routing," *IEEE Transactions on Dependable and Secure Computing*, vol. 16, no. 6, pp. 930–943, 2018.

456 A. Yazdinejad, R. M. Parizi, A. Dehghantanha, and K.-K. R. Choo, "Blockchain-enabled authentication handover with efficient privacy protection in SDN-based 5G networks," *IEEE Transactions on Network Science and Engineering*, vol. 8, no. 2, pp. 1120–1132, 2019.

457 K. D. Joshi and K. Kataoka, "pSMART: A lightweight, privacy-aware service function chain orchestration in multi-domain NFV/SDN," *Computer Networks*, vol. 178, p. 107295, 2020.

458 H. Mun, M. Seo, and D. H. Lee, "Secure privacy-preserving V2V communication in 5G-V2X supporting network slicing," *IEEE Transactions on Intelligent Transportation Systems*, vol. 23, no. 9, pp. 14439–14455, 2021.

459 S. Kuebler-Wachendorff, R. Luzsa, J. Kranz, S. Mager, E. Syrmoudis, S. Mayr, and J. Grossklags, "The right to data portability: Conception, status quo, and future directions," *Informatik Spektrum*, vol. 44, pp. 264–272, 2021.

460 D. Butin and D. Le Métayer, "A guide to end-to-end privacy accountability," in *2015 IEEE/ACM 1st International Workshop on TEchnical and LEgal aspects of data pRivacy and SEcurity*. IEEE, 2015, pp. 20–25.

461 C. Ilten, I. Kroener, D. Neyland, and H. Postigo, *Managing privacy through accountability*. Springer, 2012.

462 A. Al Omar, M. Z. A. Bhuiyan, A. Basu, S. Kiyomoto, and M. S. Rahman, "Privacy-friendly platform for healthcare data in cloud based on blockchain environment," *Future Generation Computer Systems*, vol. 95, pp. 511–521, 2019.

463 W. Du, Z. Teng, and Z. Zhu, "Privacy-MaxEnt: integrating background knowledge in privacy quantification," in *Proceedings of the 2008 ACM SIGMOD International Conference on Management of Data*, 2008, pp. 459–472.

464 M. J. Culnan, "Protecting privacy online: Is self-regulation working?" *Journal of Public Policy & Marketing*, vol. 19, no. 1, pp. 20–26, 2000.

465 C. L. Miltgen and H. J. Smith, "Exploring information privacy regulation, risks, trust, and behavior," *Information & Management*, vol. 52, no. 6, pp. 741–759, 2015.

466 S. Listokin, "Industry self-regulation of consumer data privacy and security," *UIC John Marshall Journal of Information Technology & Privacy Law*, vol. 32, p. 15, 2015.

467 S. G. Hoffman, *Regulation of cloud services under US and EU antitrust, competition and privacy laws*. Peter Lang International Academic Publishers, 2016.

468 J. Campbell, A. Goldfarb, and C. Tucker, "Privacy regulation and market structure," *Journal of Economics & Management Strategy*, vol. 24, no. 1, pp. 47–73, 2015.

469 C. Taylor and L. Wagman, "Consumer privacy in oligopolistic markets: Winners, losers, and welfare," *International Journal of Industrial Organization*, vol. 34, pp. 80–84, 2014.

470 J. Chi, E. Owusu, X. Yin, T. Yu, W. Chan, Y. Liu, H. Liu, J. Chen, S. Sim, V. Iyengar, and P. Tague, "Privacy partition: A privacy-preserving framework for deep neural networks in edge networks," in *2018 IEEE/ACM Symposium on Edge Computing (SEC)*. IEEE, 2018, pp. 378–380.

471 H. Cheng, W. Wang, and C. Rong, "Privacy protection beyond encryption for cloud big data," in *2014 2nd International Conference on Information Technology and Electronic Commerce (ICITEC)*. IEEE, 2014, pp. 188–191.

472 Y. Zhen and H. Liu, "Distributed privacy protection strategy for MEC enhanced wireless body area networks," *Digital Communications and Networks*, vol. 6, no. 2, pp. 229–237, 2020.

473 W. Lin, X. Zhang, Q. Cui, and Z. Zhang, "Blockchain based unified authentication with zero-knowledge proof in heterogeneous MEC," in *2021 IEEE International Conference on Communications Workshops (ICC Workshops)*. IEEE, 2021, pp. 1–6.

474 J. Lee, D. Kim, J. Park, and H. Park, "A multi-server authentication protocol achieving privacy protection and traceability for 5G mobile edge computing," in *2021 IEEE international conference on consumer electronics (ICCE)*. IEEE, 2021, pp. 1–4.

475 K. Kaur, S. Garg, G. Kaddoum, M. Guizani, and D. N. K. Jayakody, "A lightweight and privacy-preserving authentication protocol for mobile edge computing," in *2019 IEEE Global Communications Conference (GLOBECOM)*. IEEE, 2019, pp. 1–6.

476 J. N. Okoye, "Privacy by design," Master's thesis, NTNU, 2017.

477 A. Cavoukian and M. Chibba, "A regulator's perspective: Leading the way with privacy by design," *Cyber Security in Future Internet, Security and Privacy by Design. OUTLOOK, Visions and Research for the Wireless World*, no. 11, 2014.

478 J.-H. Hoepman, "Privacy design strategies," in *IFIP International Information Security Conference*. Springer, 2014, pp. 446–459.

479 F. Kemmer, C. Reich, M. Knahl, and N. Clarke, "Software defined privacy," in *2016 IEEE International Conference on Cloud Engineering Workshop (IC2EW)*. IEEE, 2016, pp. 25–29.

480 B. Liang, *Mobile edge computing*. Cambridge University Press, 2017.

481 J. Singh, T. Pasquier, J. Bacon, H. Ko, and D. Eyers, "Twenty security considerations for cloud-supported Internet of Things," *IEEE Internet of Things Journal*, vol. 3, no. 3, pp. 269–284, 2016.

482 P. Trakadas, N. Nomikos, E. T. Michailidis, T. Zahariadis, F. M. Facca, D. Breitgand, S. Rizou, X. Masip, and P. Gkonis, "Hybrid clouds for data-intensive, 5G-enabled IoT applications: An overview, key issues and relevant architecture," *Sensors*, vol. 19, no. 16, p. 3591, 2019.

483 X. Li, S. Liu, F. Wu, S. Kumari, and J. J. Rodrigues, "Privacy preserving data aggregation scheme for mobile edge computing assisted IoT applications," *IEEE Internet of Things Journal*, vol. 6, no. 3, pp. 4755–4763, 2018.

484 J. Zhang, Q. Zhang, S. Ji, and W. Bai, "PVF-DA: Privacy-preserving, verifiable and fault-tolerant data aggregation in MEC," *China Communications*, vol. 17, no. 8, pp. 58–69, 2020.

485 A. A. Barakabitze, A. Ahmad, R. Mijumbi, and A. Hines, "5G network slicing using SDN and NFV: A survey of taxonomy, architectures and future challenges," *Computer Networks*, vol. 167, p. 106984, 2020.

486 M. Condoluci and T. Mahmoodi, "Softwarization and virtualization in 5G mobile networks: Benefits, trends and challenges," *Computer Networks*, vol. 146, pp. 65–84, 2018.

487 K. V. Cardoso, C. B. Both, L. R. Prade, C. J. Macedo, and V. H. L. Lopes, "A softwarized perspective of the 5G networks," *arXiv preprint arXiv:2006.10409*, 2020.

488 ETSI, "Network Functions Virtualisation (NFV)," last accessed 18 September 2019. [Online]. Available: https://www.etsi.org/technologies/nfv.

489 ETSI, "ETSI Multi-access Edge Computing starts second phase and renews leadership team," ETSI News, last accessed 21 May 2020. [Online]. Available: https://www.etsi.org/newsroom/news/1180-2017-03-news-etsi-multi-access-edge-computing-starts-second-phase-and-renews-leadership-team.

490 5G Americas, "Understanding Information-Centric Networking and Mobile Edge Computing," 5G Americas, Dec. 2016, last accessed 16 May 2020. [Online]. Available: http://www.5gamericas.org/.

491 F. Giust, G. Verin, K. Antevski, J. Chou, Y. Fang, W. Featherstone, F. Fontes, D. Frydman, A. Li, A. Manzalini, and D. Purkayastha, "MEC deployments in 4G and evolution towards 5G," *ETSI White Paper #24*, vol. 24, pp. 1–24, 2018.

492 T. X. Tran, A. Hajisami, P. Pandey, and D. Pompili, "Collaborative mobile edge computing in 5G networks: New paradigms, scenarios, and challenges," *Communications Magazine*, vol. 55, no. 4, pp. 54–61, 2017.

493 M. Series, "IMT Vision–Framework and overall objectives of the future development of IMT for 2020 and beyond," *Recommendation ITU*, vol. 2083, pp. 1–21, 2015.

494 B. Yi, X. Wang, K. Li, and M. Huang, "A comprehensive survey of network function virtualization," *Computer Networks*, vol. 133, pp. 212–262, 2018.

495 R. Mijumbi, J. Serrat, J.-L. Gorricho, N. Bouten, F. De Turck, and R. Boutaba, "Network function virtualization: State-of-the-art and research challenges," *IEEE Communications Surveys & Tutorials*, vol. 18, no. 1, pp. 236–262, 2015.

496 K. Antevski, C. J. Bernardos, L. Cominardi, A. de la Oliva, and A. Mourad, "On the integration of NFV and MEC technologies: Architecture analysis and benefits for edge robotics," *Computer Networks*, vol. 175, p. 107274, 2020.

497 ETSI, "Network Functions Virtualisation (NFV); Management and Orchestration," ETSI GS NFV-MAN 001 V1.1.1, Dec. 2014. [Online]. Available: https://www.etsi.org/deliver/etsi.gs/NFV-MAN/001.60/gs.NFV-MAN001v010101p.pdf.

498 L. Bing, Z. Yunyong, and X. Lei, "An MEC and NFV integrated network architecture," *ZTE Communications*, vol. 15, no. 2, pp. 19–25, 2019.

499 R. A. Mishra, A. Kalla, K. Shukla, A. Nag, and M. Liyanage, "B-VNF: Blockchain-enhanced architecture for VNF orchestration in MEC-5G networks," in *2020 IEEE 3rd 5G World Forum (5GWF)*. IEEE, 2020, pp. 229–234.

500 ETSI, "Mobile Edge Computing (MEC); Deployment of Mobile Edge Computing in an NFV Environment," ETSI Industry Specification Group (ISG) White Paper, Feb. 2018. [Online]. Available: http://www.etsi.org/deliver/etsi.gr/MEC/001.099/017/01.01.01.60/gr.MEC017v010101p.pdf.

501 I. Ahmad, T. Kumar, M. Liyanage, J. Okwuibe, M. Ylianttila, and A. Gurtov, "Overview of 5G security challenges and solutions," *IEEE Communications Standards Magazine*, vol. 2, no. 1, pp. 36–43, 2018.

502 A. Kalla and S. K. Sharma, "A constructive review of in-network caching: A core functionality of ICN," in *2016 International Conference on Computing, Communication and Automation (ICCCA)*. IEEE, 2016, pp. 567–574.

503 Named Data Networking Project, last accessed 14 May 2020. [Online]. Available: http://named-data.net/.

504 SAIL Project, last accessed 14 May 2020. [Online]. Available: https://sail-project.eu/.

505 PSIRP Project, last accessed 14 May 2020. [Online]. Available: http://www.psirp.org/.

506 FP7 PURSUIT Project, last accessed 14 May 2020. [Online]. Available: https://www.fp7-pursuit.eu/.

507 COMNET Project, last accessed 14 May 2020. [Online]. Available: http://www.comet-project.org/.

508 D. Mansour and T. Braun, "Survey in Service-Centric Networking," 2014.

509 Q. Wu, Z. Li, J. Zhou, H. Jiang, Z. Hu, Y. Liu, and G. Xie, "SOFIA: Toward service-oriented information centric networking," *IEEE Network*, vol. 28, no. 3, pp. 12–18, 2014.

510 R. Ravindran, A. Chakraborti, S. O. Amin, A. Azgin, and G. Wang, "Realizing ICN in 3GPP's 5G NextGen core architecture," *arXiv preprint arXiv:1711.02232*, 2017.

511 L. Lei, X. Xiong, L. Hou, and K. Zheng, "Collaborative edge caching through service function chaining: Architecture and challenges," *IEEE Wireless Communications*, vol. 25, no. 3, pp. 94–102, 2018.

512 A. Azgin, R. Ravindran, and G. Wang, "Location-driven mobility support architecture for information centric networks," in *2018 International Conference on Computing, Networking and Communications (ICNC)*, 2018, pp. 905–911.

513 M. Hussaini, M. A. Naeem, B.-S. Kim, and I. S. Maijama'a, "Efficient producer mobility management model in information-centric networking," *IEEE Access*, vol. 7, pp. 42032–42051, 2019.

514 D. Kutscher, S. Eum, K. Pentikousis, I. Psaras, D. Corujo, D. Saucez, T. Schmidt, and M. Waehlisch, "Information-centric networking (ICN) research challenges," *Internet Research Task Force (IRTF)*, pp. 1–38, 2016.

515 X. Li, M. Samaka, H. A. Chan, D. Bhamare, L. Gupta, C. Guo, and R. Jain, "Network slicing for 5G: Challenges and opportunities," *IEEE Internet Computing*, vol. 21, no. 5, pp. 20–27, 2017.

516 N. Alliance, "Description of network slicing concept," *NGMN 5G P*, vol. 1, p. 1, 2016.

517 X. Foukas, G. Patounas, A. Elmokashfi, and M. K. Marina, "Network slicing in 5G: Survey and challenges," *IEEE Communications Magazine*, vol. 55, no. 5, pp. 94–100, 2017.

518 ETSI, "Multi-access Edge Computing (MEC); Support for network slicing," ETSI Group Report MEC 024, V2.1.1, Tech. Rep., Nov. 2019.

519 A. Ksentini and P. A. Frangoudis, "Toward slicing-enabled multi-access edge computing in 5G," *IEEE Network*, vol. 34, no. 2, pp. 99–105, 2020.

520 A. Vaswani, N. Shazeer, N. Parmar, J. Uszkoreit, L. Jones, A. N. Gomez, L. Kaiser, and I. Polosukhin, "Attention is all you need," in *Proceedings of the 31st International Conference on Neural Information Processing Systems*, ser. NIPS'17. Red Hook, NY, USA: Curran Associates Inc., 2017, pp. 6000–6010.

521 A. Radford, J. Wu, R. Child, D. Luan, D. Amodei, and I. Sutskever, "Language models are unsupervised multitask learners," *OpenAI blog*, vol. 1, no. 8, p. 9, 2019.

Index

a

academic researchers 6
acceleration techniques 78
acceleration training 75
access control lists (ACL) 101, 158
access network (AN) 146
access stratum (AS) 160
ACL *see* access control lists (ACL)
advanced coding/modulations 147
advanced driver assistance systems (ADAS) 24, 130
advanced metering infrastructure (AMI) 20
advanced persistent threat (APT) 147
advanced video coding 34
AirFrame open edge server 122
air pollution monitoring 93
Alliance for Telecommunications Industry Solutions (ATIS) 143
Amazon 6, 123–124
Amazon Web Services (AWS) 124, 135
  AWS Greengrass 135
AMI *see* advanced metering infrastructure (AMI)
AN *see* access network (AN)
anti-quantum cryptographic methods 161
API *see* application programming interface (API)
application-aware performance optimization 164
application descriptor (AppD) 197
application developers 6
application functions (AF) 158
application programming interface (API) 18
application-specific integrated circuits (ASIC) 37, 130, 132–133, 136
architectural/real estate visualizations 25
artificial intelligence (AI) 71, 97
  AI accelerators 118
  models 116
  technologies 83
artificial intelligence (AI)-assistive extreme communications (AEC) 114–115
attribute-based encryption (ABE) 101
audio video (AV) monitoring service 64
augmented reality (AR) 4, 5, 10, 25–26, 57, 93, 113
  based remote robotic monitoring 59–60
  functionality 46
  navigation system 44
  technology 40
authentication and key agreement (AKA) 162
authorization mechanisms 9
authorized adversaries 177
automation 21
automotive edge devices 130
autonomous driving 4
Azure IoT 135

Mobile Edge Computing and Communications: Driving Forces, Technology Foundation, and Application Areas, First Edition. Aaron Yi Ding, Chamitha De Alwis, and Madhusanka Liyanage.
© 2025 John Wiley & Sons Ltd. Published 2025 by John Wiley & Sons Ltd.

b

base stations (BS) 9, 58
Bayesian game theory 100
Beelink devices 129
Bitbucket 156
blockchain 108–109
bluetooth 23, 148
bluetooth low energy (BLE) 23, 148
brain-computer interfaces (BCI) 28, 142

c

cache data 102
caching technique 78
California Consumer Privacy Act (CCPA) 178
Cambricon M1300 Series 133
capital expenditures (CapEx) 29
centralized mode 74
central processing units (CPU) 37, 102
challenge-response procedures (CRP) 155, 167
clickjacking 54
cloud based architecture 43–45
cloud computing (CC) 7, 9, 171, 190
cloud data transfer 44
cloudlets 7, 28
cloud of things (CoT) 28
cloud-oriented offloading 83
cloud service provider (CSP) 6, 11, 155, 172, 190
cloud-to-edge 9
code division multiple access (CDMA) 13, 146
collaborative robots (Cobots) 143
commercial-off-the-self (COTS) network equipment 194
common public radio interface (CPRI) interface 121
communication network evolution 110–115
communications, computing, control, localization, and sensing (3CLS) 139
communication service providers (CSP) 121, 123
communication technologies
  blockchain 108–109
  federated learning 109
  LIS 110
  quantum computing 110
  THz communication 108
  VLC 110
  ZSM 108
compression 77–78
computation offloading 83, 177
  assessing metrics 88
  defined 87–88
  procedures 88
  representative solutions 89
connected and autonomous vehicle (CAV) 101, 143
constrained Markov decision process (CMDP) 182
consumer regulation 181
containerization 116
content-centric networking (CCN) 104
content delivery networks (CDN) 4, 6, 7, 90, 103
content stores (CS) 105
continuous approach 38
convolutional neural networks (CNN) 79, 131
cost metric 88
crowdsourcing 153
CSP *see* communication service providers (CSP)
curious adversaries 177
customer facing service portal (CFSP) 13–14, 153, 155
customer facing services (CFS) 13
custom software 116
cyber-physical systems (CPS) 145
cyber-security 145

d

data aggregation 74
data center as a service (DCaaS) 136
data center manager 122
data compression 74
data fusion 34
data handling 179–180
data-level security 200
data offloading 83
  defined 84–85
  evaluation metrics 86
  procedures 85
  representative solutions 86–87
data portability 179
data structures 105
data transferring 22
decentralization 115
decentralized information sharing 76
decentralized mode 74
dedicated core networks (DCN) 157
dedicated short range communication (DSRC) technology 24
deep learning (DL) 93–94, 141, 166
deep learning Amazon machine images 124
deep neural network (DNN) 74, 94, 134
deep packet inspection (DPI) 158, 194
deep post-decision state (PDS) learning process 182
deep Q-network (DQN) 182
Dell Technologies cloud 135
denial of service (DoS) 164
Dev Board Mini 127
device-driven data offloading 84
device-to-device (D2D) 22
  communication 103
  connectivity 148
DHCP *see* dynamic host configuration protocol (DHCP)
differential privacy 186
digital signal processors (DSP) 102, 131
distributed denial of service (DDoS) 147
distributed ledger technologies (DLT) 108
domain name system (DNS)
  amplification 151
  proxy/server 16
drone 92
dynamic adaptive streaming over HTTP (DASH) 34
dynamic host configuration protocol (DHCP) 194
dynamic voltage and frequency scaling (DVFS) technology 55

e

early-exit technique 78
edge accelerators
  ASIC 132–133
  FPGA 130–131
  future of 133
  TPU 131–132
edge artificial intelligence (AI) 5
  edge inference
    architectures 77
    importance 76–77
  edge training
    architectures 74–75
    importance 73–74
    techniques 75–76
  overview 71–73
  in practice
    applications 79–80
    energy-efficient edge AI 79
  techniques 77–79
edge-assisted multi-vehicle perception (EMP) framework 39
edge automation tool 122–123
edge-based analytics 58
edge-based video analytics (EVA) 33, 36
edge caching
  caching, defined 102
  ICN 103, 104
    data packets 104
    data structures 105
    in-network caching 105
    interest packets 104

edge caching (*contd.*)
named-content 104–105
overview 102–103
edge-centric computing 28
edge cloud (EC) 28
edge computing 4
bandwidth 4
energy saving 5
frameworks 117
latency 4
mobility and collaboration 5
personalisation 4–5
privacy and sovereignty 5
proliferation 4–5
stakeholders and models 6–7
sustainability 5
vision of edge 4–6
edge computing landscape 115, 116
hardware development timeline 117, 118
early development of, edge hardware (2000s-2010s) 118
future of 119
IoT *vs.* edge computing (2010s-2020s) 118
latest development (2020s-present) 118–119
software development timelines
early stages (2000s-2010s) 116
edge software trends (2020s-present) 117
future of 117
IoT *vs.* edge computing (2010s-2020s) 116
edge computing system solutions
application oriented systems
Amazon 123–124
enclave device blueprint 120–121
IBM 124–125
Nokia 121–123
scalys 120
legacy systems 119–120
edge device management 117
edge-enabled augmented reality (AR)
MAR
advancement towards 40–42
in 5G 42
MEC for 42–43
overview 40
edge gateways 118
edge gateway software 116
edge hardware and devices 125, 126
Beelink devices 129
coral platforms 127
Intel devices 128–129
Jetson family 126, 127
NXP 129–130
Raspberry-Pi Series 127–128
edge integration
AI/ML based communication technologies 140–141
IOT *vs.* IOE 139–140
small data 140
technologies and applications
CAV 143
cobots 143
Industry 5.0 141–142
smart and intelligent healthcare 142
UAV 142
XR toward holographic teleportation 142–143
EdgeLock secure enclave 130
Edgent 134–135
edge offloading
AI-driven edge offloading
DL 93–94
RL 94
edge computation offloading 90–91
edge data offloading 89–90
enabling technologies 92
motivation and fundamental concept
computation offloading 87–89
data offloading 84–87
rationale for 83–84
programming models 94
safety and reliability 94–95

service (target) discovery 94
use cases 92–93
edge-optimized operating systems 117
edge-oriented offloading 83
edge privacy
future 184
hybrid cloud approach 185
privacy-preserving data aggregation (PPDA) 185–186
MEC privacy 174–176
impacts 176–177
issues 177
privacy in, 5G networks 172, 173
data privacy 173
identity privacy 174
location privacy 173
privacy preservation 178–180
privacy protection mechanisms 180
privacy by design 183–184
privacy preservation, in mobile offloading 182
regulation 181–182
SDP 184
strong encryption and strong authentication 182–183
edge security 117
MEC security 146–147
possible security solutions
AI with, machine learning and deep learning 166
blockchain 167–168
context awareness 167
modern authentication 168
NS 166–167
PUF 167
SECaaS 165
VMI 168–169
VPLS 169
security requirements, of MEC
classical security requirements 160–163
MEC specific 163–165
security threats, in MEC 146
threat vectors
vs. access network 146–150
vs. MEN 150–153
edge servers 129
edge storage
data robustness in 100
data security and privacy 101
design 98, 99
edge-to-cloud integration 117
education applications 25
efficient routing 200
e-health 35
electricity-generating plants 20
electromagnetic pulse (EMP) 152
embedded privacy enhancing mechanisms 180
embedded systems 118
encapsulation security payload (ESP) 161
enclave device blueprint 120–121
enclaves 121
end-to-edge offloading 91
end-to-end (E2E)
encryption strategy 161
latency 42
end user device (device-driven) 85
energy consumption metric 88
energy-efficient edge AI 79
enhanced connectivity 58–59
enhanced dedicated core networks (eDCN) 157
enhanced mobile broadband (eMBB) 21, 42, 56, 58, 112, 190
EPS integrity algorithm (EIA) 161
e-trading 25
EU General Data Protection Regulation (GDPR) 66
European General Data Protection Regulation 177
European 5G Infrastructure Public Private Partnership (5G-PPP) 18
European Telecommunication Standards Institute (ETSI) 3, 18, 31, 58, 108, 143, 181, 192

event-driven approach 38
evolved common public radio interface (eCPRI) 122
evolved packet system (EPS) 160
eXtended reality (XR) application 94
extensible authentication protocol (EAP) 162
extremely low-power communication (ELPC) 113
extremely reliable low-latency communication (eRLLC) 113

f

fairness metric 88
Federal Aviation Administration (FAA) 26
federated learning (FL) 109, 114, 186
FemtoClouds 28
field of view (FoV) 25, 38
field-programmable gate arrays (FPGA) 37, 118, 128–131, 136
fifth-generation (5G) mobile technology 10
5G-based applications 10
5G-based edge computing paradigms 172
5G based gNodeB (gNB) 9
5G-based real-time applications 27
5G information-centric networking (ICN) model 157
5G infrastructure public private partnership (5G PPP) 181
fixed networks 13
FlashLinQ 148
fog computing 7, 27–28
fog-to-fog connections 28
forensic security 145
forwarding information base (FIB) 105
fourth generation (TPU v4) 132
4G long term evolution (LTE) networks 42, 112
further enhanced mobile broadband (FeMBB) 112

g

gaming applications 25
general-purpose hardware platforms 29
global system for mobile communication (GSM) 13
glucometers 23
Google 6, 37
Google Lens 41
governmental or continental-level regulation 181
graphics processing units (GPU) 37, 102, 118, 131
Greengrass 135

h

Hay Day 4
head mounted display (HMD) 25
healthcare 80
Health Insurance Portability and Accountability Act (HIPAA) 178
high altitude platforms (HAP) 26
high area traffic capacity 114
high-efficiency video coding 34
high spectrum efficiency 113–114
homomorphic encryption 186
hospital to home (H2H) services 142
Huawei 18
human-machine-interface (HMI) 25
human-to-human (H2H) 23
human-to-machine (H2M) 23
hybrid IoT-edge-cloud mode 75
hybrid mode 75
hyrax 28

i

IBM 18, 124–125
ICT services 6
image compression/coding 44
image segmentation 34
implantable cardiac defibrillators (ICD) 23
i.MX processors 129

industrial augmented reality (IAR) devices 48
industrial edge devices 130
industrial experts 6
industrial internet of things (IIoT) 11, 97
industry frameworks
  Azure IoT 135
  DCaaS 136
  KubeEdge 133–135
  private edge service providers 135–136
industry self-regulation 181
Industry Specification Group (ISG) 3, 18
inferentia 124
information centric networking (ICN) 92
  MEC integration
    challenges 201
    mutual benefits of 200–201
  named content 199
information-centric networking (ICN) 102, 103
infotainment systems 130
infrastructure as a service (IaaS) 11
  cloud customer 184
infrastructure-driven data offloading 84
in-network caching 105, 200
integrating hypervisor-based technologies 16
Intel 18
Intel Celeron N3450 processor 129
intelligent reflecting surfaces (IRS) 113
intelligent transportation systems (ITS) 23–24, 101
intelligent wearable devices (IWD) 142
Intel NUC (next unit of computing) 128
Intel Xeon scalable processors 123
interference-aware receivers 147
intermediary access points (IAP) 65
International Telecommunication Union (ITU) 181
internet of everything (IoE) 139
internet of medical things (IIoMT) 142
internet of things (IoT) 139
  IoT connectivity solutions 130
  IoT greengrass 123–124
internet of vehicles (IoV) 143
internet protocol (IP) 104
Internet Protocol television (IPTV) 19
internet service providers (ISP) 5, 11, 89, 185
internet services 35
intrusion detection and prevention systems (IDPS) 158, 166
intrusion detection system (IDS) 152, 168, 194
in-vehicle systems 19

k

Kernel-VM (KVM) 168
Kilo virtual machine (KVM) 30
KubeEdge 133–135

l

large intelligent surfaces (LIS) 110, 141
large language models (LLM) 207
"last-mile" network 6
legacy systems 119–120
legacy telecommunication systems 185
license shared access (LSA) 147
light amplification by stimulated emission of radiation (LASER) diodes 110
light emitting diode (LED) 110
lightweight virtualization technologies 16
line-of-sight (LoS) transmission 26
live/retrospective approach 38
local area data network (LADN) 151
localized architecture 46–48
location-based services (LBS) 173, 176
long range (LoRa) 23
low altitude platforms (LAP) 26
low-latency data processing 58
low power wide area (LPWA) 23

m

machine learning (ML) 97, 166
machine-to-machine (M2M) communication 11, 22, 29

machine type communication (MTC) 10–11
machine type communication devices (MTCD) 22
machine type communication gateways (MTCG) 23, 150
macro base stations 102
management and orchestration (MANO) 194
man-in-the-middle (MitM) 145
market regulation 181
Markov decision process (MDP) 159, 182
massive machine-type communications (mMTC) 22–23, 42, 190
message queuing telemetry transport (MQTT) 134
metaverse 97
metropolitan advanced delivery network (MADNet) 87
micro data centers (MDC) 28
microservers 118
Microsoft HoloLens 41
migrating services 177
miligram's reality-virtuality continuum 40
mini PC 118
mist 7
mist computing 28
MitM attacks 147
mixed reality (MR) 10, 25–26, 40
mobile ad hoc network (MANET) 148
mobile augmented reality (MAR) 33, 40
  architecture 50
    cloud based architecture 43–44
    hybrid architecture 48–49
    localized architecture 46–48
    MEC based architecture 45–46
  cloud based MAR system 44
mobile broadband and low-latency (MBBLL) 114
mobile cloud computing (MCC) 4, 7, 26, 27
mobile computing environments 27
mobile edge applications (ME Apps) 16, 62
mobile edge computing (MEC) 18, 57, 139
  classical security requirements
    authentication and authorization 162–163
    availability 161–162
    confidentiality 160–161
    integrity 161
  deployment challenges and solutions
    communication issues 65
    legal adaptation and ethics 66
    scalability limitation 66
    security and privacy issues 65
  MEC application orchestrator (MEAO) 203
  specific security requirements
    mitigating service impeding threats 163–164
    robustness 164
    secure charging and billing 163
    secure mobile offloading schemes 164
    secure virtualized environment and orchestration 164–165
  use cases
    AR assisted remote surgery 62
    AR based remote robotic monitoring 59–60
    patient monitoring and treatment 60–62
    realization 62–65
    smart medical data aggregation 60–62
mobile edge computing (MEC) based MAR applications
    automotive 53
    healthcare 51–52
    IIoT 51
    smart city and home 52–53
    travel and tourism 52
  technical aspects of
    communication 54

energy management 55
mobility management 55–56
privacy 54
security 53–54
service migration 55
service offloading 55
Mobile Edge Computing Industry Specification Group (MEC ISG) 192
mobile edge computing (MEC) integration, in 5G backhaul
challenges 193
needs 191–193
technologies for
ICN 199–201
network slicing 201–203
NFV 194–197
software defined networking 197–199
mobile edge host (MEH) 13, 15–16, 62, 151–152
MEH 15–16
MEPM 16
VIM 16–17
mobile edge orchestrator (MEO) 13, 14–15, 153, 154
mobile edge platform (MEP) 15, 16, 62, 150
mobile edge platform manager (MEPM) 16, 58
mobile edge service (MES) 17
mobile edge services and interfaces
Mm-management interfaces 17–18
Mp-MEC platform interfaces 17
Mx-external interfaces 18
mobile edge system level 13
CFSP 13–14
MEO 14–15
OSS 14
UALCMP 14
UE applications (UE Apps) 13
mobile equipment 38
mobile gaming applications 21
mobile network operators (MNO) 6, 10, 43, 192, 195
mobile networks 121
mobile offloading process 164
mobile virtual network operators (MVNO) 190
mobility as service 80
mobility management 55–56, 159–160
ModelIO design 94
model search technique 78
motion-to-photon (MTP) 114
movidius neural compute stick 128
multi-access edge computing (MEC) 10, 19, 42, 58, 102, 122, 190
AR 25–26
computing paradigms 26
cloudlets 28
comparisons 28–30
fog computing 27–28
MCC 27
defined 9
evolution of 11–12
ITS 23–24
mMTC applications 22–23
MR 25–26
needs 9–11
reference architecture
mobile edge host level 15–17
mobile edge services and interfaces 17–18
mobile edge system level 13–15
standardization 18–19
use cases and applications
UAV 26
use cases and applications
critical infrastructure 19–21
ITS 23–24
mMTC 22–23
UHD video streaming 21–22
VR 25–26
multi-input approach 38
multi-path transport control protocol (TCP) 92

multiple-input-multiple-output (MIMO) 113, 141, 147
multi-protocol label switching (MPLS) 27, 169
multi Radio Access Technology (multi-RAT) 19
MX Industrial Edge 123

n

narrowband IoT (NB-IoT) 23, 148
National EarlyWarning Score (NEWS2) 60
National Paediatric EarlyWarning System(PEWS) 60
near field communication (NFC) 148
net metering 20
network address translation (NAT) 158, 194
network-as-a-service (NaaS) 108
network exposure function (NEF) 159
network function virtualization (NFV) 13, 19, 131, 145, 190
- MEC integration
  - challenges 197
  - mutual benefits 195–197
- NFVI 154
- NFVO 154, 194

network function virtualization infrastructure (NFVI) 154
network function virtualization orchestration (NFVO) 154, 194
network service part includes NFV infrastructure (NFVI) 194
network slice instances (NSI) 157, 167, 203
network slice managers (NSM) 157, 167
network slicing (NS) 145, 157, 166–167, 190
- MEC integration
  - challenges 203
  - mutual benefits of 202–203
- NSM 157

network softwarization, in 5G networks 190–191
neural architecture search (NAS) 78
neural processing units (NPU) 132
neuro-stimulators 23
new radio (NR) 13
Next G Alliance 143
Next Generation Mobile Networks (NGMN) Alliance 143, 181
next generation NodeB (gNB) 46
Nokia Networks 18
non-access stratum (NAS) 160
non-3GPP technologies 23
non-orthogonal multiple access (NOMA) 114, 141
NTT DOCOMO 18
Nvidia Jetson 126

o

object detection 34
object tracking 34
offloading overhead 86
offshore plants 21
open edge computing (OEC) platform 28
open edge fronthaul gateway (FHGW) 121–122
OpenFog Consortium 27
Open Networking Foundation (ONF) 181
OpenVINO toolkit 129
operating expense (OpEx) 29
operation and business support systems (OSS/BSS) 194
operation support system (OSS) 14, 153, 154
optical wireless communication (OWC) technologies 141
original equipment manufacturers (OEM) 25, 181
overloaded communication channels 21
over the top (OTT)
- providers 190
- streaming channels 22

over-the-top (OTT) 185
oximeters 23

p
pacemakers 23
packet data network gateway (PGW) 194
partitioning model 78
patient health record (PHR) 64
peer-to-peer (P2P) 6, 103
  networks 4
  streaming 22
pending interest table (PIT) 105
Personal Data Protection Act (PDPA) 178
personal digital assistant (PDA) 29
Personal Information Protection Electronic Documents Act (PIPEDA) 178
petroleum extraction 20
physical unclonable functions (PUF) 167, 176
Pi 3 Series 128
Pi 4 Series 128
Pi Zero 128
platforms as a Service (PaaS) 11
Pokemon Go 41
policy control function (PCF) 157, 163
polymerase chain reaction (PCR) 60
pre-copy technique 89
privacy compliance 179
privacy officers 184
privacy-preserving data aggregation (PPDA) 185–186
privacy-preserving training 75–76
privacy preserving video 39
privacy protection as a service (PPaaS) 185
Private Branch eXchang (PBX) 29
private edge service providers 135–136
proof of concept (PoC) framework 18
protocol data unit (PDU) sessions 157, 158
Proximity Services (ProSe) 148
public switched telephone networks (PSTN) 13
pull-based communications 200

q
quality of experience (QoE) 14, 27, 86, 88, 102
quality of service (QoS) 14
quantum computing 110, 115
quantum resistance (QR) method 161
QUIC 92

r
radio access network (RAN) 9, 42, 191, 203
radio access technologies (RAT) 13
radio resource control (RRC) connection 161
radio units (RU) 121
Raspberry-Pi Series 127–128
real-time transport protocol (RTP) 34
recurrent neural networks (RNN) 79, 131
reduction of quality (RoQ) attacks 158
reinforcement learning (RL) 94
remotely operated vehicles (ROV) 60
remote robotic surgeries 25
resource facing services (RFS) 14
resource optimization 59
return on investment (RoI) 193
road safety 93
robot-assisted surgery (RAS) 62
robot ethics 66
robotic appliances 57
robust and secure edge storage (RoSES) model 101
role-based access control (RBAC) 101
round-trip time (RTT) processing 25

s
SafeShareRide 39
Sagemaker 124
satellite communication (SATCOM) 152
scalability optimization 59

second generation (TPU v2) 132
secure function evaluation 187
secure multi-party computation (SMC) 175, 186
security as a service (SECaaS) 165
service function chaining (SFC) 158
service functions (SF) 158
service level agreements (SLA) 14, 36, 175
service migration 55, 158–159
session management function (SMF) 157
sewage governance techniques 20
side channel attacks (SCA) attacks 168
single-board computers (SBC) 118
single-input approach 38
6 degree of freedom (6DoF) video 42
size, weight, and power (SWAP) 26
slice/service type (SST) values 203
small base stations 102
small cell BS (SCBS) 159
small data 140
smart cities 35, 79–80
smart devices 80
smart grids 20, 153
smart home applications 53
smart metering 20
SmartStore 100
software as a service (SaaS) 11
software defined mobile networks (SDMN) 171
software defined networking (SDN) 145, 163, 190
  MEC integration
    challenges for 198–199
    mutual benefits 198
  SDMN 171
software defined privacy (SDP) 184
software defined VPLS (Soft-VPLS) approach 169
softwarization 190
specialized hardware 118
split computing (SC) 78
static/dynamic configurator 135
stationary equipment 38
store edge networked data (SEND) 99
superfluid clouds 28
supervisory control and data acquisition (SCADA) systems 19, 120
system-on-module (SoM) 127

t

tactile internet 10
tensor processing units (TPU) 118, 130, 131–132, 136
terahertz (THz) frequency band 108
thin clients 44
third generation (TPU v3) 131
3rd Generation Partnership Project (3GPP) 146, 181
threat vector (TV) 146
  vs. access network 146, 147
    ad hoc connectivity b/w base station 148–149
    ad hoc connectivity b/w user equipment 148–149
    UE 149–150
  architectural
    mobility management 159–160
    NS 157
    service migration 158–159
    traffic steering 157–158
  vs. core network
    CFSP 155
    5G core network 156
    MEO 154
    MEO *vs.* 5G core network 155–156
    OSS 154
    UALCMP 153, 154
  vs. MEN
    connectivity b/w edge and core 152–153
    connectivity b/w mobile edge hosts 152
    MEH 151–152
    MEH and cloud servers 153
    MEPM 150–151

Mobile Edge Apps 153
VIM 151
time-critical cyber-physical systems (CPS) 203
traffic steering 157–158
transport control protocol (TCP) 104
Transport Systems Catapult 24
TrustBox Edge Family 120, 121
trusted execution environments (TEE) 121
Trusted Platform Module (TPM) 2.0 120
trusted third party (TTP) 171
trust-oriented data access (TODA) method 101

u

UAV-to-Everything (U2X) network 142
UE App life-cycle management proxy (UALCMP) 153, 154
ultra high definition (UHD) video streaming 10
ultra-low latency 24, 47, 54
ultra-reliable and low-latency communication (URLLC) 25, 42, 56, 57, 190
ultraviolet (UV) emission 60
ultra wideband (UWB) 23
unmanned aerial vehicles (UAV) 26, 92
USB accelerator 127
user application lifecycle management proxy (UALCMP) 14, 18
user equipments (UE) 13, 58, 174
user plane function (UPF) 151, 159

v

VAP optimizer 36
VDU *see* virtualized distributed unit (VDU)
vehicle-to-everything (V2E)
adaptation 23
communication 130
Vehicular Ad hoc NETworks (VANET) 143
vehicular engines 20
vehicular networks (VN) 24
vehicular value-added services 24
versatile video coding 34
video analytics
applications of 34–35
components 34
EVA
definition and technologies 35–37
in practice 37–39
pipeline 33–34
video decoding 34
video encoding 34
video storage 37
virtualization infrastructure (VI) 15, 16, 151
virtualization infrastructure manager (VIM) 16–18, 62, 151, 194
virtualized distributed unit (VDU) 122
virtualized network functions (VNF) 13
virtual machine introspection (VMI) 168–169
virtual machines (VM) 15, 62
virtual network function descriptor (VNFD) 197
virtual network function managers (VNFM) 194
virtual network functions (VNF) 194
virtual private local area network service (VPLS) 169
virtual private network (VPN) 169, 194
virtual radio access network (vRAN) 122
virtual reality (VR) 4, 5, 10, 25–26, 40, 57
visible light communication (VLC) 110, 141
vision processing unit (VPU) 128
vital sign monitors 23
VM *see* virtual machines (VM)
Vodafone 18

w

wearable devices 80
web-based AR (web AR) 26
web browsers 102

web real-time communication (WebRTC) 34
Wi-Fi access points (AP) 159
Wi-Fi direct 148
wireless body area network (WBAN) 22
wireless communications technologies 42
wireless electroencephalogram (EEG) headsets 28
wireless local area networks (WLAN) 13, 146
World Health Organization (WHO) 57
Worldwide Interoperability for Microwave Access (WiMAX) 13

y
YouTube 22

z
zero-knowledge proofs (ZKP) 187
Zero touch network and Service Management (ZSM) 108
ZigBee 23, 148